THE STRENGTH IN
NUMBERS

THE FUNDAMENTALS OF
MATHEMATICS AND STATISTICS

TIMOTHY C. KEARNS

authorHOUSE

AuthorHouse™
1663 Liberty Drive
Bloomington, IN 47403
www.authorhouse.com
Phone: 833-262-8899

Published by AuthorHouse   01/21/2021

ISBN: 978-1-6655-1443-9 (sc)
ISBN: 978-1-6655-1442-2 (e)

To those that feel the sense of the mysterious.

Table of Contents

(4) FUNDAMENTALS OF TRIGONOMETRY
   AND THE CALCULUS

***** TRIGONOMETRY *****

***** THE CALCULUS *****

Part 2   PROBABILITY

(5) COMBINATORICS

(6) FUNDAMENTALS OF PROBABILITY

Part 3   STATISTICS

Introduction

This book is an informal overview of mathematical and statistical fundamentals, and is written for students that are interested in pursuing studies and work in the mathematical sciences, but seek something like a textbook that is more friendly and readable. The book consists of three sections: basic mathematics, probability, and statistics. Despite the fact that the part dealing with mathematics includes the basics of single variable calculus, the remainder of the book focuses on probability theory and statistics in an algebraic, less formal, and more accessible way.

The first section of the book which deals with mathematics covers basic logic, set theory, the various number systems, algebra, geometry, trigonometry, and an introduction to the single variable calculus.

The second section of the book deals with probability and covers combinatorics, the basics of probability and random variables, probability mass functions, expected value and distribution parameters.

The third section deals with statistics. We discuss simple random samples and statistics, the law of large numbers, the discrete distributions known as the Bernoulli, Binomial, and the Poisson, the continuous distributions known as the Normal, Chi-Square, and the T-distributions, along with the Central Limit Theorem. In chapters 9 and 10, we focus on large sample hypothesis testing and confidence intervals.

Then, in Chapter 11 we cover small sample hypothesis testing and confidence intervals. Finally, chapter 12 is a brief introduction to linear regression and correlation.

When I think about who would want to read this book, I must admit that it would be perhaps someone like many former students that I have tutored. Many of these people needed a rigorous treatment of statistics, but lacked a good foundation in mathematics. I have a Bachelor of Science degree (with distinction) in Statistics from Va. Tech, with a strong minor in math, including applied and theoretical mathematics. I feel that most of my education has really been from my reading extensively in math and science after college graduation, and also from my many years of math tutoring. The book is not meant for those among us that are math challenged, but I do not feel that it is excessively rigorous or unnecessarily difficult for most that would read it. It is something in between these two extremes. In fact, the material in the probability and statistics parts is almost completely of an algebraic nature. Perhaps an aspiring math teacher before or after college graduation, or a technically minded person returning to graduate school after being away from math and statistics for a while would find the book most beneficial.

Students can pick and choose to study only those parts of the book that they feel they need to. Some will go on to take more advanced courses in math and statistics, so this book is meant to prepare them with what they will most likely need to know, to give them a good foundation from which to build upon. Included are a few sections about the two types of infinity associated with the real numbers, which is actually not an introductory topic, but I think that it is important enough to include. It is my hope that the reader finds the material about different types of infinity to be both enlightening and actually interesting. It should help the reader in their understanding of math and stats.

So, this book is an affordable and comprehensive overview and primer for those needing to know about the basics of mathematics and statistics, two very important subjects to the technically and scientifically minded. Most of the book includes some exercises meant to help students learn the material as they work their way through. So, those serious students that seek a good treatment of the basics of math and stats should find this book to be a good supplemental reading and a great resource.

Timothy C Kearns
Fairfax County, Va. USA
January, 2021

PART 1

<u>MATHEMATICAL</u>
<u>FUNDAMENTALS</u>

In the first four chapters we present the fundamental knowledge of math that one needs to more completely understand applied math and statistics. Chapters 1 and 2 will present the logic, set, and number theory that should help the student of applied mathematics. Then chapters 3 and 4 present the indispensable tools of more advanced mathematics: algebra, geometry, trigonometry and the single variable calculus. These branches of math form a foundation that the reader can confidently build upon in the higher math and statistics courses that many will take.

(1) <u>MATHEMATICAL REASONING</u>

(1.1) <u>Mathematics as a Basis for Science</u>

Mathematics is a language that we use in describing and understanding the world we live in. The subjects of logic, set theory, algebra, geometry, and further developments in mathematics are a large part of the foundation upon which science rests. In fact, mathematical reasoning may not be just the foundation of science, it may very well be the foundation of much more. The ancient Pythagoreans were a school of Greek mathematicians that believed that everything could be reduced to numbers, that the entire world had been built using mathematics. They believed that the truth behind the everyday reality we experience lies in numbers.

Here in this chapter we will explore some of the most basic elements of logic, and in the next we will explore some of the most basic elements of modern set theory, especially infinite sets. Logic and set theory are for mathematicians the greater part of the foundation of their subject.

We live in a world of constraints, which are mathematical constraints. Physical theory tells us that the cosmic speed limit for all material objects and electromagnetic radiation,

5

is a fundamental constant which we commonly call c (the speed of light in a vacuum), and it is approximately 300,000 kilometers per second. Nothing can travel faster than this. The description and behavior of atoms, molecules, and all the elementary particles of which they are made are subject to mathematical laws, rules, and formulas which cannot be violated. For example, it appears that we live in a three dimensional space; and we can't draw an object of four dimensions. In arithmetic, we are constrained by the fact that $3 + 4 = 7$, and it cannot be 900. And as well, we must live with the fact that $3 \times 4 = 12$. It simply is not 87. If we want to place 30 tennis balls among 5 bins in such a way that there are an odd number of balls in each bin, some reflection tells us that it is quite impossible when we realize that the sum of 5 odd numbers cannot be even. In all of math and nature there are rules, laws, formulas, and so on which cannot be violated, and all of these constraints have been for the most part determined by mathematical types of reasoning. It seems that mathematics determines what is possible and what is not. The existence of anything in this world requires some kind of logical or mathematical basis.

We use the two subjects of mathematics and statistics, that one can refer to as deductive and inductive reasoning respectively, to build our growing knowledge of all the various parts of the world. These two subjects are very much connected to form our science and our scientific method. It is these two fundamental and very important subjects which we use to prove theorems and other things in mathematics and in science, allowing us to describe and master the physical world. Our scientific method, which is based on reason and experiment requires that the laws of science be accessible to everyone. When we have some hypothesis or theory which makes predictions, it's necessary that in principle everybody should be able to do their own experimental investigations, in order to verify or disregard the hypothesis or theory for themselves, in order to seek the truth about things. It is this dependence on reason and experiment that forms the scientific way of thinking.

So, as best we can figure, we exist in a world of logical consequences and order. If this were not the case, and there was only chaos, there would not be any patterns or inviolable laws from which anything would form. It seems that the very existence of our universe itself requires some such patterns and inviolable laws in order to exist. This kind of structure makes it possible for observing entities such as ourselves to exist. This is the kind of world that we find ourselves in. Mathematics and the languages of the many sciences have turned out to be the most effective tools for describing and for explaining the natural world in which we find ourselves. Here in this first chapter, we will become familiar with some basic logic and how logic helps us to construct certain lines of reasoning that we can use to establish mathematical and scientific truths. If you find the truth tables and material of this chapter to appear a bit exotic looking, bear with it. It is important material in mathematics.

***** BASIC LOGIC *****

Propositions and Truth Value

In this treatment of logic, we will consider propositions, or statements, and some logical operators or relations between these propositions that are useful for us in mathematics. Propositions will usually be denoted by p, q, r, s, t, and so on. When we state a proposition, we say that it has a truth value. By this we mean to say that it is either true (T) or false (F), a very clear cut binary situation. If it is true, then the proposition states something that we would agree is exactly in accordance with what we observe and know. If it is false, it is precisely not in accordance with what we observe and know.

Using an example in math, suppose x = 3. Then suppose the propositions p and q are as follows:

p: x is greater than or equal to 2
q: x is less than 0

From the knowledge of mathematics that we all collectively have, we would agree that proposition p is true (T), and the proposition q is false (F).

Logical Relations and Truth Tables

Let's turn our attention to the five basic logical relations that we call: NOT (~), OR (∨), AND (∧), IMPLICATION (→), and BI-IMPLICATION (↔).

We will use so-called truth tables to define these relations. A truth table shows all the possible combinations of the truth values of the proposition(s), and the truth value of the logical relation under consideration. If we are dealing with only one propostion p, then there are only two distinct possibilities for the truth value of p. If we are dealing with two propositions p and q, then there are four different possible combinations for the truth values of p and q. If we are dealing with three propositions p, q, and r, then there are eight different possible combinations for the truth values of p, q, and r. And so on it goes.

(1) For a proposition p, the truth value of ~p is defined by the following truth table: (this is just negation, similar to the idea of complements that we will see in set theory, or to negative numbers that we encounter in arithmetic)

Proposition p	~p
T	F
F	T

(2) For two propositions p and q, the truth value of p ∨ q is defined by the truth table: (p ∨ q is true whenever at least one of the propositions is true – this is p OR q)

Proposition p	Proposition q	p ∨ q
T	T	T
T	F	T
F	T	T
F	F	F

(3) For two propositions p and q, the truth value of p ∧ q is defined by the truth table: (p ∧ q is true only when both of the propositions are true – this is p AND q)

Proposition p	Proposition q	p ∧ q
T	T	T
T	F	F
F	T	F
F	F	F

(4) For two propositions p and q, the implication p ⟶ q is usually thought of as (if p, then q) or (p implies q). The truth value of p ⟶ q is defined by the truth table:

Proposition p	Proposition q	p ⟶ q
T	T	T
T	F	F
F	T	T
F	F	T

The only time that an implication is false is when we try to say that a true statement implies something false. Let's consider a little bit more about implications.

The following is a True statement:
10 is a whole number between 9 and 11, therefore
$2(9) < 2(10) < 2(11)$. This is a valid line of reasoning
from a true premise to a true conclusion.

The following is a False statement:
10 is a whole number between 9 and 11, therefore
$2 \cdot (10)$ is a negative number. This is not a valid line of
reasoning, where a true premise implies something false.

The following is a True statement:
A car is an animal, therefore $2 \cdot (10) = 20$. This is considered
to be a valid line of reasoning. It is considered valid to make
a correct conclusion even if the premise is false.

The following is a True statement:
A car is an animal, therefore $2 \cdot (10) = 73$. This is considered
to be a valid line of reasoning. It is considered valid to make
a false conclusion from a false premise.

With regard to the last two True statements, sometimes in
mathematics, things are simply defined to be a certain way
if it doesn't cause difficulties or make for an inconsistency
in the larger scheme of things. An example of this, for those
familiar with factorials, is the fact that 0! is simply defined
to be 1, and everything works out just fine. The idea of
implication, the idea of one thing leading to some other
thing as just presented is central to many theorems and
propositions. Equally important is the defining of this idea

in two directions. This leads us to the final logical relation that we consider here, called bi-implication.

(5) For two propositions p and q, the relation p ⟷ q is usually thought of as bi-implication or (p if and only if q) or (p iff q), which means that (p ⟶ q and q ⟶ p). The truth value of p ⟷ q is defined by the truth table:

Proposition p	Proposition q	p ⟷ q
T	T	T
T	F	F
F	T	F
F	F	T

The two-way implication is true when both propositions p and q are true or when both propositions p and q are false.

Now, let's show with a couple examples how we can use truth tables to analyze some more complicated logical propositions.

Example (1):
Construct the truth table for (p → r) ∨ (q → r) which involves the propositions p, q, and r. Since we have a statement that

12

involves three propositions, there are 8 possibilities for the truth values of the three propositions. Let's define s and t: s = (p → r) and t = (q → r). The truth table is:

p	q	r	s	t	s ∨ t
T	T	T	T	T	T
T	T	F	F	F	F
T	F	T	T	T	T
T	F	F	F	T	T
F	T	T	T	T	T
F	T	F	T	F	T
F	F	T	T	T	T
F	F	F	T	T	T

The last column shows the truth values for the statement (p → r) ∨ (q → r). Since this statement is on the large scale an OR statement, then the statement is always true except for the case where neither of the two statements p → r or q → r is true. Assume that there are only three ways for "something to imply r," and they are p or q or something else. Then this result makes sense.

Example (2):
Construct the truth table for (p ↔ q) ∧ (p → q) which involves the propositions p and q. We will let s = (p ↔ q) and t = (p → q). Since we have two propositions in this case, there are only four truth value combinations that would be possible. The truth table is:

13

p	q	s	t	s ∧ t
T	T	T	T	T
T	F	F	F	F
F	T	F	T	F
F	F	T	T	T

So, the last column shows the truth values for the statement (p ↔ q) ∧ (p → q). This statement is true only when both parts are true because it is largely an AND statement, where one side is the two-way implication. Two-way implication is true only in two cases, so it makes sense that the larger statement would be true in at most two cases.

(1.2) Logical Arguments

Implication - Contrapositive Pairs
Suppose we have an implication p → q. Then this is logically equivalent to (~q) → (~p). This is called the contrapositive for the implication. Let's verify that these two statements are logically equivalent using truth tables:

p	q	~p	~q	p → q	(~q) → (~p)
T	T	F	F	T	T
T	F	F	T	F	F
F	T	T	F	T	T
F	F	T	T	T	T

We can see that the last two columns are identical. Therefore, an implication $p \rightarrow q$ is logically equivalent to $(\sim q) \rightarrow (\sim p)$. We call these two statements an implication – contrapositive pair.

When we have the statement $p \rightarrow q$, we call the statement $q \rightarrow p$ the converse. Therefore, the statement $\sim p \rightarrow \sim q$ is the contrapositive for the converse, and we call it the inverse. So, the converse and the inverse form a second implication – contrapositive pair.

Logical Arguments
We are all familiar with the old syllogism:

All men are mortal.
Socrates is a man.
Therefore, Socrates is mortal.

This syllogism is called a logical argument. We call the first two statements the hypotheses, and then we have the conclusion. There are many other types of logical arguments that could be considered, but it's only the basic idea of arguments that we want to consider in this book. If we label the two hypotheses, statements (1) and (2), and label the conclusion statement (3), then this type of logical argument can be codified to the following form:

$[(1) \wedge (2)] \rightarrow (3)$.

When we have a three part logical argument of this form, we can evaluate it using truth tables. Let's consider a few examples of logical arguments which involve implications, which we have discussed above. Let's start by showing how the simple implication $p \rightarrow q$ can be codified into our logical argument form shown above and evaluated using truth tables.

Example (1):
For the statement $p \rightarrow q$, write it in the form:

If   (1) $p \rightarrow q$
and   (2) p
Then (3) q

Then we can use truth tables to evaluate the simple implication by checking to see if $[(1) \wedge (2)] \rightarrow (3)$ is always true, or in other words, if $[(p \rightarrow q) \wedge (p)] \rightarrow q$ is always true. Firstly, let $s = p \rightarrow q$, $r = s \wedge p$, and then let $t = r \rightarrow q$. So,

p	q	s	r	t
T	T	T	T	T
T	F	F	F	T
F	T	T	F	T
F	F	T	F	T

The last column represents $[(p \rightarrow q) \wedge (p)] \rightarrow q$, and it is true in all cases. When a logical argument is always true, we call it a tautology. A valid logical argument always turns out to be a tautology. We can see that when one or more of the hypotheses of a valid argument is false, this doesn't change a valid argument's truth value from being true. This is what guarantees for us that in the formal language of math, when we use a valid argument, we will always arrive at a valid conclusion.

We will use the following two statements p and q for both of the next two examples: p: $10 < x < 20$ and q: $0 < x < 30$. These next two examples show invalid logical arguments. We call the first "Denying the Antecedent." We call the second "Denying the Consequent."

Example (2):
Analyze the following argument:

If   (1) $p \rightarrow q$
and   (2) $\sim p$
Then (3) q

Using truth tables:

Let s = $p \rightarrow q$, r = s $\wedge$ ($\sim p$), and then let t = r $\rightarrow$ q .

p	q	~p	s	r	t
T	T	F	T	F	T
T	F	F	F	F	T
F	T	T	T	T	T
F	F	T	T	T	F

The last column represents: $[(p \rightarrow q) \wedge (\sim p)] \rightarrow q$. Since its truth values are not always true, it is not a valid argument. For the statements p and q above, we cannot just say that (x is not between 10 and 20) implies (0 < x < 30). That is, we cannot just replace statement p with statement ~p. This is an example of an incorrect line of reasoning, a fallacious argument.

Example (3):
Analyze the following argument:

If   (1) p → q
and   (2) p
Then (3) ~q

Using truth tables:
Firstly, let s = p → q, r = s ∧ p, and then let t = r → ~q .

p	q	~q	s	r	t
T	T	F	T	T	F
T	F	T	F	F	T
F	T	F	T	F	T
F	F	T	T	F	T

The last column represents: $[(p \rightarrow q) \wedge (p)] \rightarrow \sim q$. Since it's truth values are not always true, it is not a valid argument. To say (x is between 10 and 20) implies (x is not between 0 and 30) is just not true. Nor in general is anything other than our given statement q. The lesson of this example is that in an implication $p \rightarrow q$, we cannot just assume that $\sim q$ will also follow from statement p. That is invalid reasoning.

For the next example, we will use the statements:
p: 0 < x < 10,   q: x is positive,   r: x > -5.

Example (4):
Analyze the following argument:

If   (1) $p \rightarrow q$
and   (2) $q \rightarrow r$
Then (3) $(\sim r) \rightarrow (\sim p)$

This argument can be changed into the statement:
$[(p \rightarrow q) \wedge (q \rightarrow r)] \rightarrow [(\sim r) \rightarrow (\sim p)]$.

To analyze it using a truth table, use the following definitions: Let $s = p \rightarrow q$, $t = q \rightarrow r$, $u = s \wedge t$,
and $v = ((\sim r) \rightarrow (\sim p))$.

p	q	r	~p	~r	s	t	u	v	u → v
T	T	T	F	F	T	T	T	T	T
T	T	F	F	T	T	F	F	F	T
T	F	T	F	F	F	T	F	T	T
T	F	F	F	T	F	T	F	F	T
F	T	T	T	F	T	T	T	T	T
F	T	F	T	T	T	F	F	T	T
F	F	T	T	F	T	T	T	T	T
F	F	F	T	T	T	T	T	T	T

The last column represents the logical argument:
[(p → q) ∧ (q → r)] → [(~r) → (~p)] , and we can see that it is a valid argument because it is a tautology. So it turns out that we can conclude:

[(0 < x < 10) → (x is positive) and (x is positive) → (x > -5)]
→ [(x ≤ -5) → (x is not between 0 and 10)]

This statement presents no logical difficulties. It is a correct line of reasoning.

In mathematics, we spend a lot of energy deriving results and proving theorems. We are always using the tools of logic presented above, though we may not be aware of it in our demonstrations. The study of the subject of logic (which of course we have just scraped the surface of) is important for us so that we can have correct arguments and draw to correct conclusions. This is so-called deductive reasoning, which is what so much of mathematics is really all about.

################ Exercises ################

(1) Construct the truth table for: ~(p ∧ (~q)) .

(2) Construct the truth table for: ~(p ∧ q) → [(~p) ∧ (~q)] .

(3) Are the two statements always true?
     (a)   ~(p ∨ q) ↔ [(~p) ∧ (~q)]
     (b)   ~(p ∧ q) ↔ [(~p) ∨ (~q)]

(2) <u>SETS AND NUMBERS</u>

(2.1) <u>Sets and Cardinality</u>

The use of set theory and its intrinsic logic is one of the most fundamental aspects of math. Set notions are used in practically every branch of mathematics and are quite indispensable. When discussing sets, we make the distinction between sets with a finite number of members and those containing an infinite number of members. The sets that will prove to be of the greatest interest are infinite sets, and these will almost always be some set of numbers. However, finite sets are often very useful for illustrating certain operations with sets and certain properties of sets. It will be assumed that the reader is already familiar with many important number sets. However, the sets of rational, irrational, real, and complex numbers will be studied in much more detail later in this chapter.

Let me start by defining a set as a well-defined collection of objects. As examples, consider the following sets:

A_1 = {apple, pear, pineapple, orange}
A_2 = {cat, dog, bird, elephant}
A_3 = {Ford, Volvo, Chrysler, Mercedes}

Each of these sets clearly specifies a collection of objects.

If we want to say that an object is a member of, or is not a member of a set, we have a notation for that. We would write:

"apple" $\in A_1$ to say that "apple" is an element of A_1,
or "apple" is a member of A_1, and

"bat" $\notin A_2$ to say that "bat" is not an element of A_2,
or "bat" is not a member of A_2.

Now, what do the three sets A_1, A_2, A_3 have in common? Clearly, they all consist of four different things, four types of fruit, four types of animals, and four different car manufacturers. They all have the property of "four-ness." What this suggests is that some numbers (whole numbers) are themselves a property of sets. The idea of the symbol 4, used in our common base 10 number system, or what I have called "four-ness" is a property that all sets of four objects possess. All sets containing ten members possess the property of "ten-ness." All sets containing one billion members possess the property of "one billion-ness," and so on. So we will say that for all sets, a fundamental property of the set is the quantity of things that are in the set.

The number of objects that are in a set is called the cardinality of the set, abbreviated "card." For example, for the sets above we would write card(A_1) = 4, card(A_2) = 4, and the card(A_3) = 4. The cardinality of any set with a finite number of elements or members is a whole number. Note, for example, that the cardinality of a set could not be an irrational number, like $\sqrt{17}$. At least in this book, we will not consider any such set, if in fact it exists.

We can speak of the cardinality of infinite sets as well. As we will see, the number of elements in an infinite set is denoted by what is called a transfinite number. We will learn that there are an infinite number of types of infinity, so we will have a systematized way of denoting the cardinality of different types of infinite sets.

(2.2) <u>Specifying the Members of a Set</u>

Let me denote the set B_1 in the following two ways:

B_1 = {a, e, i, o, u}
B_1 = { x | x is a vowel in the English alphabet}

We call the first way the "roster" method for specifying the members of the set B_1. The roster method can be most useful when the number of members in the set is small, or the members of the set can easily be listed. The second way is the "property" method. The way that we would say the above is something like: "The set B_1 contains things x, where x is a vowel in the English alphabet." We would usually use the property method when the number of members is large or infinite, or where it is just not easy or possible to list the members in a roster fashion. However, the roster method or a variation of it can also be used for infinite sets as well, and we will do this quite frequently. For example, consider the set B_2:

$B_2 = \{1, 2, 3, \ldots\}$

It is clear that set B_2 is the set of positive whole numbers even though we can't list them all. Note that we use braces often when describing sets. We need to mention that the order that we write down the members of a set does not matter (when we can write them all down in the first place) and each member of a set need be listed only once. For example: {A, C, R, Z} and {Z, A, R, C} specify the same set.

Many times we use a special symbol to refer to a specific set, or we can just provide a verbal description of the members of a set (consider the descriptions of some of the number sets below). In this book, we will reserve the following symbols for certain sets of importance in mathematics:

N = {1, 2, 3, . . .} will refer to the Natural numbers, also called the Counting numbers, and also called the set of positive Integers.

W = {0, 1, 2, 3, . . .} will refer to the Whole numbers. The only difference between N and W is that W includes 0. We can also refer to W as the non-negative Integers. The number 0 is thought of as neither negative or positive.

Z = {. . . , -3, -2, -1, 0, 1, 2, 3, . . .} will refer to the Integers.

$Q = \{$all numbers of the form $\frac{a}{b}$, where a, b $\in$ Z, and b $\neq$ 0$\}$ will be called the rational numbers. Note that we have been constructing sets from previous ones: $N \subset W \subset Z \subset Q$. (Here we are using subset notation which is discussed in the next section). This pattern of building up sets from previous ones stops with Q.

IRR = {all irrational numbers}. IRR is the set of all real numbers that are not rational. The sets Q and IRR have no overlap or intersection.

R = {the collection of all rational and irrational numbers}. This is the set of Real numbers.

C = {all numbers of the form a + bi | a, b $\in$ R, $i = \sqrt{-1}$}. This is the set of Complex numbers. The number i is called a purely imaginary number, and also called the imaginary unit.

There is another set that is very important and is referred to with a special symbol. We call it the null set or the empty set. We denote it with the symbol $\oslash$. The null set is the set that contains nothing. It is somewhat analogous to the number 0 in our familiar real number system.

(2.3) Subsets

Consider the following two sets:
$D_1 = \{2, 4, 6, 8\}$ and $D_2 = \{2, 4, 6, 8, 10, 12, 14\}$.

We say that set D_1 is a subset of the set D_2 or that the set D_2 contains the set D_1, since every member of D_1 is also a member of D_2, and we denote this by writing $D_1 \subset D_2$. Note that $D_2 \not\subset D_1$ since the elements 10, 12, and 14 are not members of set D_1. We sometimes call D_1 a proper subset of D_2, since D_1 does not contain all the elements of D_2.

Any set is a subset of itself, so $D_1 \subset D_1$ and $D_2 \subset D_2$. The null set is a subset of any set. So, for any set A, $\emptyset \subset A$. This of course implies that $\emptyset \subset \emptyset$.

################## Exercises ##################

Suppose we start with the sets A = {w,t,s,u,h,f},
B = {w,t,s,u,h,f}, C = {t,s,u,f}, and D = {w,t,r,s,u,f}.

(1) (a) Is $C \subset A \subset B$? (b) Is C a proper subset of A, why or why not? (c) Is A a proper subset of B, why or why not?

(2) (a) Is $C \subset D \subset A$? (b) Is C a proper subset of D, why or why not? (c) Is D a proper subset of A, why or why not?

(2.4) Unions, Intersections, Complements

Consider the following sets:
$E_1 = \{10, 20, 30, 40, 50\}$
$E_2 = \{60, 70, 80, 90\}$
$E_3 = \{40, 50, 60, 70\}$.

Unions

The union of two sets A and B is the set which contains all the things that are in set A or in set B (which of course also includes all the things that are in both), and we write this as $A \cup B$. For example,

$$E_1 \cup E_2 = \{10, 20, 30, 40, 50, 60, 70, 80, 90\}$$
$$E_1 \cup E_3 = \{10, 20, 30, 40, 50, 60, 70\}$$
$$E_2 \cup E_3 = \{40, 50, 60, 70, 80, 90\}$$
$$E_1 \cup E_2 \cup E_3 = \{10, 20, 30, 40, 50, 60, 70, 80, 90\}$$

For any sets A and B,

$$A \subset (A \cup B) \text{ and } B \subset (A \cup B).$$
$$A = (A \cup A), \quad A = (A \cup \emptyset), \text{ and } \emptyset = (\emptyset \cup \emptyset).$$
If $B \subset A$, then $(A \cup B) = A$.

Intersections

The intersection of two sets A and B is the set which contains all the things that are in set A and in set B, that is, the overlap of the two sets, and we write this as $A \cap B$. Using the three sets E_1, E_2, and E_3 from above:

$$E_1 \cap E_2 = \emptyset$$
$$E_1 \cap E_3 = \{40, 50\}$$
$$E_2 \cap E_3 = \{60, 70\}$$
$$E_1 \cap E_2 \cap E_3 = \emptyset, \text{ since there are no elements}$$
common to all three of these sets.

Two sets whose intersection is the null set, such as the two sets E_1 and E_2 above, are said to be mutually exclusive or disjoint. The intersection of E_1, E_2, and E_3 is the empty set, since there is no element common to all three sets, but they are not a mutually exclusive collection of sets, since some pairs of these three sets are not disjoint. For any two sets A and B,

$$(A \cap B) \subset A \text{ and } (A \cap B) \subset B.$$
$$A = (A \cap A), \quad (A \cap \emptyset) = \emptyset, \text{ and } \emptyset = (\emptyset \cap \emptyset).$$
If $B \subset A$, then $(A \cap B) = B$.

Complements
Let U be a universal set, a set which contains everything, in some context, and let A be a subset of U. The complement of set A, written A^c, is the set of all things in U that are not in A. Therefore, $(A \cup A^c) = U$ and $(A \cap A^c) = \emptyset$. We say that A and A^c are together a collection of mutually exclusive and exhaustive sets, because all things in U are either in A or in A^c, but not in both because $A \cap A^c = \emptyset$.

Sometimes in mathematics, the set U is partitioned into $n > 2$ pairwise disjoint sets $\{A_1, A_2, \ldots, A_n\}$, where the universal set $U = (A_1 \cup A_2 \cup \ldots \cup A_n)$. Pairwise disjoint means that we have $A_i \cap A_j = \emptyset$ when $i \neq j$. To have the sets pairwise disjoint guarantees that the entire collection is a collection of mutually exclusive sets. To see this, suppose that the intersection of some collection of m sets, where $3 \leq m \leq n$ is not empty. Then we could select three of those m sets, say A_i, A_j, and A_k, where i, j, and k are different numbers. Then since $(A_i \cap A_j \cap A_k)$ is not the empty set, there would be some element x common to all three sets. But $A_i \cap A_j = A_i \cap A_k = A_j \cap A_k = \varnothing$, since the sets in the collection are pairwise disjoint. But at the same time, these three intersections must also contain the element x. This is clearly a contradiction. So, the entire collection of n sets is mutually exclusive, and we call $\{A_1, A_2, \ldots, A_n\}$ a collection of mutually exclusive and exhaustive sets, exhaustive meaning that their union is U.

There are a couple other useful results involving complements that I would like to mention:

(A) The first useful result is that if we partition a universal set U into the two sets A and A^c, and if E is any other set within U, then we could express the set E in this way:

$$E = (A \cap E) \cup (A^c \cap E)$$

What this says is that E is the part of E that is in A along with the part of E that is in A^c. The two parts are disjoint since A and A^c are. We could also do this if the universal set U were partitioned into n mutually exclusive and exhaustive sets $\{A_1, A_2, \ldots, A_n\}$. In this case we would have:

$$E = (A_1 \cap E) \cup (A_2 \cap E) \cup \ldots \cup (A_n \cap E)$$

where the n sets in this union are mutually exclusive.

(B) The second useful result is DeMorgan's Laws for Sets. If we have any collection of n sets $\{A_1, A_2, \ldots, A_n\}$ within a universal set U, then it is true that:

$$(A_1 \cup A_2 \cup \ldots \cup A_n)^c = A_1^c \cap A_2^c \cap \ldots \cap A_n^c \quad \text{and}$$
$$(A_1 \cap A_2 \cap \ldots \cap A_n)^c = A_1^c \cup A_2^c \cup \ldots \cup A_n^c$$

Let's prove this for the case of two sets. The proof for n sets is completely analogous. We must say at this point that to prove that two sets A and B are equal, we must prove that $A \subset B$ and that $B \subset A$. We will use this fact to prove the first law for two sets. Then once the first Law is proven, we will use some obvious set manipulations to easily prove the second Law. For the case of two sets, the first law says that:

<u>Law 1</u>: $(A_1 \cup A_2)^c = A_1^c \cap A_2^c$
<u>Proof</u>:
(1) Let $x \in (A_1 \cup A_2)^c$. Then x is not a member of A_1 or A_2. So, $x \in A_1^c$ and $x \in A_2^c$. So, $x \in A_1^c \cap A_2^c$. Therefore $(A_1 \cup A_2)^c \subset A_1^c \cap A_2^c$.
(2) Now Let $x \in A_1^c \cap A_2^c$. Then $x \notin A_1$ and $x \notin A_2$. So, $x \in (A_1 \cup A_2)^c$. Therefore $A_1^c \cap A_2^c \subset (A_1 \cup A_2)^c$. Therefore, $(A_1 \cup A_2)^c = A_1^c \cap A_2^c$.

<u>Law 2</u>: $(A_1 \cap A_2)^c = A_1^c \cup A_2^c$
<u>Proof</u>:
Law 1 says that $(A_1 \cup A_2)^c = A_1^c \cap A_2^c$. Since for any set A, $A = (A^c)^c$, we have $(A_1^c \cap A_2^c)^c = (A_1 \cup A_2)$. Now, interchange the roles of A_1 and A_1^c, and A_2 and A_2^c, to get $(A_1 \cap A_2)^c = A_1^c \cup A_2^c$, which proves Law 2.

################# Exercises #################

Suppose we have a universal set U and the sets
A, B ,C, D, E, F and G:

U = {0,1,2,3,4,5,6,7,8,9,10,11,12,13,14,15,16,17,18,19,20}
A = {0,2,4,6,8,10}
B = {6,7,8,9,10,11,12,13}
C = {4,6,8,12}
D = {15,17,20}
E = {0,7,9,13,18,19}
F = {1,2,3,5,10,11,14,16}
G = {0,2,4,6,7,8,9,10,11,12,13,15,17,18,19,20}

(1) What is (a) $(A \cup B)$?,   (b) $(A \cup C)$?,   (c) $(A \cup D)$?
       (d) $(A \cup E)$?   (e) $(B \cup C)$?,   (f) $(B \cup D)$?,
       (g) $(B \cup E)$?   (h) $(C \cup D)$?   (i) $(C \cup E)$?,
       (j) $(D \cup E)$?,   (k) $(C \cup D \cup E)$?

(2) What is (a) $(A \cap B)$?,   (b) $(A \cap C)$?,   (c) $(A \cap D)$?
       (d) $(A \cap E)$?   (e) $(B \cap C)$?,   (f) $(B \cap D)$?,
       (g) $(B \cap E)$?   (h) $(C \cap D)$?   (i) $(C \cap E)$?,
       (j) $(D \cap E)$?

(3) What are (relative to the universal set U) the sets
   (a) A^c (b) B^c (c) C^c (d) D^c (e) E^c

(4) Which pairs of sets from {A,B,C,D,E} are pairwise
  disjoint?

(5) (a) Are sets {A,B,C,D,E} a collection of pairwise disjoint sets?

(b) Are they a collection of exhaustive sets?

(6) (a) What is $(A \cup B \cup C \cup D \cup E)^c$?

(b) Are the sets $\{G, (A \cup B \cup C \cup D \cup E)^c\}$ a collection of mutually exclusive and exhaustive sets?

(7) (a) Are {C,D,E} a set of mutually exclusive sets?

(b) Are they a collection of exhaustive sets?

(8) Are $\{C \cup D \cup E), F\}$ a collection of mutually exclusive and exhaustive sets?

(9) If we have the universal set U = {0,1,2,5,6,7,9,10,11}, and set A = {1,2,5,6,7} and set E = {6,7,9,10}.

(a) What is A^c?

(b) What is $(A \cap E)$?

(c) What is $(A^c \cap E)$?

(d) What set is $(A \cap E) \cup (A^c \cap E)$?

(2.5) The Cartesian Product of two Sets

If we have two sets A and B, then the Cartesian Product of A and B is (written $A \times B$):

$$A \times B = \{\text{the set of all ordered pairs (a,b)} \mid a \in A, b \in B\}.$$

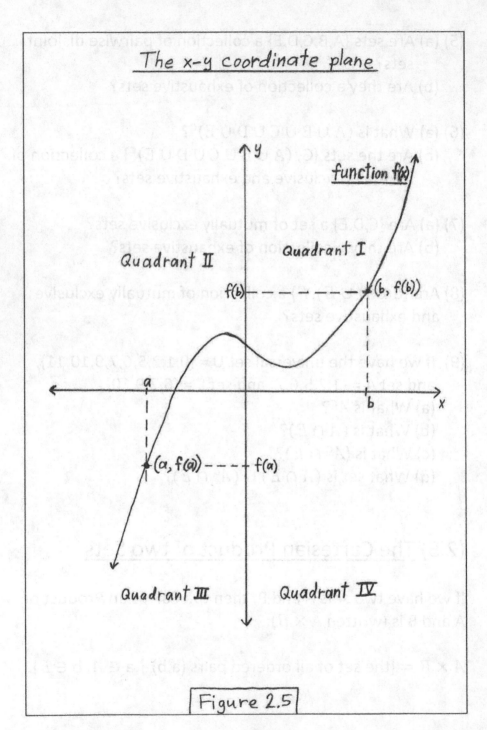

The x-y coordinate plane

function f(x)

Quadrant II

Quadrant I

f(b) — — — — — — — (b, f(b))

a

b

x

(a, f(a)) — — — — f(a)

Quadrant III

Quadrant IV

Figure 2.5

This can be generalized to n sets. The main use of this concept in applied mathematics is in defining so-called n-dimensional (Euclidean) space R^n:

$R^n = R \times R \times \cdots \times R$ (n sets of real numbers)
   = {the set of all n-tuples $(a_1, a_2, \ldots, a_n)$, | all $a_i \in R$}

R^1 is the real number line (Real numbers).
R^2 is the two-dimensional coordinate plane, what we call
   the x-y coordinate plane. Every point in the plane can be
   represented by an ordered pair of real numbers. See
   Figure 2.5 on the previous page.
R^3 is the three-dimensional coordinate space, where every
   point in three dimensional space can be represented by
   a 3-tuple of real numbers, called its coordinates.

(2.6) <u>Rational Numbers and Countable Infinity</u>

<u>The Rationals</u>
The rational numbers comprise a collection of numbers that
we could call "ordinary numbers," numbers that we are all
familiar with in the everyday world. We are all familiar
intuitively with 10 chairs, $25.15 for dinner at our favorite
restaurant, $\frac{1}{4}$ of a pumpkin pie, and a 20% discount on a new
shirt. All of these things involve rational numbers, or what I
have called the set of "ordinary numbers." We are not so in
tune with $\sqrt[3]{947}$ acres, or 15π to 3 odds of winning a bet.
These things involve irrational numbers, which are not what

we have called "ordinary numbers." We do need to know about irrational numbers to more fully understand the real numbers, but that will come later.

Rational numbers are all numbers which can be expressed as a ratio of two integers. That is, if x is a rational number, then there exists two integers a and b such that $x = \frac{a}{b}$. The integer 923 can be written as $\left(\frac{923}{1}\right)$, 45.173 can be written as $\left(\frac{45,173}{1000}\right)$, and 0.0035 can be written as $\left(\frac{35}{10,000}\right)$.

Rational numbers always have a decimal expansion which repeats some pattern after a certain point, even if the number repeats an infinite number of zeroes, as is the case with 2.45 = (2.450000 . . .). To see why this must be true, consider the fraction $\frac{257}{99}$. To find its corresponding decimal expansion we can do the long division. In this case, at each step of the division, the remainder is one of the numbers {0, 1, 2, . . . , 98}. There are 99 possible remainders, a large number of remainders, but still a finite number of them. So inevitably, in the long division, we must get a remainder which was a previous remainder and a repeating pattern would develop. In the case of $\frac{257}{99}$, we would find that $\frac{257}{99}$ is (2.59595959 . . .), which we write as $2.\overline{59}$. The bar is over the part which repeats. This is how we find the repeating decimal expansion for any ratio of two integers. Now, we need to show how to find a ratio of two integers that corresponds to a given repeating decimal expansion.

Example (1):
Consider the decimal expansion $(61.371371371\ldots) = 61.\overline{371}$. Let $n = 61.\overline{371}$. Then $1000n = 61,371.\overline{371}$. Subtracting n from 1000n, we would have:
$999n = (61,371.\overline{371} - 61.\overline{371}) = 61310$, or
$999n = 61310$. So, $n = 61.\overline{371} = \left(\frac{61310}{999}\right)$.

Example (2):
Consider the decimal expansion $(5.2939393\ldots) = 5.2\overline{93}$.
Let $n = 5.2\overline{93}$. Then $10n = 52.\overline{93}$, $1000n = 5293.\overline{93}$.
So $1000n - 10n = 990n = (5293.\overline{93} - 52.\overline{93}) = 5241$.
So $n = 5.2\overline{93} = \left(\frac{5241}{990}\right)$.

Countable Infinity
There is a very important distinction to be made between the rational numbers and the real numbers, and it concerns infinity. It turns out that they constitute two different types of infinity. The rational numbers are countably infinite and the real numbers are uncountably infinite. We will now discuss this further.

The simplest type of infinity is the infinity of the natural numbers $N = \{1, 2, 3, \ldots\}$. We call this countable infinity. A countably infinite set is one which can be put into a 1-1 (one-to-one) correspondence with this set N. Consider sets:

$M_1 = \{5, 10, 15, \ldots\}$
$M_2 = \{1000, 2000, 3000, \ldots\}$

M_1 is the set of all positive multiples of 5. M_2 is the set of all positive multiples of 1000. It is easy to see that there is a 1-1 correspondence between M_1 and N and between M_2 and N by considering the functions f(n) and g(n):

f(n) = 5n, for (n = 1, 2, 3, . . .) and
g(n) = 1000n, for (n = 1, 2, 3, . . .).

What these 1-1 functions show is that there are just as many numbers in M_1 as there are in N, and there are just as many numbers in M_2 as there are in N. We have $M_2 \subset M_1 \subset N$, but the card($M_1$) = card($M_2$) = card(N).

We can also set up a 1-1 function h(n) from M_1 to M_2 in the following way: h(n) = 200n, for (n = 5, 10, 15, . . .). This demonstrates that there are the same number of members in the two sets M_1 and M_2.

We can set up a 1-1 correspondence between N and W using the function j(n): $j(n) = \begin{cases} 0 & \text{for } n = 1 \\ (n-1) & \text{for } n = 2, 3, 4, ... \end{cases}$
So, W has the same cardinality as N, even though $N \subset W$.

As for the integers Z = {0, 1, -1, 2, -2, . . .}. It is easy to see the 1-1 correspondence between Z and N using k(n):

$$k(n) = \begin{cases} 0, & \text{for } n = 1 \\ (^n/_2) & \text{for } n = 2, 4, 6, ... \\ -(^{(n-1)}/_2) & \text{for } n = 3, 5, 7, ... \end{cases}$$

We will denote the cardinality of any countably infinite set with the transfinite number α_1. A countably infinite set can in principle be listed. When two sets have the same number of members, we say that they are equivalent. This is to be distinguished from equality, which means that both sets have the exact same members. Something interesting to note from above is that an infinite set can be equivalent to a subset of itself. It is easy to see this for countably infinite sets. Later, we will see that it is true for uncountably infinite sets also. We can let this be a definition of an infinite set.

We have now arrived at what may be a surprising result, and is the focus of this section. The rational numbers are countably infinite also. How could this be? All possible fractions? There are an infinite number of them between any two?

We will now show how this is actually true. We will show that the rationals are countably infinite through a procedure that in principle allows us to list them. To make things easier, we will represent the rational number $\frac{a}{b}$ with the ordered pair (a,b), where of course a and b are integers with $b \neq 0$.

By following the indicated path through the array shown on the next page, we can in principle list all possible fractions. We know that for every fraction there are an infinite number of other fractions that represent the same rational number. For example, the set of fractions $\left\{\frac{2}{3}, \frac{4}{6}, \frac{6}{9}, \cdots\right\}$ all represent the same rational number $\frac{2}{3}$, which is in lowest terms. So, when we follow the indicated path through this array and discard repeats, we have a listing of all the rational numbers Q:

$(0,1) \rightarrow (1,1) \quad (-1,1) \rightarrow (2,1) \quad (-2,1) \rightarrow (3,1) \quad (-3,1) \quad \cdots$
$\swarrow \qquad \nearrow \qquad \swarrow \qquad \nearrow \qquad \swarrow$
$(0,-1) \quad (1,-1) \quad (-1,-1) \quad (2,-1) \quad (-2,-1) \quad (3,-1) \quad (-3,-1) \quad \cdots$
$\downarrow \nearrow \qquad \swarrow \qquad \nearrow \qquad \swarrow$
$(0,2) \quad (1,2) \quad (-1,2) \quad (2,2) \quad (-2,2) \quad (3,2) \quad (-3,2) \quad \cdots$
$\swarrow \qquad \nearrow \qquad \swarrow$
$(0,-2) \quad (1,-2) \quad (-1,-2) \quad (2,-2) \quad (-2,-2) \quad (3,-2) \quad (-3,-2) \quad \cdots$
$\downarrow \nearrow \qquad \swarrow$
$(0,3) \quad (1,3) \quad (-1,3) \quad (2,3) \quad (-2,3) \quad (3,3) \quad (-3,3) \quad \cdots$
$\swarrow$
$(0,-3) \quad (1,-3) \quad (-1,-3) \quad (2,-3) \quad (-2,-3) \quad (3,-3) \quad (-3,-3) \quad \cdots$
$\downarrow \qquad \cdot \qquad \cdot \qquad \cdot \qquad \cdot \qquad \cdot \qquad \cdot$

So, the rationals Q = {0, 1, -1, 2, $\frac{1}{2}$, $-\frac{1}{2}$, -2, 3, $\frac{1}{3}$, . . .}.

The cardinality of the rationals is also α_1. It has exactly the same number of members as any of the other countably infinite sets that we have encountered in this section. Such is the quite different nature of entities that we call infinite sets. Now, we will show that there is yet a higher type of infinity, and we will call it uncountable infinity. Countable and uncountable infinity are the only types of infinity that we will discuss (or need to discuss in this book) because they are the only types of infinity that are involved with the real and complex numbers.

(2.7) Real Numbers and Uncountable Infinity

Irrational Numbers

Before we get to the real numbers, we have to first discuss the irrational numbers, a set of numbers which we will denote as IRR. Irrational numbers are numbers that cannot be written as the ratio of two integers. I want to reproduce in modern form a classical proof that the $\sqrt{2}$ is irrational. It was the ancient Greeks that first discovered the existence of irrational numbers with a proof similar to this one. The Greeks believed that all quantities were what we (or they) called commensurate. This means that if we had two lengths a and b, that there existed two whole numbers m and n such that m lengths of size "a" would equal n lengths of size "b". Using our modern algebra, this means that:

$$ma = nb.$$

So, the ratio of the lengths $\frac{a}{b}$ would be the rational number $\frac{n}{m}$, after dividing both sides of the previous equation by b and then m. The Greeks ran into a problem when they considered a square with sides of length 1 and they wanted to determine the length of the diagonal. Using the result called the Pythagorean theorem, which was known to them at the time of this proof, the length of the diagonal is:

$$\sqrt{1^2 + 1^2} = \sqrt{2}\,.$$

They came up with a proof that the $\sqrt{2}$ was not rational, using a revolutionary technique attributed to Greek (or some mediterranean mathematicians of those times) which we now call "Proof by Contradiction." We also refer to it as "Reductio ad Absurdum," which means a Reduction to Absurdity. It is a proof technique that is well known today and is commonly used by all undergraduate math majors. In a Proof by Contradiction, you assume that the negation of the result that you want to prove is true and show that this leads to a contradiction of your hypotheses (starting points that you are assuming to be true). Therefore the result that you want to prove is true. The proof goes like this:

<u>Theorem</u>: The $\sqrt{2}$ is irrational.

<u>Proof</u>: Assume that the $\sqrt{2}$ is rational, that is $\sqrt{2} = \frac{a}{b}$, where a and b are positive integers that are in lowest terms. Then squaring both sides, we get $2 = \frac{a^2}{b^2}$, or $a^2 = 2b^2$.

*****

At this point in the proof we need to explain that all whole numbers have a unique factorization into prime numbers. Prime numbers are positive whole numbers whose only divisors are 1 and themselves, like 7 or 31, for example. The first few primes are: 2,3,5,7,11, . . . and so on. There are an infinite number of primes. Numbers that are not prime are called composite, like 24 or 100, for example. The prime factorization of 24 is $24 = (2 \cdot 2 \cdot 2 \cdot 3)$, and the prime factorization of 100 is $100 = (2 \cdot 2 \cdot 5 \cdot 5)$.

*****

Now, back to the proof. We see that a^2 is divisible by 2. So, 2 is in the prime factorization of a^2, which means that 2 is in the prime factorization of "a" (think about why this would be true). Therefore, 2 divides "a" evenly. So, there is a whole number "h" such that a = 2h. This means then that $a^2 = 4h^2$. But from above, $a^2 = 2b^2$, so $4h^2 = 2b^2$, which means that $b^2 = 2h^2$. So, it follows that 2 divides b^2 and hence 2 divides "b" evenly, for the same reason that 2 divides "a" evenly (by referring to prime factorizations). So, we have that 2 divides "a" and 2 divides "b." This is a contradiction of our hypothesis that a and b were two whole numbers in lowest terms. Therefore, we can conclude that the $\sqrt{2}$ is not rational. So, we call $\sqrt{2}$ an irrational number.

This is how it was proved that irrational numbers exist. This was a result that was shocking to the Greeks, but nowadays we take it more in stride. To us, what is important to know about irrationals is that they have decimal representations which never repeat any pattern. Examples of irrational numbers do not roll off the tongue as easily as rationals do. However, with the help of a calculator, some examples are:

$\sqrt{3} = 1.73205080\ldots$
$\sqrt[5]{83} = 2.42000140\ldots$

$\pi = 3.141592654\dots$
$e = 2.71828182\dots$

π is of course the ratio of the circumference of any circle to its diameter, and "e" is the base of the natural logarithms, which are very important numbers and are discussed later.

Real Numbers

The Real numbers are the union of the two disjoint sets: the rationals Q and the irrationals IRR. In set notation,

$$R = (Q \cup IRR)$$

These are all of the numbers that are on the real number line. Sometimes mathematicians just refer to it as the real line. The real numbers are an ordered set of numbers, meaning that if we have any two reals x and y, then only one of the following must be true: $x < y$, $x = y$, or $x > y$. The real numbers constitute what we call in algebra, an ordered field. We will now explain the properties of a field. The Complex numbers, which we will discuss in the next section are a field of numbers also, but they are not ordered.

The properties that are satisfied by an algebraic field are listed below as properties (1a) through (3). They will not seem earth shattering to many readers, but they present the basic ideas and principles that our ordinary algebra is built upon. At the very least, it should give a name to many of the properties of the real numbers that are already familiar to most readers.

The set of real numbers along with the operations of addition and multiplication form an algebraic field. The properties pertaining to addition will be listed first, then the properties pertaining to multiplication, and then finally we state the distributive property of real numbers:

(1a) If x and y are any two real numbers, then (x + y) is also a real number. This means that the real numbers are closed with respect to addition.

(1b) If x and y are any two real numbers, then (x + y) = (y + x). This is the commutative property of addition.

(1c) If x, y, and z are any three real numbers, then (x + y) + z = x + (y + z). This is the associative property of addition.

(1d) For any real number x, (x + 0) = (0 + x) = x. The number 0 is called the additive identity.

(1e) For any real number x, there exists a real number (-x) such that (x + (-x)) = ((-x) + x) = 0. (-x) is called the additive inverse of x.

(2a) If x and y are any two real numbers, $(x \cdot y)$ is another real number. This means that the real numbers are closed with respect to multiplication.

(2b) If x and y are any two real numbers, $(x \cdot y) = (y \cdot x)$. This is the commutative property of multiplication.

(2c) If x, y, and z are any three real numbers, then $(x \cdot y) \cdot z = x \cdot (y \cdot z)$. This is the associative property of multiplication.

(2d) For any real number x, $(x \cdot 1) = (1 \cdot x) = x$. The number 1 is called the multiplicative identity.

(2e) For any real number x ≠ 0, there exists a real number $\left(\frac{1}{x}\right)$ such that $(x) \cdot \left(\frac{1}{x}\right) = \left(\frac{1}{x}\right) \cdot (x) = 1$. $\left(\frac{1}{x}\right)$ is called the multiplicative inverse of x.

(3) For any three real numbers x, y, and z, $x \cdot (y + z) = (x \cdot y) + (x \cdot z)$. This is called the distributive property. We say that multiplication distributes over addition.

These 11 properties of the real numbers, along with some other definitions and notations developed along the way, can be used to develop all of our modern algebraic methods and results that we usually call "High School Algebra."

Uncountable Infinity

The main result of this section is to show that the real numbers constitute an infinite set of numbers which are believed to be of the next higher type of infinity, with cardinality α_2, that we call uncountable infinity. There has been much energy expended to find a type of infinity in between countable and uncountable infinity, but all attempts to do that have failed. The uncountably infinite collection of real numbers is believed to be a continuum. For every point on the number line there corresponds only one real number, and for every real number there is only one corresponding point on the number line. The set of rationals do not form a seamless continuum. If the number line consisted only of rationals, the number line would be, figuratively speaking, like swiss cheese, with holes in it. An entirely greater infinity of numbers, the irrationals, would be missing.

We will now very easily show that this is true by using another Proof by Contradiction. Let's restrict ourselves to real numbers on the interval [0,1], that is all numbers x such that $0 \leq x \leq 1$.

Theorem: The real numbers in [0,1] are not countable.

Proof: Assume that they are countable. Then they could be listed. We will assume that $0 = (0.0000000000 \ldots)$ and that $1 = (0.9999999999 \ldots)$. Let's write this list as:

$$0. a_1 a_2 a_3 a_4 \ldots$$
$$0. b_1 b_2 b_3 b_4 \ldots$$
$$0. c_1 c_2 c_3 c_4 \ldots$$
$$0. d_1 d_2 d_3 d_4 \ldots$$
$$\vdots \qquad \vdots$$
$$\vdots \qquad \vdots$$

where each of the subscripted letters is one of our numeric digits {0, 1, 2, 3, 4, 5, 6, 7, 8, 9}.

Now, construct a number $z = (0. z_1 z_2 z_3 z_4 \ldots)$ in the following way:

Choose $z_1 \neq a_1$, $z_2 \neq b_2$, $z_3 \neq c_3$, $z_4 \neq d_4$, and so on. Then z will differ from every number on the list in at least one decimal place. In other words, z is not on the list. This is a contradiction of our hypothesis that all of the numbers in [0,1] could be listed. We could add z to the list and do the same thing over and over again. It then becomes clear to us that the infinity of the real numbers in [0,1] is quite different from the infinity of the rationals in [0,1]. Mathematicians call this higher type of infinity so-called uncountable infinity.

Since all real numbers between 0 and 1 are uncountably infinite, then clearly the entire set of real numbers is also uncountably infinite, and they have cardinality α_2. To show this we can construct a 1-1 function from the interval (0, 1) to the real numbers by letting $y = f(x) = \tan\left(-\frac{\pi}{2} + (x)(\pi)\right)$, where x is in the interval (0,1) and y ranges from $-\infty$ to ∞. The reader may find it helpful to see figure 5.3 with the graph of the tangent function in Section 4.1 in order to understand this last result.

It turns out that there is an infinite hierarchy of types of infinity. There is an infinity of objects that is greater than the infinity of the real numbers, and it has cardinality α_3. Then there is a next higher infinity of objects with cardinality α_4, and so on indefinitely. In this book, we are concerned mainly with the real numbers, so we do not need to concern ourselves with these orders of infinity that are higher than countable and uncountable infinity.

Measure

We should mention that the rational numbers are dense in the real numbers. This means that every real number is either rational or it is the limit of a sequence of rationals. So, every irrational number is the limit of a sequence of rationals. Recall that π was the irrational number written (3.141592654 . . .) The irrational number π is the limit of the sequence:

$$\left\{3, \frac{31}{10}, \frac{314}{100}, \frac{3141}{1,000}, \frac{31415}{10,000}, \frac{314159}{100,000}, \ldots\right\}$$

where each of the members of this sequence is a rational number. (In fact, π is the limit of many such similar rational sequences that converge to π). Nevertheless, it is true that there actually is in some sense (thinking in terms of infinity) many more irrational numbers than there are rational numbers. Mathematicians have developed a branch of mathematical analysis known as measure theory, where they use more advanced math than we want to use here, to define the measure of a real number set. It is like the measure of the collective width of the numbers in the set. The measure of the real numbers in an interval of real numbers [a,b] is (b − a). For example, the measure of the real number interval [3,13] is (13 - 3) = 10 units. This uses our usual measure of distance. What can be shown quite easily is that the measure of a countable set within the real numbers is always 0. This says that the measure of the rational numbers within any interval [a,b] is 0. In fact, the measure of the rational numbers in (-∞, ∞) is 0. In other words, from a measure theoretic point of view, it is as if the rational numbers are not even there. That is the reason that we said earlier that the real number line with only rational numbers on it would be like swiss cheese, with holes in it. The irrational numbers make up all the difference, and the measure of the irrational numbers in [a,b] is (b − a)!

One way that this idea can be applied is in probability theory. We could draw a part of the real number line on the floor and drop a nickel on the line and ask what is the probability that the nickel lands on a rational number? If we could figure with infinite precision the exact number that the nickel lands on, we would discover that the probability is 0 (it is impossible) that it would land on a rational number, and the probability is 1 (it is certain) that it would land on an irrational number!

(2.8) The Complex Numbers

As stated in a previous section, the complex numbers C are all numbers of the form (a + bi), where a and b are real numbers and i is the imaginary unit; $i = \sqrt{-1}$. A complex number (a + bi) can be thought of as a number in a plane, which we call the complex plane. Even though (a + bi) is located at what we might call the point with coordinates (a,b), this complex number is not generally thought of as a two dimensional ordered pair of coordinates in the x-y plane (two-dimensional Euclidean space), but rather just the point (a,b) itself. Nevertheless, we can still think of it as a two-dimensional number. We can consider the real numbers (all complex numbers with b = 0) to be along a horizontal number line with the number 1 as the unit, and all the purely imaginary numbers (all the complex numbers except the origin with a = 0) to be along a vertical number line with the number $i = \sqrt{-1}$ as the unit. The number (a + bi), is "a" units in the real direction and "b" units in the imaginary direction, just like in the usual coordinate plane.

The complex numbers are not ordered, but they satisfy all the same field properties that the real numbers do. The arithmetic of complex numbers is very much like the way that we deal with binomial expressions in our ordinary algebra (see section 3.3 for the expansion of binomial expressions). The only thing that is different in complex number arithmetic is simply that $i^2 = \left(\sqrt{-1}\right)^2 = -1$.

We will now consider the arithmetic of the complex numbers: That is sums, differences, products, and quotients of complex numbers. We will also consider the so-called magnitude |z| for the complex number z along with the so-called complex conjugate $\bar{z}$ of z. The complex conjugate is important for considering the quotient of two complex numbers. One of the major differences in the complex arithmetic is that we can take the square root of negative numbers. For example,

$$\sqrt{-36} = \sqrt{36} \cdot \sqrt{-1} = \pm 6i .$$

(1) If we take the sum or difference of two complex numbers $z_1 = (a + bi)$ and $z_2 = (c + di)$, we have:

$$z_1 + z_2 = (a + bi) + (c + di)$$
$$= a + c + bi + di$$
$$= (a + c) + (b + d)i, \qquad \text{and}$$

$$z_1 - z_2 = (a + bi) - (c + di)$$
$$= a - c + bi - di$$
$$= (a - c) + (b - d)i$$

which can be thought of as the point with coordinates $((a + c), (b + d))$ in the complex plane when we consider the sum, or as the point $((a - c), (b - d))$ when we consider the difference.

(2) If we take the product of two complex numbers $z_1 = (a + bi)$ and $z_2 = (c + di)$, we have:
$$z_1 \cdot z_2 = (a + bi) \cdot (c + di) = ac + bci + adi + bdi^2$$
$$= ac + (bc + ad)i + bd(-1)$$
$$= (ac - bd) + (bc + ad)i$$

which can be thought of as the point with coordinates $((ac - bd), (bc + ad))$ in the complex plane.

(3) The magnitude of a complex number $z = (a + bi)$ is the distance from the origin to the number and it is denoted by $|z| = \sqrt{a^2 + b^2}$. Note that there are an infinite number of complex numbers that have this particular magnitude; they all lie on a circle centered at the origin.

The complex conjugate for the complex number $z = (a + bi)$ is $\bar{z} = (a - bi)$, and the product of complex conjugates is: $z \cdot \bar{z} = (a + bi) \cdot (a - bi) = a^2 - b^2(i^2)$, which is $(a^2 + b^2)$. Note that the product of complex conjugates is the square of the magnitude of the complex number z, and is a real number. That is: $z \cdot \bar{z} = |z|^2$. So, $|z| = \sqrt{(z \cdot \bar{z})}$.

(4) If we take the quotient of two complex numbers $z_1 = (a + bi)$ and $z_2 = (c + di)$, we have to multiply the numerator and the denominator by the conjugate of the denominator: (of course z_2 is not 0)

$$z_1 \div z_2 = \frac{(a+bi)}{(c+di)} = \frac{(a+bi)}{(c+di)} \cdot \frac{(c-di)}{(c-di)}$$

$$= \left(\frac{ac + bd}{c^2 + d^2}\right) + \left(\frac{bc - ad}{c^2 + d^2}\right) i$$

which can be thought of as the point with coordinates $\left(\left(\frac{ac + bd}{c^2 + d^2}\right), \left(\frac{bc - ad}{c^2 + d^2}\right)\right)$ in the complex plane.

The complex numbers with their imaginary parts, square roots of negative numbers, and all that, at first sound sort of unusual and exotic, and we can question their relevance in the real world. However, they are of great use to those mathematicians and scientists working with math at a high level. They are simply a more general type of number that contains the real numbers as a special case, which allows us to solve problems and gain new insights into the world of math and science that we otherwise would not be able to do. The terms imaginary and complex are just unfortunate historical misnomers. The complex and imaginary numbers are just as "real" as the real numbers are. After all, they are just a set of symbols which can be manipulated according to certain rules, and they form a consistent whole, an algebraic field.

Now, suppose we want to solve the following equation

$$x^2 = 2.$$

If we are a child in elementary school and only have the rational numbers in our mathematical language, we would not be able to find a solution to that problem. The question would seem meaningless. However, the middle school student has learned that the real number line also contains the so-called irrational numbers which have non-repeating infinite decimal expansions. The exact value of the irrational number $\sqrt{2}$ can be determined to any desired degree of accuracy, and $\left(\sqrt{2}\right)^2 = \left(-\sqrt{2}\right)^2 = 2$. So, our equation has these as its two solutions.

Now if we ask that middle school student to find the solution to

$$x^2 = -2,$$

that student would not be able to find a solution. The question would seem meaningless. How can something squared be negative? But the high school student has learned that we can define a more general type of number, the complex numbers, which contain the real numbers with all their familiar rules of association as a special case, and that we can now speak of the square root of a negative number. So, the last equation above can be solved. It has the two solutions $\sqrt{2}i$ and $-\sqrt{2}i$ in this expanded number system that we call the complex number system.

The complex numbers are of use to us because they allow us to solve certain problems that we cannot solve otherwise. As a more general type of number, they provide us with a more general understanding of the world. If they did not provide meaningful solutions to certain math problems of interest to us, then we would simply put them away somewhere and not bother ourselves with them. But that is not the case. The set of complex numbers can be shown to satisfy exactly the same rules of association as any other algebraic field, like the real numbers do (once again simply a historical misnomer). Once we get over our reluctance to accept that the square root of a negative number is a "real" entity in a mathematical sense, then we can expand our minds and accept the reality and usefulness of the complex numbers.

The main result that is worth mentioning about the complex numbers, in the algebra encountered in this book, is that the complex field is algebraically complete. What this means is that all polynomials of degree n with real coefficients always have n complex roots. Often some of these roots are real, with complex roots always occurring in complex conjugate pairs. For example, the polynomial equation:

$$x^3 - 3x^2 + x - 3 = 0 \text{ can be factored into the form:}$$

$$(x - 3)(x^2 + 1) = (x - 3)(x - i)(x + i) = 0 .$$

This has only one real root, x = 3, and the two complex conjugate roots are x = i, and x = $-i$. We have a total of n = 3 complex roots when we remember that a real number is just a special type of complex number.

################ Exercises ##################

Perform the following complex number calculations:

(1) $\left(17 - \frac{1}{2}i\right) + \left(2 - \frac{1}{4}i\right)$ (2) $\left(9 + \frac{1}{3}i\right) + \left(8 - \frac{2}{9}\right)i$

(3) $(10 - 7i) - \left(11 - \frac{1}{7}i\right)$ (4) $(1 + i) - (1 - i)$

(5) $(4 + 2i)(4 - 3i)$ (6) $(2 + 3i)(3 - 9i)$

(7) $\frac{(1-3i)}{(2+6i)}$ (8) $\frac{(-7i)}{(18i)}$ (9) $\frac{1}{7i}$

(3) FUNDAMENTALS OF ALGEBRA & GEOMETRY

***** ALGEBRA *****

(3.1) Introduction

The Power of Algebra

Before the 17th Century, mathematicians did not have much algebra to work with. The mathematicians of the early 17th century developed most of what we would recognize as modern algebra along with analytic geometry. By the end of that century, Isaac Newton, Gottfried Leibniz, and many others had developed calculus, primarily because they had the algebra to work with. The calculus introduced a lot of new mathematical notation, much of which survives to the present day.

If one should wonder just how important of an advance the development of algebra is, they should think carefully about the following problem:

(3) <u>FUNDAMENTALS OF ALGEBRA & GEOMETRY</u>

***** ALGEBRA *****

(3.1) <u>Introduction</u>

<u>The Power of Algebra</u>
Before the 17th Century, mathematicians did not have much algebra to work with. The mathematicians of the early 17th century developed most of what we would recognize as modern algebra along with analytic geometry. By the end of that century, Isaac Newton, Gottfried Leibniz, and many others had developed calculus, primarily because they had the algebra to work with. The calculus introduced a lot of new mathematical notation, much of which survives to the present day.

If one should wonder just how important of an advance the development of algebra is, they should think carefully about the following problem:

Suppose we are told that John is one half of what Laura's age will be 14 years from now. Laura is 12 years younger than Edgar will be 5 years from now. And, finally Edgar is 10 years older than John will be next year.

If you are asked to find John's, Laura's, and Edgar's current ages, most people with no proficiency in mathematics would probably say: "Oh No!!! Just forget it!" In fact, it would be a head scratcher even for mathematicians if they were told that they could not introduce any algebraic symbolism in trying to solve it. But if these mathematicians were allowed to use algebra, they would probably do something like this:

Let J, L, and E represent John, Laura, and Edgar's present ages. Reading the statement of the problem, they would probably write:

$$J = \frac{1}{2}(L + 14) \qquad\qquad J = \frac{1}{2}L + 7$$

$$L = (E + 5) - 12 \quad\Longrightarrow\quad L = E - 7$$

$$E = (J + 1) + 10 \qquad\qquad E = J + 11$$

Now substituting $(J + 11)$ for E in the second equation, we have the two equations:

$$J = \frac{1}{2}L + 7 \qquad\qquad J - \frac{1}{2}L = 7$$

$$\Longrightarrow$$

$$L = J + 4 \qquad\qquad -J + L = 4$$

Adding the last two equations together, we get, $\frac{1}{2}L = 11$, which says that L = 22. Going Back to L = J + 4, it is clear that J = 18. Then going back to E = J + 11, we see that E = 29. So, the present ages of John, Laura, and Edgar are 18, 22, and 29 years respectively. All so easy! All solved within a minute or two!

What I am trying to say is that algebra is one of the greatest innovations in the history of mathematics and science. The solving of mathematical problems took a giant leap forward with the advent of algebra. In the case above, a little algebra makes solving the problem, which many people would not even care to try without it, quite trivial.

Some Fundamental Ideas about Algebra
To understand probability and statistics, and anything higher in mathematics, you will not go far if you don't understand algebra. It is so central to applied math. So, we include a review of the basic concepts and methods of this important subject. Despite the fact that in this chapter we will cover a lot of things that will not be used later in the book, the reader is encouraged to give most of it a thorough reading and to work the exercises that are included. You can never get too much exercise with and understanding of the many aspects of algebra. It should help your mathematical ability in general.

As one can see from the example above, concerning the

ages of {John, Laura, and Edgar}, algebra involves the simple representation of an unknown quantity with a letter, which we call a variable. In addition, we have various rules for manipulating these variables. The purpose of solving an equation is usually to find a value for the unknown variable which makes the equation a true statement. In a particular setting, quantities which do not vary are called constants. Constants are numbers, or symbols that represent specific numbers. Constants can take different values in different applications of the same equation. In a lot of the equations that we will encounter, the unknown variables are often denoted by x, y, or z, and constants are often denoted with a, b, or c, and so on. When we want to know the solution of an equation like the following:

$$3x - 40 = -20$$

we are looking for a number x such that when we multiply x by 3, and then subtract 40 from that result, we get -20. When we want to know the solution of an equation like the following:

$$x^2 - x + 3 = 0$$

we are looking for a number x such that when we multiply it by itself, then subtract the number x from that result, and then add 3 to that result, we get 0.

Learning the techniques for solving equations like these, and of course many more things in algebra, is indispensable for the subjects of probability and statistics and for so much more if the student needs to go further in mathematics and science. For the following equation:

$$y = ax^2 + bx + c$$

the letters a, b, and c would usually represent three given real numbers. We may want to determine the nature of the solution set:

{all points (x, y) in the plane | $y = ax^2 + bx + c$}.

The solution set here is a parabolic curve in a coordinate plane, which we call the x-y coordinate plane. Things such as this are very important problems for students of the various sciences, engineering, and technology.

The reader should understand that a major advantage that we get from algebra is that the solution of an equation involving variables and constants can be applied to many different problems of the same general kind. We don't need to solve different problems of the same general form with a completely new line of reasoning.

For example, many problems in scientific applications involve a linear relationship between two variables, which we can call x and y. A symbolic formulation of this relationship can always be written in the following way: y = mx + b, for some constants m and b. No matter what x and y may represent, if there is a linear relationship between x and y, we can always represent it in this way. Only the two constants m and b will be different in each individual case. Anything that we can say about one linear relationship between variables x and y, when it is expressed symbolically like this, is the same as what we can say about any other linear relationship between two variables x and y.

As another example, many problems in scientific applications involve a quadratic relationship between two variables, that we can call x and y. A symbolic formulation of this relationship can always be written in the following way: $y = ax^2 + bx + c$, for some constants a, b, and c. It doesn't matter what x and y may represent, if there is a quadratic relationship between x and y, we can always represent it in this way. Only the three constants a, b, and c will be different in each individual case. Anything that we can say about one quadratic relationship between variables x and y, when it is expressed symbolically like this, is the same as what we can say about any other quadratic relationship between two variables x and y. So, when we develop a technique of solution for a quadratic equation that involves only symbols, then we can use this technique to solve many different quadratic equations. A simple and concise algebraic expression or equation can be applied to finding the solution of many different problems of the same form. This is important because in applied problems that scientists and mathematicians encounter, there are many different patterns that commonly repeat themselves.

Another major advantage of algebra is that the ways that symbols relate to each other and combine with each other is always the same and based on logic. Mathematics is a language with logical relations built in, because the logical relations that are present between the symbols in an algebraic equation are the same as the logical relations which exist between numbers in arithmetic.

Finally, we should realize that algebraic expressions and equations easily facilitate our ability to work with and see relationships between two or more quantities (variables). We had no ability to do this before it was invented. This aspect of mathematics has lead to the rapid expansion of science and technology that has occurred in the relatively short span of only the last three or four centuries.

(3.2) Algebraic Notation

Multiplication and Quotients

One thing that is new in algebra to the uninitiated is that we quite often use the juxtaposition of two quantities to mean multiplication. For example, even though we can certainly write $2 \cdot H$ or $2 \cdot (H)$ or $(2) \times (H)$ to indicate 2 times H, the way that this is commonly written in algebra is 2H. The product of 36 and x is usually written 36x, and so on. The product of $(44 - y)$ and $(y + 3)$, depending on the context, and possibly for emphasis, can be written as $(44 - y) \cdot (y + 3)$ or as $(44 - y) \times (y + 3)$, but we usually write $(44 - y)(y + 3)$.

To indicate the quotient of 100 and $(2x + 6)$, we sometimes use the notation $100 \div (2x + 6)$. It depends once again on the context and a need for emphasis. It would usually be expressed as $\frac{100}{(2x+6)}$. In algebra, we like to use fractions.

To indicate the quotient of the two fractions $\left(\frac{2}{x}\right)$ and $\frac{(4x+1)}{33}$, we may sometimes write $\left(\frac{2}{x}\right) \div \frac{(4x+1)}{33}$, but in algebra we often need to simplify expressions further, so we would usually say:

$$\left(\frac{2}{x}\right) \div \frac{(4x+1)}{33} = \left(\frac{\frac{2}{x}}{\frac{(4x+1)}{33}}\right) = \left(\frac{2}{x}\right) \cdot \left(\frac{33}{(4x+1)}\right)$$

$$= \frac{66}{(x)(4x+1)} = \frac{66}{4x^2 + x}.$$

Precedence of Operations

Note that multiplcations and divisions are normally done before additions and subtractions, unless we have parentheses to make clear what the precedence of operations is. For example, if we have $32 \cdot x + 4$, then this means that 32 is multiplied by x first and then 4 is added to that result. So we would normally write this as $(32x + 4)$ using the juxtaposition of 32 and x for multiplication. If we have $32 \cdot (x + 4)$, then this means $32(x + 4)$, where the sum of x and 4 in the parentheses is to be done first and then that result is to be multiplied by 32. Note the use here of the juxtaposition of 32 and the expression (x + 4) for the multiplication of these two quantities. We can simplify this further with the distributive law:

32(x + 4) = (32)·(x) + (32)·(4) which is 32x + 128.

As another example, if we have $9 \cdot x \cdot y + 4 \cdot x - 90 \cdot \left(\frac{3}{y}\right)$, without any other parentheses, then this would usually be expressed as $9xy + 4x - 90\left(\frac{3}{y}\right)$, where the multiplications are indicated by juxtaposition. The last product could be simplified, resulting in $9xy + 4x - \left(\frac{270}{y}\right)$.

Absolute Value
The absolute value of a number, or its magnitude, is the distance from the origin to the number. We use x within vertical lines such as |x| to denote the absolute value of x. For example, the |34| = 34, and the |-12| is 12. A more technical way of defining the absolute value of a quantity x is that the $|x| = \begin{cases} x & , & if\ x \geq 0 \\ -x & , & if\ x < 0 \end{cases}$.

Inequality
If x is less than y, we write x < y.
If x is less than or equal to y, we write x ≤ y.
If x is greater than y, we write x > y.
If x is greater than or equal to y, we write x ≥ y.

We can also have a multi-part inequality, such as $x < y < z$. This means that x is less than y and y is less than z, or that y is strictly between x and z (y cannot equal x or z).

For example, $(1 < x < 14)$ says that x is greater than 1 and less than 14, or x is between 1 and 14 (but x cannot be equal to 1 or 14).

If $(3 \leq x \leq 9)$, this says that x is greater than or equal to 3 and x is less than or equal to 9, or it could be said that x is between 3 and 9 inclusive ("inclusive" meaning that x could be 3 or 9, one of the endpoints).

If $(5 < x \leq 11)$, this says that x is greater than 5 and x is less than or equal to 11. In this case it is possible that x could equal 11 but x is definitely greater than 5.

Absolute Value and Inequality

If we have $|x| < 7$, this says that x is a quantity that is less than 7 units from the origin. When thinking of distance from the origin, think of $|x|$ as $|x - 0|$. So, $-7 < x < 7$.

If $|x| \geq 2$, then x is greater than or equal to 2 units from the origin. This can also be written: $x \leq -2$ or $x \geq 2$.

If $|x - 10| < 3$, then x is all numbers which are less than 3 units from 10 (Note: we are not considering distance from the origin in this case). In terms of inequalities, this means $7 < x < 13$.

If $|x - 25| \geq 5$. This says that x is a number greater than or equal to 5 units from 25. So, this means $x \leq 20$ or $x \geq 30$.

Exponential Notation

When we multiply a quantity by itself a certain number of times, there is a special notation for this. For example, $(x)(x) = x^2$, and this notation is pronounced "x-squared." $(x)(x)(x) = x^3$, and this notation is pronounced "x-cubed." $(x)(x) \cdots (x) = x^n$, (n factors of x) is "x to the nth power."

Laws of Exponents

(1) If we multiply $x^2 \cdot x^3 = (x)(x) \cdot (x)(x)(x)$
$$= x^{2+3} = x^5.$$
This prompts the law of exponents that says:
$x^n \cdot x^m = x^{n+m}$, where n and m are real numbers.

(2) If we have $(x^2)^3 = (x)(x) \cdot (x)(x) \cdot (x)(x)$
$$= x^{2 \cdot 3} = x^6.$$
This prompts the law of exponents that says:
$(x^n)^m = x^{n \cdot m}$, where n and m are real numbers.

(3) Dividing x^3 by x^5, we have $\dfrac{x^3}{x^5} = \dfrac{(x)(x)(x)}{(x)(x)(x)(x)(x)} = \dfrac{1}{x^2}$
$$= x^{-2} = x^{3-5}.$$
This prompts the law of exponents that says:
$\dfrac{x^n}{x^m} = x^{n-m}$, where n and m are real numbers.

(4) As a special case of law (3), $x^{-n} = x^{0-n} = \dfrac{x^0}{x^n} = \dfrac{1}{x^n}$.

It follows also that $x^n = \dfrac{1}{x^{-n}}$, by taking reciprocals.

(5) The nth root of x is denoted $x^{\frac{1}{n}}$. As special cases, we have the $\sqrt{x} = x^{\frac{1}{2}}$, the $\sqrt[3]{x} = x^{\frac{1}{3}}$, and so on.

(6) The notation $x^{\frac{m}{n}} = \sqrt[n]{(x^m)} = \left(\sqrt[n]{x}\right)^m$. Both of these forms give the same result. As an example,

$(8)^{\frac{7}{3}} = \left(\sqrt[3]{8}\right)^7 = (2)^7 = 128$, and

$(8)^{\frac{7}{3}} = \sqrt[3]{(8)^7} = \sqrt[3]{(2,097,152)} = 128$.

Many times, one form is easier to evaluate than the other.

Logarithmic Notation
A logarithm is an exponent. If "a" is an allowable base, then the notation $\log_a x$ represents an exponent y such that $a^y = x$. Allowable bases are numbers "a", where a > 0, but a ≠ 1.

If a is 10, then we are using common logarithms. If a is "e," then we are using natural logarithms. We mentioned the number "e" in the section on irrationals. $e \approx 2.71828$. "e" is one of those numbers that crops up a lot in mathematics like the number π does. We have a special notation for natural logarithms. We use the notation ln(x) or lnx to indicate the $\log_e x$.

<u>Laws of Logarithms</u>
For real numbers x and y, where x > 0, and y > 0,

(1) $\log_a(xy) = \log_a x + \log_a y$.

(2) $\log_a\left(\frac{x}{y}\right) = \log_a x - \log_a y$.

(3) $\log_a(x)^r = r \log_a x$.

(4) $a^{(\log_a x)} = x$ and $\log_a(a^x) = x$, for all bases "a".

(This is true because $f(x) = a^x$ and $g(x) = \log_a x$ are inverse functions when the base is the same, which means that $f(g(x)) = g(f(x)) = x$). We will discuss functions and inverse functions in more detail in a later section.

(5) $\log_a b = \frac{\log_c b}{\log_c a}$, for all allowable bases "a" and "c."
As a special case of law (5), $\log_a b = \frac{\ln b}{\ln a}$. This allows you to evaluate a logarithm like $\log_7(3.2)$ as $\frac{\ln(3.2)}{\ln(7)}$. If you want to use common logarithms, then $\log_7(3.2) = \frac{\log_{10}(3.2)}{\log_{10}(7)}$.
Either way, $\log_7(3.2)$ can be evaluated using your pocket calculator, because most scientific calculators have keys for common and natural logarithms.

<u>Square Roots</u>
The notation $\sqrt{x}$ represents a number such that:

$(\sqrt{x})(\sqrt{x}) = x$, where x is a non-negative number.
We call $\sqrt{x}$ the square root of x.

The specialized notation around x is called radical notation.

Some examples,
If $(\sqrt{36})(\sqrt{36}) = 36$, then $\sqrt{36} = 6$, since $6^2 = 36$, and
If $(\sqrt{81})(\sqrt{81}) = 81$, then $\sqrt{81} = 9$, since $9^2 = 81$.

When we speak of the $\sqrt{N}$, where $N \geq 0$, we usually refer to the positive square root. The number $(-\sqrt{N})$, when squared, also equals N. So, a positive number N has the two square roots $-\sqrt{N}$ and $+\sqrt{N}$. So, when we consider the solution of $x^2 = N$, where $N \geq 0$, we say that the equation has the two solutions x = $\pm\sqrt{N}$.

It is important to note that $\sqrt{abc} = \sqrt{a} \cdot \sqrt{b} \cdot \sqrt{c}$ and that $\sqrt{\dfrac{a}{b}} = \dfrac{\sqrt{a}}{\sqrt{b}}$. These are actually laws of exponents:

(A) $(abc)^{\frac{1}{2}} = (a)^{\frac{1}{2}} \cdot (b)^{\frac{1}{2}} \cdot (c)^{\frac{1}{2}}$ (B) $\left(\dfrac{a}{b}\right)^{\frac{1}{2}} = \dfrac{(a)^{\frac{1}{2}}}{(b)^{\frac{1}{2}}}$.

Cube Roots
The notation $\sqrt[3]{x}$ represents a number such that:
$(\sqrt[3]{x})(\sqrt[3]{x})(\sqrt[3]{x}) = x$, where x is any real number.
We call $\sqrt[3]{x}$ the cube root of x.

Some examples,

$(\sqrt[3]{64})(\sqrt[3]{64})(\sqrt[3]{64}) = 64$, so $\sqrt[3]{64} = 4$, since $4^3 = 64$.
$(\sqrt[3]{8})(\sqrt[3]{8})(\sqrt[3]{8}) = 8$, so $\sqrt[3]{8} = 2$, since $2^3 = 8$.
$(\sqrt[3]{-27})(\sqrt[3]{-27})(\sqrt[3]{-27}) = -27$, so $\sqrt[3]{-27} = -3$,
   since $(-3)^3 = -27$.

There is only one cube root $\sqrt[3]{N}$ for a given real number N, whether N is positive or negative. So, when we consider the solution of $x^3 = N$, there is only one real solution.

It is important to note that $\sqrt[3]{abc} = \sqrt[3]{a} \cdot \sqrt[3]{b} \cdot \sqrt[3]{c}$ and that $\sqrt[3]{\dfrac{a}{b}} = \dfrac{\sqrt[3]{a}}{\sqrt[3]{b}}$. These are actually laws of exponents:

(a) $(abc)^{\frac{1}{3}} = (a)^{\frac{1}{3}} \cdot (b)^{\frac{1}{3}} \cdot (c)^{\frac{1}{3}}$ (b) $\left(\dfrac{a}{b}\right)^{\frac{1}{3}} = \dfrac{(a)^{\frac{1}{3}}}{(b)^{\frac{1}{3}}}$.

n^{th} roots

The situation is very similar with n^{th} roots where n > 3. If n is even (n = 4, 6, 8, . . .), the equation $x^n = N$ has only the two solutions x = $\pm\sqrt[n]{N}$ (when considering only the real solutions). If n is odd (n = 5, 7, 9, . . .), the equation $x^n = N$ has only the one solution x = $\sqrt[n]{N}$ (when considering only the real solutions). We call these kinds of quantities the fourth root of x, or the seventh root of x, and so on, depending on what n is. It is also important to notice that:

$$\sqrt[n]{abc} = \sqrt[n]{a} \cdot \sqrt[n]{b} \cdot \sqrt[n]{c} \text{ and that } \sqrt[n]{\dfrac{a}{b}} = \dfrac{\sqrt[n]{a}}{\sqrt[n]{b}}.$$

Thinking of these as laws of exponents:

(a) $(abc)^{\frac{1}{n}} = (a)^{\frac{1}{n}} \cdot (b)^{\frac{1}{n}} \cdot (c)^{\frac{1}{n}}$ (b) $\left(\frac{a}{b}\right)^{\frac{1}{n}} = \frac{(a)^{\frac{1}{n}}}{(b)^{\frac{1}{n}}}$.

We can actually expand the laws of exponents further:

(a) $(abc)^{\frac{m}{n}} = (a)^{\frac{m}{n}} \cdot (b)^{\frac{m}{n}} \cdot (c)^{\frac{m}{n}}$ (b) $\left(\frac{a}{b}\right)^{\frac{m}{n}} = \frac{(a)^{\frac{m}{n}}}{(b)^{\frac{m}{n}}}$.

################# Exercises ##################

Find the solution set with inequality notation:
(1) $|x - 13| < 9$ (2) $|x| \le 30$ (3) $|x - 2| \ge 4$
(4) $|x - 5| > h$ (h > 0)   (5) $|x + 20| > \varepsilon$ $(\varepsilon > 0)$
(6) $|x - 11| \ge 3$ (7) $|x - 25| < 10$

Simplify the following expressions (without radicals):

(8) $\dfrac{x^8 y^3}{x^2 y}$ (9) $\dfrac{x^{-2} y^{-3}}{x^{-6} y^2}$ (10) $(x^3 y^8 z^4)^{\frac{1}{5}}$ (11) $\sqrt[3]{\dfrac{x^2}{y^9}}$

(12) $x^{\frac{1}{2}} \cdot x^2 \cdot x^{\frac{7}{2}}$ (13) $(x^2 y^3 z^4)^3$ (14) $\sqrt{x^{-2} y^{-4} z^4}$

(15) $e^{\ln(2x)}$ (16) $\ln\left(e^{(2x)}\right)$ (17) $5^{\log_5 x} \cdot 6^{\log_6 x}$

(18) $2^x 16^{(x^2 + 2)}$ (19) $\ln(3x) + \ln(4x) + \ln(5x)$

(20) $\log_{10} 30x^2 - \log_{10} 6x$ (21) $\log_4(x)^7$ (22) $\ln(x)^3$

Solve for x in each of the following equations:

(23) $x^2 = 9$ (24) $x^2 = 25$ (25) $x^2 = 27$

(26) $x^3 = -125$ (27) $x^5 = 32$ (28) $x^4 = 81$

What number does each of the following represent?

(29) (a) $\sqrt{(3)^4}$ (b) $\sqrt{(100)^2}$ (c) $\sqrt[6]{(4)^3}$

　　(d) $\sqrt[3]{27,000}$ (e) $\sqrt[5]{1024}$ (f) $\sqrt[7]{13}$

(3.3) <u>Expansions and Factorizations</u>

Two of the most important algebraic manipulations of expressions are based on the fact that multiplication distributes over addition for real numbers. If we have the product $a(b + c)$, then we can expand the product $a(b + c) = ab + ac$, or going in the opposite direction we can factor the sum $ab + ac = a(b + c)$.

<u>EXPANSIONS</u>
Let a, b, c, d, and e be real numbers. Then:

<u>Example (1)</u>:
$$(a + b)^2 = (a + b)(a + b)$$
$$= a(a + b) + b(a + b)$$
$$= a^2 + ab + ba + b^2$$
$$= a^2 + 2ab + b^2$$

Some examples:

(A) $(2x + 5y)^2 = (2x + 5y)(2x + 5y)$
$$= (2x)(2x + 5y) + (5y)(2x + 5y)$$
$$= 4x^2 + 10xy + 10xy + 25y^2$$
$$= 4x^2 + 20xy + 25y^2$$

(B) $\left(8\sqrt{x} + \sqrt{y}\right)^2 = \left(8\sqrt{x} + \sqrt{y}\right)\left(8\sqrt{x} + \sqrt{y}\right)$
$$= (8\sqrt{x})(8\sqrt{x} + \sqrt{y}) + (\sqrt{y})(8\sqrt{x} + \sqrt{y})$$
$$= 64x + 8\sqrt{x}\sqrt{y} + 8\sqrt{x}\sqrt{y} + y$$
$$= 64x + 16\sqrt{xy} + y$$

(C) $(3x^2 + 13y^2)^2$
$$= (3x^2)(3x^2 + 13y^2) + (13y^2)(3x^2 + 13y^2)$$
$$= 9x^4 + 39x^2y^2 + 39x^2y^2 + 169y^4$$
$$= 9x^4 + 78(xy)^2 + 169y^4$$

Example (2):
$(a - b)^2 = (a - b)(a - b)$
$$= a(a - b) - b(a - b)$$
$$= a^2 - ab - ba + b^2$$
$$= a^2 - 2ab + b^2$$

Some examples:
(A) $(4x - 10y)^2 = (4x - 10y)(4x - 10y)$
$$= (4x)(4x - 10y) + (-10y)(4x - 10y)$$
$$= 16x^2 - 40xy - 40xy + 100y^2$$
$$= 16x^2 - 80xy + 100y^2$$

(B) $\left(5\sqrt{x} - 3y\right)^2 = \left(5\sqrt{x} - 3y\right)\left(5\sqrt{x} - 3y\right)$

$$= \left(5\sqrt{x}\right)\left(5\sqrt{x} - 3y\right) + (-3y)\left(5\sqrt{x} - 3y\right)$$
$$= 25x - 15\sqrt{x}(y) - 15(y)\sqrt{x} + 9y^2$$
$$= 25x - 30\sqrt{x}(y) + 9y^2$$

(C) $(x^3 - 2y^2)^2 = (x^3 - 2y^2)(x^3 - 2y^2)$

$$= (x^3)(x^3 - 2y^2) + (-2y^2)(x^3 - 2y^2)$$
$$= x^6 - 2x^3y^2 - 2y^2x^3 + 4y^4$$
$$= x^6 - 4x^3y^2 + 4y^4$$

Example (3):

$(a + b)(c + d) = a(c + d) + b(c + d)$
$$= ac + ad + bc + bd$$

Some examples:

(A) $(4x - 9)(2x + 5)$

$$= (4x)(2x + 5) + (-9)(2x + 5)$$
$$= 8x^2 + 20x - 18x - 45$$
$$= 8x^2 + 2x - 45$$

(B) $\left(\sqrt{x} - \sqrt{2}\right)\left(\sqrt{x} + \sqrt{2}\right)$

$$= \left(\sqrt{x}\right)\left(\sqrt{x} + \sqrt{2}\right) + \left(-\sqrt{2}\right)\left(\sqrt{x} + \sqrt{2}\right)$$
$$= x + \left(\sqrt{x}\right)\left(\sqrt{2}\right) - \left(\sqrt{2}\right)\left(\sqrt{x}\right) - 2$$
$$= x - 2$$

(C) $(3x - y)(x + y)$

$$= (3x)(x + y) + (-y)(x + y)$$
$$= 3x^2 + 3xy - yx - y^2$$

76

$$= 3x^2 + 2xy - y^2$$

The distributive law can be generalized.

Example (4):
$$(a)(b + c + d + e) = ab + ac + ad + ae$$

Some examples:

(A) $\left(\sqrt{x}\right)(2 + 2x + 2x^2 + 2x^3)$
$$= 2\left(x^{\frac{1}{2}}\right) + 2\left(x^{\frac{1}{2}}\right)(x) + 2\left(x^{\frac{1}{2}}\right)(x^2) + 2\left(x^{\frac{1}{2}}\right)(x^3)$$
$$= 2\sqrt{x} + 2x^{\frac{3}{2}} + 2x^{\frac{5}{2}} + 2x^{\frac{7}{2}}$$

(B) $(2x)(4 + 3y - 7z^2 - 2h)$
$$= 8x + 6xy - 14xz^2 - 4xh$$

(C) $(9x)(3 + 2y - 3x^3) = (27x + 18xy - 27x^4)$

Example (5):
$$(a + b)(c + d + e)$$
$$= (a)(c + d + e) + (b)(c + d + e)$$
$$= ac + ad + ae + bc + bd + be$$

Two examples:

(A) $(2x + 3)(4 - x + 3y^2)$
$$= (2x)(4 - x + 3y^2) + (3)(4 - x + 3y^2)$$
$$= 8x - 2x^2 + 6xy^2 + 12 - 3x + 9y^2$$
$$= 5x - 2x^2 + 6xy^2 + 9y^2 + 12$$

(B) $(9x - 1)(10x + 4y - 3z)$
$$= (9x)(10x + 4y - 3z) - (1)(10x + 4y - 3z)$$
$$= (90x^2 + 36xy - 27xz - 10x - 4y + 3z)$$

Sometimes we employ the following algebraic technique where it can be used to advantage. We show two examples:

Example (6):
(A) $(a + b)^2 = (a + c - c + b)^2$
$$= \big((a + c) + (b - c)\big)^2$$
$$= (a + c)^2 + (b - c)^2 + 2(a + c)(b - c)$$

(B) $(2x + 7y)^2 = (2x + \varepsilon - \varepsilon + 7y)^2$
$$= \big((2x + \varepsilon) + (7y - \varepsilon)\big)^2$$
$$= (2x + \varepsilon)^2 + (7y - \varepsilon)^2 + 2(2x + \varepsilon)(7y - \varepsilon)$$

FACTORIZATIONS

Example (1):
$$a^2 - b^2 = (a + b)(a - b)$$

Some examples:
(A) $x^2 - 9 = (x + 3)(x - 3)$
(B) $16x^2 - 36y^2 = (4x + 6y)(4x - 6y)$
(C) $4x^2 - 100z^2 = (2x + 10z)(2x - 10z)$

Example (2):
For an expression like $x^2 + ax + b$, we look for two

78

numbers c and d such that $(c \cdot d) = b$ and $(c + d) = a$. Then the factorization would be:

$$x^2 + ax + b = (x + c)(x + d).$$

This is many times a matter of trial and error, and it is not always possible to find two such whole numbers c and d.

Some examples:
(A) $x^2 - 3x - 18 = (x + 3)(x - 6)$
(B) $x^2 + 11x + 28 = (x + 4)(x + 7)$
(C) $x^2 + 5x + 4 = (x + 1)(x + 4)$
(D) $x^2 - 7x + 10 = (x - 2)(x - 5)$

Example (3):
Factoring higher order polynomials:
Two examples:

(A) $x^3 + 2x^2 - 99x = (x)(x^2 + 2x - 99)$
$$= (x)(x - 9)(x + 11)$$

(B) $x^4 - 3x^3 + x^2 - 3x = (x)(x^3 - 3x^2 + x - 3)$
$$= (x)(x - 3)(x^2 + 1)$$
$$= (x)(x - 3)(x - i)((x + i)$$

Note that $(x^2 + 1)$ cannot be factored into two linear factors involving only real numbers.

################# Exercises #################

Expand each of the following (simplify where necessary):

(1) $(2x - 4)^2$ (2) $(x + 12)^2$ (3) $(3x + 3)^2$

(4) $(4x - 10)^2$ (5) $(x)(3x - 2y + 4z)$

(6) $(\sqrt{x})\left(15\sqrt{x} - \frac{1}{\sqrt{x}} + 2\sqrt{x}(x)\right)$ (7) $(x^2 + 3)(9 - x)$

(8) $(x + 3)(x - 3)$ (9) $(x^3 - x^5)(x + x^4)$

(10) $(3x + 4)(5x + 7)$

Factor each of the following expressions:

(11) $(x^2 + x - 12)$ (12) $(x^2 - 16x + 63)$

(13) $(x^3 + 11x^2 + 30x)$ (14) $(x^4 - 10x^3 + 16x^2)$

(15) $(x^2 - 10{,}000)$ (16) $(16x^2 - 4y^2)$

(17) $(25x^2 - 125x - 150)$

(18) $(x^3 - x^2 - x + 1)$ (Hint: Divide this by $(x - 1)$)

(3.4) Expressions and Equations with Fractions

Dealing with fractions is very important in algebra.
We will consider some examples of such expressions, and
then some equations. Note that just as with fractions
involving numbers, when we deal with sums or differences
of fractions involving variables we also need a common
denominator. The first three examples deal with this point.
Then we will include some more examples dealing with
simplification of fractions:

Examples:

(1) $\dfrac{1}{x} + \dfrac{1}{y} = \left(\dfrac{y}{y}\right)\left(\dfrac{1}{x}\right) + \left(\dfrac{x}{x}\right)\left(\dfrac{1}{y}\right) = \dfrac{y}{xy} + \dfrac{x}{xy} = \left(\dfrac{x+y}{xy}\right)$

(2) $\dfrac{1}{(x+a)} - \dfrac{1}{(x-b)} = \left(\dfrac{x-b}{x-b}\right)\left(\dfrac{1}{(x+a)}\right) - \left(\dfrac{x+a}{x+a}\right)\left(\dfrac{1}{(x-b)}\right)$

$= \dfrac{(x-b)-(x+a)}{((x+a)(x-b))} = \dfrac{-(a+b)}{(x+a)(x-b)}.$

(3) $\dfrac{\left(\frac{1}{x}\right)}{\left(\frac{1}{y}+\frac{1}{z}\right)} = \dfrac{\left(\frac{1}{x}\right)}{\left(\left(\frac{z}{z}\right)\left(\frac{1}{y}\right)+\left(\frac{y}{y}\right)\left(\frac{1}{z}\right)\right)} = \dfrac{\left(\frac{1}{x}\right)}{\left(\frac{y+z}{yz}\right)}$

$= \left(\dfrac{1}{x}\right) \cdot \left(\dfrac{yz}{y+z}\right) = \dfrac{yz}{(x)(y+z)} = \dfrac{yz}{xy+xz}.$

(4) Remove square roots from the numerator of
    the fraction $\left(\dfrac{\sqrt{x+h}-\sqrt{x}}{h}\right).$

81

$$\left(\frac{\sqrt{x+h}-\sqrt{x}}{h}\right) = \left(\frac{\sqrt{x+h}-\sqrt{x}}{h}\right)\left(\frac{\sqrt{x+h}+\sqrt{x}}{\sqrt{x+h}+\sqrt{x}}\right)$$

$$= \frac{(x+h)-x}{(h)(\sqrt{x+h}+\sqrt{x})} = \frac{(h)}{(h)(\sqrt{x+h}+\sqrt{x})}$$

$$= \left(\frac{1}{(\sqrt{x+h}+\sqrt{x})}\right).$$

(5) Simplify: $\dfrac{(x-7)^4(x+3)(x-2)^3}{(x-2)^2(x+3)^5(x-7)^2}$.

The fraction $\dfrac{(x-7)^4(x+3)(x-2)^3}{(x-2)^2(x+3)^5(x-7)^2}$ simplifies to

$(x-7)^{4-2} \cdot (x+3)^{1-5} \cdot (x-2)^{3-2}$ which equals

$(x-7)^2 \cdot (x+3)^{-4} \cdot (x-2) = \dfrac{(x-7)^2(x-2)}{(x+3)^4}$.

(6) Simplify: $\dfrac{x^{-3}y^4z^{-6}}{x^5y^2z^{-4}}$ (eliminating negative exponents)

The fraction $\dfrac{x^{-3}y^4z^{-6}}{x^5y^2z^{-4}}$ simplifies to

$(x)^{-3-5} \cdot (y)^{4-2} \cdot (z)^{-6-(-4)}$ which equals

$(x)^{-8} \cdot (y)^2 \cdot (z)^{-2}$ which equals $\dfrac{y^2}{x^8 \cdot z^2}$.

We will now deal with equations that involves fractions of algebraic expressions. This is of course a very important type of equation that we need to give some consideration.

<u>Example (7)</u>:

Solve for x in the equation: $\sqrt{x} = \dfrac{2x + 12}{\sqrt{x}}$:

Starting with $\sqrt{x} = \dfrac{2x + 12}{\sqrt{x}}$, multiply both sides by the $\sqrt{x}$ to get: $x = 2x + 12$. Then, subtract x and 12 from both sides of the equation to rearrange things. We get the solution x = -12.

<u>Example (8)</u>:

Solve for x in the equation: $x = \dfrac{3x^2 - 5}{x}$.

Multiplying both sides of the equation by x leads to: $x^2 = 3x^2 - 5$. Now subtract x^2 and add 5 to both sides of the equation to rearrange things. Then we have:

$2x^2 = 5$ or $x^2 = \dfrac{5}{2} \rightarrow x = \pm\sqrt{\dfrac{5}{2}} = \pm\sqrt{2.5} = \pm 1.58$,

rounding to two decimal places.

<u>Example (9)</u>:

If p = 0.30, solve for n (where n is a positive integer) in the inequality:

$$\left(\dfrac{(1.96)\sqrt{(p)(1-p)}}{\sqrt{n}}\right) \le 0.03$$

Square both sides and substitute 0.30 for p to get:

$$\frac{(3.8416)(0.30)(0.70)}{n} \le 0.0009$$

Multiplying both sides by n and doing the calculations we get: $n \ge 896.37$. Since n must be a positive integer, then n should be chosen to be greater than or equal to 897 to ensure that the inequality above is satisfied.

Example (10):

Solve for x in the equation: $\dfrac{3-x}{x^2} = \dfrac{10}{5+x}$.

Multiplying both sides of the equation by $(x^2)(5+x)$, we get:

$$(3-x)(5+x) = 10x^2 \rightarrow 15 - 5x + 3x - x^2 = 10x^2$$

Moving all terms to one side and collecting like terms we get the quadratic equation (see section (3.7)):

$$11x^2 + 2x - 15 = 0 .$$

Using the quadratic formula, a topic of section (3.7), leads to the two solutions:

$$x = \frac{-2 \pm \sqrt{2^2 - (4)(11)(-15)}}{2(11)} = \frac{-2 \pm \sqrt{4+660}}{22} = \{-1.262, 1.080\},$$

after rounding to three decimal places.

############### Exercises ###############

Simplify the following fractions:

(1) $\left(\dfrac{1}{2x}\right) + \left(\dfrac{1}{4x}\right)$
(2) $\left(\dfrac{1}{x-1}\right) - \left(\dfrac{3}{x+1}\right)$

(3) $\left(\dfrac{x^2+x}{x} + \dfrac{2x+x^3}{x^2}\right)$
(4) $\left(\dfrac{x^2-4}{x+2} + \dfrac{x^2-9}{x+3}\right)$

(5) $\left(\dfrac{16x}{5x^2} - \dfrac{3x+2}{5x}\right)$
(6) $\left(\dfrac{14x}{\frac{1}{x^3} + \frac{1}{x^3}}\right)$

Solve the following equations for x:

(7) $\dfrac{x-2}{x} = \dfrac{x+1}{x+2}$
(8) $\dfrac{2\sqrt{x}}{x+1} = \dfrac{x+3}{\sqrt{x}}$

(9) $\dfrac{1}{\sqrt{x}} = \dfrac{\sqrt{x}-5\sqrt{x}}{x^2}$
(10) $\dfrac{x}{20} = \dfrac{36}{x+2}$

(11) $\sqrt{x} = \dfrac{2x+3}{\sqrt{x}}$
(12) $\dfrac{x}{4-x} = \dfrac{4+x}{x-2}$

(3.5) Relations and Functions

We will now discuss relations and a special type of relation that we call a function in more detail.

Relations:
A relation is any mapping from a set X to a set Y, called the Domain and Codomain respectively. In most of applied

85

mathematics it would consist of an arbitrary collection of isolated points or an arbitrary curve in the coordinate plane, or a mixture of the two. For relations, if we have a member x in X, it is possible that it be mapped to more than one element in Y. One important relation that we encounter a lot in mathematics is a circle in the x-y plane with equation:

$$(x - h)^2 + (y - k)^2 = r^2$$

where the center is the point with coordinates (h, k) and the radius is r. Note that for most of a circle, there are two y-values for every x-value. When we discuss functions, the reader should understand that the top and the bottom of the circle can be considered as functions when considered separately.

Another important relation is an ellipse with equation:

$$\frac{(x - h)^2}{a^2} + \frac{(y - k)^2}{b^2} = 1$$

where the center is the point with coordinates (h, k) and where a and b are two positive numbers that are not equal (which gives the ellipse its oval shape). Note that for most of an ellipse there are two y-values corresponding to an x-value, just as with circles. When we discuss functions, the reader should realize that the top and bottom of the ellipse can be considered as functions when we consider them separately.

Functions:
Now we will discuss functions, a topic of more importance for us in this book and in a lot of applied mathematics. A function is a special type of relation. Most functions that we will encounter in this book are continuous curves that are familiar to us, like lines, parabolas, exponential curves, and so forth.

A function f is a mapping from a set X (called the Domain) to another set Y (called the Codomain), where every element of X (which we call x) is mapped to only one element y in Y, and we write this as y = f(x). We say that f maps X to Y and denote this as f: $X \rightarrow Y$. Every element of X is mapped to some element in Y. However, not every element y in Y necessarily has a pre-image x in X such that f(x) = y. The set of elements in Y that do have a pre-image x in X is called the range of f, written f(X). In general f(X) $\subset Y$, and we would say that f is an into function. If f(X) = Y, then we say that f is an onto function. If for any two different x-values x_1 and x_2, $f(x_1) \neq f(x_2)$, then we say that f is one-to-one (or 1-1).

Inverse Functions:
If a function f is 1-1 and onto, then f has an inverse function $f^{-1}: Y \rightarrow X$ such that $f^{-1}(f(x)) = x$ for every x $\in X$, and $f(f^{-1}(y)) = y$ for every y $\in Y$. The next two examples deal with inverse functions. For the graphs of the following function-inverse function pairs, the reader should see the figure on the next page.

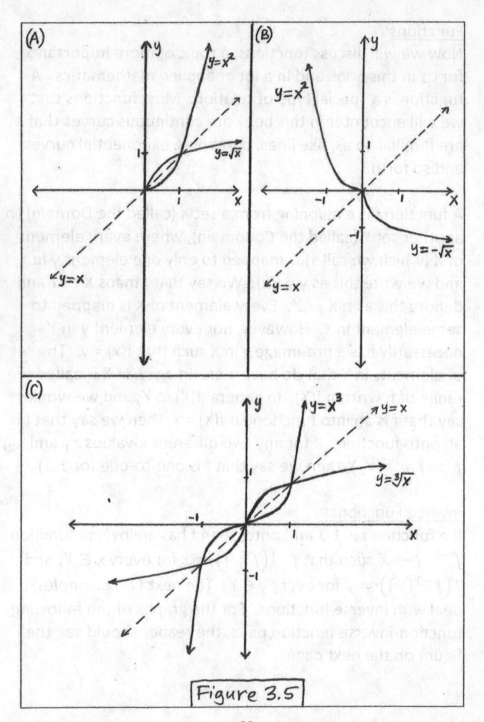

Figure 3.5

<u>Example (1):</u>
Let $y = f(x) = x^2$. It is a parabola that has a vertex at the origin of the x-y coordinate plane and opens upward on either side of the origin. The Domain X is all real numbers, the Codomain Y is all real numbers, the range f(X) is the set of non-negative real numbers. This function f: X → Y is a mapping from X into Y since no member x of X maps to a negative real number. So, it is not a mapping from X onto Y. So, considering its entire domain, $f(x) = x^2$ does not have an inverse function $f^{-1}: Y \rightarrow X$. However, from the function $y = x^2$, we can come up with two pairs of functions that are inverses of each other. When we want to find an inverse function for a function f(x), interchange the roles of x and y, and then solve for y. In this case, we would start by writing $x = y^2$, so then $y = \pm\sqrt{x}$. So we have the two functions: $y = -\sqrt{x}$ and $y = \sqrt{x}$. Note that a function and its inverse are reflections across the line y = x.

Let's not worry about the point (0,0) for the moment, to make the wording easier. It is easier to talk of the positive real numbers and the negative real numbers, rather than having to speak of the non-negative real numbers and the non-positive real numbers, which we have to do when we include that single point.

(a) If we restrict ourselves to x > 0, the function $y = x^2$ is a 1-1 and onto function from the positive reals to the positive reals, and also for $y = \sqrt{x}$. So, if $f(x) = x^2$ and $g(x) = \sqrt{x}$, then $f(g(x)) = (\sqrt{x})^2 = x$ and $g(f(x)) = \sqrt{x^2} = x$.

Now, go back and include the point (0,0) in the discussion,
and we have $\begin{cases} f(x) = x^2, & x \geq 0 \\ g(x) = \sqrt{x}, & x \geq 0 \end{cases}$ as a function-inverse function pair.

(b) If we restrict ourselves to x < 0, the function $y = x^2$ is a 1-1 and onto function from the negative reals to the positive reals. For x > 0, the function $y = -\sqrt{x}$ is a 1-1 and onto function from the positive reals to the negative reals.

So, consider $f(x) = x^2$ and $g(x) = -\sqrt{x}$.

For x > 0, $f(g(x)) = \left(-\sqrt{(x)}\right)^2 = x$, and

For x < 0, $g(f(x)) = -\sqrt{(x)^2} = -\sqrt{(-x)^2} = -(-x) = x$.

Therefore, $f(g(x)) = x$, when x is in the domain of $g(x)$, and $g(f(x)) = x$, when x is in the domain of $f(x)$.

Including the point (0,0) in the discussion, we have:

$\begin{cases} f(x) = x^2, & x \leq 0 \\ g(x) = -\sqrt{x}, & x \geq 0 \end{cases}$ as a function-inverse function pair.

Example (2): Now, consider the two functions $f(x) = x^3$ and $g(x) = \sqrt[3]{x}$. Both of these functions are 1-1 and onto functions from the entire set of real numbers to the entire set of real numbers. Also, $f(g(x)) = \left(\sqrt[3]{x}\right)^3 = x$, where x is any real number, and $g(f(x)) = \sqrt[3]{x^3} = x$, where x is any real number. Therefore, f(x) and g(x) are a function-inverse function pair.

(3.6) Linear Equations

A linear equation is one where the power on all of the variables is one. Linear equations are very important in mathematics. If a, b, c, and d are real numbers, or so-called constants, as we call things that are considered as non-variable in a given context, then:

ax + b = 0 is a linear equation in one variable. The solution is a single number on the real number line.

ax + by + c = 0 is a linear equation in two variables. The set of all solutions (x, y) of this equation is a line in the x-y coordinate plane (two-dimensional space).

ax + by + cz + d = 0 is a linear equation in three variables. The set of all solutions (x, y, z) of this equation is a plane in x-y-z coordinate space (three-dimensional space).

Example (1):

Solve the equation $\frac{1}{5}x - \frac{1}{8} = 0$ for x. We need to isolate x on one side of the equation: We have $\frac{1}{5}x = \frac{1}{8}$ after adding $\frac{1}{8}$ to both sides of the equation. Then, we have $x = \frac{5}{8}$ after

91

multiplying both sides by 5. So, $x = \frac{5}{8}$ solves the equation. If we desire to check this result, then we can substitute $\frac{5}{8}$ for x in the original equation: $\frac{1}{5}\left(\frac{5}{8}\right) - \frac{1}{8} = \frac{1}{8} - \frac{1}{8} = 0$, which is a true statement.

Example (2):
Solve the equation $6x - 24 = 18$ for x.

We need to isolate x on one side of the equation. We have $6x = 18 + 24 = 42$ by adding 24 to both sides, so that the equation has been simplified to $6x = 42$. Then, after dividing both sides by 6, we get x = 7 as the solution.

If we desire to check this result, then we substitute 7 for x in the original equation, getting the statement: $6(7) - 24 = 42 - 24 = 18$, which is a true statement.

Example (3):
Solve for y in the equation $2y - 3x = 14$.

We need to isolate y on the left side of the equation. We get $2y = 3x + 14$ after adding 3x to both sides. Then after dividing both sides by 2, we get: $y = \frac{3}{2}x + 7$.

We can say that y is a linear function of x. Sometimes we indicate this by writing: $y = f(x) = \frac{3}{2}x + 7$.

So, for example, when x = $\frac{2}{3}$, the y-value = $f\left(\frac{2}{3}\right) = \frac{3}{2}\left(\frac{2}{3}\right) + 7$, which is equal to 8. This type of linear equation, or linear function involving two variables, is the equation of a line in the two dimensional x-y coordinate plane.

The equation of a line in the x-y coordinate plane is commonly expressed as y = mx + b. "m" is called the slope and "b" is called the y-intercept. "m" tells us how much y changes for every one unit increase in x. For example, if m = 3, a one unit increase in x results in a 3 unit increase in y. The line would have a positive slope, meaning that as we go from left to right, the y-values on the line increase. If m = $-\frac{1}{3}$, then a one unit increase in x would result in a $\frac{1}{3}$ unit decrease in y. The line would have a negative slope, meaning that as we go from left to right, the y-values on the line decrease. These relationships between x and y hold no matter where we start on the line. The y-intercept "b" is the y-coordinate of the point where the line intersects the y-axis, namely the point (0,b). A horizontal line has a y-intercept of b and the slope m = 0, so the equation of a horizontal line is y = b. The slope of a vertical line where x is always a number c is undefined (one can think of it as ∞). We can write the equation of a vertical line, since every point on the line has an x-coordinate of c, as x = c.

The figure on the next page shows two lines: one is a positively line and one is a negatively sloped line in the x-y coordinate plane. y = x − 2 is positively sloped, and y = -2x + 2 is negatively sloped.

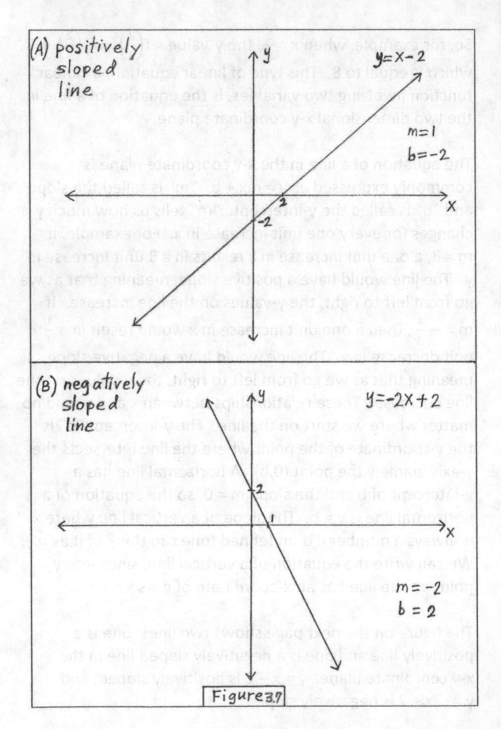

(A) positively sloped line

$y = x - 2$

$m = 1$
$b = -2$

(B) negatively sloped line

$y = -2x + 2$

$m = -2$
$b = 2$

Figure 3.7

Example (4):

Write $15x - 12y + 2x = 7x + 14$ in the form y = mx + b.

Rearranging terms: $-12y = -10x + 14$.

Then, dividing by -12, we have: $y = \frac{5}{6}x - \frac{7}{6}$.

This is the equation of a line. It passes through the y-axis at the point $\left(0, -\frac{7}{6}\right)$, and it has positive slope m = $\frac{5}{6}$.

Example (5):

Solve for z in $\frac{1}{2}x - 10y + 4z = 8$.

We need to isolate z on the left side of the equation. After subtracting $\frac{1}{2}x$ and adding 10y to both sides, we get: $4z = -\frac{1}{2}x + 10y + 8$. Then, after dividing both sides by 4 we get: $z = -\frac{1}{8}x + \frac{10}{4}y + 2$, or $z = -\frac{1}{8}x + \frac{5}{2}y + 2$.

So, z a linear function of x and y, which we can write as
$$z = f(x, y) = -\frac{1}{8}x + \frac{5}{2}y + 2.$$

The set of (x, y, z) coordinates that satisfy this function $z = f(x, y)$ constitute a plane in three dimensional space.

If we set x = 16 and y = 10, then we can determine that:
$z = f(16,10) = -\frac{1}{8}(16) + \frac{5}{2}(10) + 2 = 25$. So, the point (16, 10, 25) is a point on the plane.

Example (6):

For this example refer to the figure on the next page. Let the monthly cost y of producing x widgets depend on a certain fixed amount of overhead costs plus the cost of the human labor, in a linear way. The fixed overhead per month is $3950.00 for the costs associated with the facilities. For each widget produced, it costs $3.50 in labor costs.

(a) What is the equation relating the monthly cost y for producing x widgets, where x ≥ 0?

Once we determine the relationship between x and y, we could think of it as a line in the x-y plane. The fixed cost is the cost when x = 0, so this is the y-intercept b. For each widget produced, the labor cost is $3.50 and this remains the same for however many are produced. So, this labor cost is the slope m of the cost line. The cost y of producing x widgets is therefore:

$y = (3950.00) + (3.50)x$

where y is in dollars, and x ≥ 0.

(b) What is the monthly cost of making 80 widgets?

From our equation above, the monthly cost would be
$y = (3950.00) + (3.50)(80) = \$4230.00$.

(c) Given this cost for making 80 widgets, what price would guarantee that the widget making company breaks even?

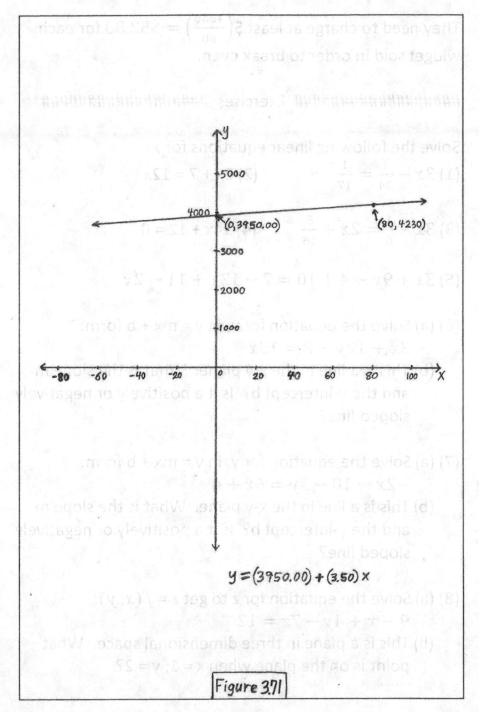

Figure 3.71

They need to charge at least $\$\left(\frac{4230}{80}\right) = \$52.88$ for each widget sold in order to break even.

################# Exercises #################

Solve the following linear equations for x:

(1) $3x - \frac{1}{34} = \frac{1}{17}$ (2) 2x + 7 = 12x

(3) $3x - \frac{1}{8} = 2x + \frac{5}{16}$ (4) 14x + 12 = 0

(5) $3x + 9x - 4 + 10 = 7 - 12x + 11 + 2x$

(6) (a) Solve the equation for y, in y = mx + b form:
$$3x + 12y - 8 = 13x$$
  (b) This is a line in the x-y plane. What is the slope m and the y-intercept b? Is it a positively or negatively sloped line?

(7) (a) Solve the equation for y, in y = mx + b form:
$$-2x - 10 - 3y = 6x + 4$$
  (b) This is a line in the x-y plane. What is the slope m and the y-intercept b? Is it a positively or negatively sloped line?

(8) (a) Solve the equation for z to get z = $f(x, y)$:
$$9 - x + 4y - 7z = 12$$
  (b) This is a plane in three dimensional space. What point is on the plane when x = 1, y = 2?

98

(3.7) Quadratic Equations

An equation of the form $ax^2 + bx + c = 0$ (where a, b, and c are real numbers) is a quadratic equation in the variable x. Since this is a second degree polynomial equation in the one variable x, it turns out that there are always two complex solutions, which may be a repeated real root or two distinct real roots, or two complex conjugate roots.

If we write $y = ax^2 + bx + c$, the set of all points (x,y) that satisfy this equation represent a parabolic curve in the x-y plane. This type of parabola opens upward or downward, depending on whether "a" is positive or negative. Firstly, we wish to find the roots (or the x-axis intercepts) of the parabolic function $y = ax^2 + bx + c$. These occur where the graph of the function goes through the x-axis, which is where y = 0. By setting y = 0, we have $ax^2 + bx + c = 0$, which is a quadratic equation in standard form in one variable x. To derive a formula for the solutions of this equation we employ an algebraic technique that we call "completing the square."

The Quadratic Formula

Starting with $ax^2 + bx + c = 0$, we first divide both sides by "a" because we want the coefficient on the x^2 term to be 1, to get:

$$x^2 + \left(\frac{b}{a}\right)x + \left(\frac{c}{a}\right) = 0, \quad \text{or} \quad x^2 + \left(\frac{b}{a}\right)x = -\left(\frac{c}{a}\right).$$

Now take the "square of one half of the coefficient" on the x-term, and then add it to both sides of the equation to get

$$x^2 + \left(\frac{b}{a}\right)x + \left(\frac{b}{2a}\right)^2 = \left(\frac{b}{2a}\right)^2 - \left(\frac{c}{a}\right).$$

Now, the three terms on the left can be combined to "complete the square," giving us:

$$\left(x + \left(\frac{b}{2a}\right)\right)^2 = \left(\frac{b}{2a}\right)^2 - \left(\frac{c}{a}\right).$$

So, $\left(x + \left(\frac{b}{2a}\right)\right)^2 = \left(\frac{b}{2a}\right)^2 - \left(\frac{c}{a}\right) = \left(\frac{b^2 - 4ac}{4a^2}\right).$

So, $x + \left(\frac{b}{2a}\right) = \pm\dfrac{\sqrt{b^2 - 4ac}}{2a}.$

Now with a bit more algebraic manipulation, we have the so-called quadratic formula, for finding the roots of a quadratic equation in one variable x:

***** $\quad x = \dfrac{-b \pm \sqrt{b^2 - 4ac}}{2a}.$ *****

The two roots are: $\left\{\left(\dfrac{-b + \sqrt{b^2-4ac}}{2a}\right), \left(\dfrac{-b - \sqrt{b^2-4ac}}{2a}\right)\right\}.$

The quantity $(b^2 - 4ac)$ is called the discriminant. If it is positive, then there are two distinct real roots. If it is zero, then there are two equal real roots (a repeated real root).

If it is negative, then there are two complex conjugate roots. Again, these are the x-coordinates of the points where the graph of the parabola $y = ax^2 + bx + c$ passes through the x-axis. The parabola may cross in two distinct places (discriminant is positive). It may bounce off the x-axis at only one point (discriminant is zero). It may not cross the x-axis at all (discriminant is negative, where we would be taking the square root of a negative number which leads to the two complex conjugate roots).

Standard Form for a Quadratic (Parabolic) Function

Now, we want to derive a standard form for a quadratic (or parabolic) function by once again completing the square, starting with $y = ax^2 + bx + c$:

So, we have $y = ax^2 + bx + c$. Divide both sides by "a."

$$\frac{y}{a} = x^2 + \left(\frac{b}{a}\right)x + \left(\frac{c}{a}\right)$$

$$\frac{y}{a} = x^2 + \left(\frac{b}{a}\right)x + \left(\frac{b}{2a}\right)^2 - \left(\frac{b}{2a}\right)^2 + \left(\frac{4ac}{4a^2}\right)$$

$$\frac{y}{a} = \left(x + \frac{b}{2a}\right)^2 + \left(\frac{4ac - b^2}{4a^2}\right). \text{ Then,}$$

$$y = a\left(x - \left(\frac{-b}{2a}\right)\right)^2 + \left(\frac{4ac - b^2}{4a}\right). \text{ So, finally, we have:}$$

$$\text{*****} \quad y - \left(\frac{4ac - b^2}{4a}\right) = a\left(x - \left(\frac{-b}{2a}\right)\right)^2 \quad \text{*****}$$

101

as our standard form for a quadratic function.

The vertex has coordinates: $\left(\dfrac{-b}{2a}, \dfrac{4ac-b^2}{4a}\right)$.

If a > 0, then the parabola opens upward. If a < 0, then the parabola opens downward.

From above, we know that the parabola goes through the x-axis at the two points:

$$\left\{ \left(\dfrac{-b + \sqrt{b^2 - 4ac}}{2a}\right), \left(\dfrac{-b - \sqrt{b^2 - 4ac}}{2a}\right) \right\}.$$

For the next two examples refer to Figure (3.8).

Example (1):
If we have the quadratic function $y = 2x^2 - 9x + 4$, the vertex is at:

$$\left(\dfrac{-b}{2a}, \dfrac{4ac-b^2}{4a}\right) = \left(\dfrac{9}{4}, \dfrac{32-81}{8}\right) = \left(\dfrac{9}{4}, -\dfrac{49}{8}\right) = (2.25, -6.125).$$

It opens upward on both sides of the vertex because a = 2 is positive. The value of -6.125 is the minimum value of this quadratic (parabolic) function.

We can find the y-intercept, where the graph goes through the y-axis, by setting x = 0. In this case, we would get:

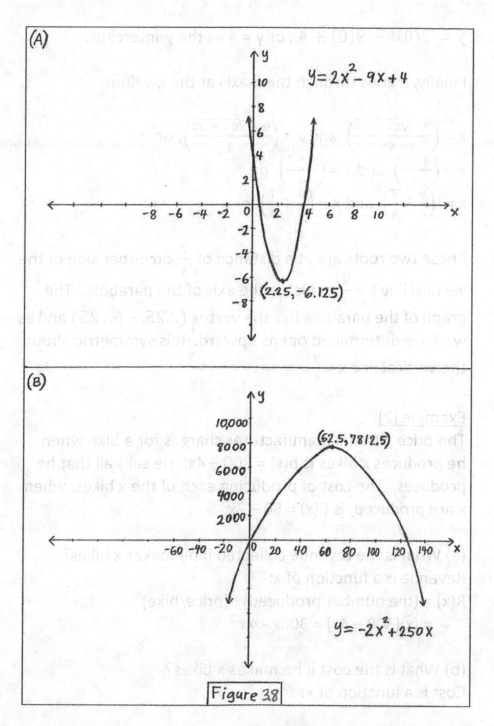

(A)

$y = 2x^2 - 9x + 4$

(2.25, -6.125)

(B)

(62.5, 7812.5)

$y = -2x^2 + 250x$

Figure 38

103

$y = 2(0)^2 - 9(0) + 4$, or y = 4 as the y-intercept.

Finally, it goes through the x-axis at the x-values:

$x = \left(\dfrac{9 - \sqrt{81 - 32}}{4}\right)$ and $x = \left(\dfrac{9 + \sqrt{81 - 32}}{4}\right)$, or

$x = \left(\dfrac{9 - 7}{4}\right)$ and $x = \left(\dfrac{9 + 7}{4}\right)$, or

$x = \left(\dfrac{9}{4} - \dfrac{7}{4}\right)$ and $x = \left(\dfrac{9}{4} + \dfrac{7}{4}\right)$, or $x = \dfrac{1}{2}$, x = 4.

These two roots are at a distance of $\dfrac{7}{4}$ on either side of the vertical line $x = \dfrac{9}{4}$, which is the axis of the parabola. The graph of the parabola has the vertex $(2.25, -6.125)$ and as we have determined opens upward. It is symmetric about the vertical line $x = \dfrac{9}{4}$.

Example (2):
The price p that a manufacturer charges for a bike when he produces x bikes is p(x) = 300 – 4x. He sells all that he produces. The cost of producing each of the x bikes, when x are produced, is C(x) = 50 – 2x.

(a) What is the revenue collected if he makes x bikes?
Revenue is a function of x:
R(x) = (the number produced) · (price/bike)
      = (x)(300 – 4x) = 300x - $4x^2$.

(b) What is the cost if he makes x bikes?
Cost is a function of x:

$C(x)$ = (the number produced) $\cdot$ (cost/bike)
= $(x)(50 - 2x)$ = $50x - 2x^2$.

(c) What is the profit from producing x bikes?

Profit is a function of x:

$P(x) = R(x) - C(x) = (300x - 4x^2) - (50x - 2x^2)$
$\quad = -2x^2 + 250x$.

If $y = P(x)$, we can say that $y = -2x^2 + 250x$. So, the profit function is a parabolic curve in the x-y plane. We know it opens downward since the coefficient on the x^2 term is negative. Therefore, the profit function will have a maximum.

(d) What is the vertex of $y = -2x^2 + 250x$, and what is the maximum (since we know obviously that the maximum is at the vertex)?

We know that the vertex is at: $\left(\frac{-b}{2a}, \frac{4ac-b^2}{4a}\right)$ = (62.5, 7812.50).
So, the profit is maximized when x = 62 or 63 bikes are made, and the maximum profit is $7812.00. (Note that the number of bikes sold must be a whole number).

(e) What is the minimum and maximum number of bikes that the company can produce to make any kind of profit at all?

We must find where $y = -2x^2 + 250x$ is zero. This happens when y = 0. So a profit of zero occurs when x = 0, or where

$x = 125$. Outside of the range [0, 125], the profit would be negative. The company must make anywhere from 1 to 124 bikes to have any profit at all.

################# Exercises #################

Find the solutions of the following quadratic equations:

(1) $x^2 - 2x - 8 = 0$

(2) $2x^2 - 10x + 2 = 0$

(3) $x^2 + 4x - 1 = 0$

(4) $3x^2 - 3x - 3 = 0$

(5) $2x^2 - 7x + 1 = 0$

(6) $x^2 - 9x - 10 = 0$

(7) $x^2 + 6x + 5 = 0$

(8) $x^2 + 11x + 1 = 0$

(9) Describe the graph of $y = \frac{1}{3}x^2 - 4x + 2$. Find the x-intercepts (where y = 0), y-intercepts (where x = 0), the vertex, and whether the graph opens up or down.

(10) Describe the graph of $y = x^2 + 2x + 1$. Find the x-intercepts (where y = 0), y-intercepts (where x = 0), the vertex, and whether the graph opens up or down.

(3.8) Higher Degree Polynomial Equations

This section is about polynomial functions of degree higher than two, that is, higher than the linear and quadratic functions that we have been working with up to this point. Things linear and quadratic constitute the larger share of what we spend our time working with in mathematics. In algebra, when we consider polynomial functions of degree greater than two, we run into issues of factoring them. Here, we will not discuss the procedures for factoring such polynomials since that is usually taught in other courses. So, in this section we will tell you how they factor if it is necessary, in both the examples and the exercises. The reader should find the material in this section interesting and should give it a read, and check out the exercises at the end of this section.

In our study of linear and quadratic equations in one variable, we were dealing with first and second degree equations respectively, where finding roots was easy, that is, we discussed functions of the two common forms: $y = ax + b$ and $y = ax^2 + bx + c$, where x and y are variables and a, b, and c are constants specific to a given application, and where of course y is a function of x.

As one could expect, we can study polynomial equations in one variable such as:

$$a_n x^n + a_{n-1} x^{n-1} + \cdots + a_2 x^2 + a_1 x + a_0 = 0 \quad (3.8.1)$$

where the $a_i's$ are real numbers, and $n \geq 3$. The so-called fundamental theorem of algebra says that an nth degree polynomial like this one always has n complex solutions, some of which are very often real numbers (note that we require that at least $a_n \neq 0$). If we write:

$$y = a_n x^n + a_{n-1} x^{n-1} + \cdots + a_2 x^2 + a_1 x + a_0 \quad (3.8.2)$$

then we have an nth degree polynomial function, the graph of which is in the x-y plane. When we set y = 0 in (3.8.2), we have equation (3.8.1). The real roots of equation (3.8.1) are the x-values where the graph of the function in equation (3.8.2) intersects the x-axis. In this section we will briefly discuss nth degree polynomial functions for $n \geq 3$.

Example (1):
For any odd degree polynomial function like
$y = ax^3 + bx^2 + cx + d$, or
$y = ax^5 + bx^4 + cx^3 + dx^2 + ex + f$, and so on,

the so-called end-behavior depends on the value of "a,"

If a is positive, then
$y \rightarrow (+\infty)$ as $x \rightarrow (+\infty)$, and $y \rightarrow (-\infty)$ as $x \rightarrow (-\infty)$.

If a is negative, then
$y \rightarrow (-\infty)$ as $x \rightarrow (+\infty)$, and $y \rightarrow (+\infty)$ as $x \rightarrow (-\infty)$.

Since complex roots always occur in complex conjugate pairs, the number of real roots for an odd power n is always one of {n, n − 2, . . . , 1}. For example:

If n = 3, the polynomial equation always has either
(a) three real roots, or
(b) one real root and two complex roots.

If n = 5, the polynomial equation always has either
(a) five real roots, or
(b) three real roots and two complex roots, or
(c) one real root and four complex roots.

This pattern is the same for odd degree n ≥ 7.

Example (2):
For any even degree polynomial function like
$y = ax^4 + bx^3 + cx^2 + dx + e$, or
$y = ax^6 + bx^5 + cx^4 + dx^3 + ex^2 + fx + g$, and so on,

the so-called end-behavior depends on the value of "a,"

If a is positive, then
$y \to (+\infty)$ as $x \to (+\infty)$, and $y \to (+\infty)$ as $x \to (-\infty)$.

If a is negative, then
$y \to (-\infty)$ as $x \to (+\infty)$, and $y \to (-\infty)$ as $x \to (-\infty)$.

Since complex roots always occur in complex conjugate pairs, the number of real roots for an even power n is always one of {n, n − 2, . . . , 0}. For example:

If n = 4, the polynomial equation always has either
(a) four real roots, or
(b) two real roots and two complex roots, or
(c) no real roots and four complex roots.

If n = 6, the polynomial equation always has either
(a) six real roots, or
(b) four real roots and two complex roots, or
(c) two real roots and four complex roots, or
(d) no real roots and six complex roots.

This pattern is the same for even degree n ≥ 8.

Some examples of cubic and quartic (fourth degree) polynomial functions are shown in the figure on the next page.

Example (3):
Find the solutions of the polynomial equation
$x^3 - 3x^2 - 18x = 0$?

Factoring, we get:
$$x^3 - 3x^2 - 18x = (x)(x^2 - 3x - 18)$$
$$= (x)(x - 6)(x + 3) = 0.$$

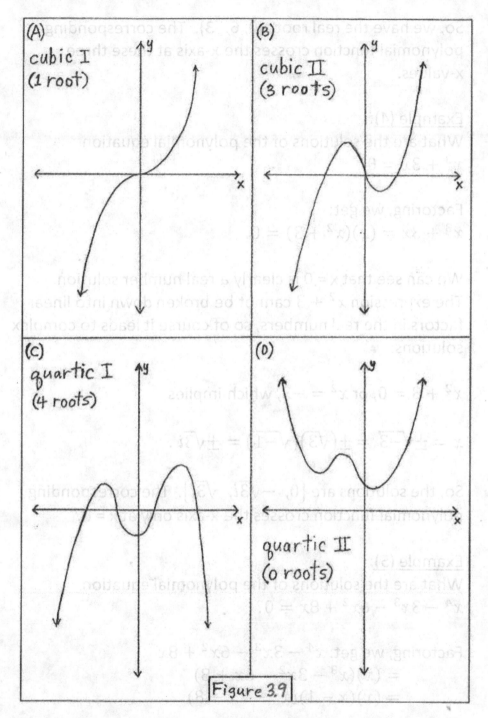

Figure 3.9

So, we have the real roots {0, 6, -3}. The corresponding polynomial function crosses the x-axis at these three x-values.

Example (4):
What are the solutions of the polynomial equation $x^3 + 3x = 0$?

Factoring, we get:
$x^3 + 3x = (x)(x^2 + 3) = 0$.

We can see that x = 0 is clearly a real number solution. The expression $x^2 + 3$ cannot be broken down into linear factors in the real numbers, so of course it leads to complex solutions:

$x^2 + 3 = 0$, or $x^2 = -3$, which implies

$x = \pm\sqrt{-3} = \pm(\sqrt{3})(\sqrt{-1}) = \pm\sqrt{3}i$.

So, the solutions are $\{0, -\sqrt{3}i, \sqrt{3}i\}$. The corresponding polynomial function crosses the x-axis only at x = 0.

Example (5):
What are the solutions of the polynomial equation $x^4 - 3x^3 - 6x^2 + 8x = 0$?

Factoring, we get: $x^4 - 3x^3 - 6x^2 + 8x$
$$= (x)(x^3 - 3x^2 - 6x + 8)$$
$$= (x)(x - 1)(x^2 - 2x - 8)$$

$$= (x)(x - 1)(x + 2)(x - 4) = 0$$

So, the solutions are {0, 1, -2, 4}. The corresponding polynomial function crosses the x-axis at these four values.

Example (6):
What are the solutions of the polynomial equation
$x^4 + 4x^2 - 5 = 0$?

Factoring, we get:
$$x^4 + 4x^2 - 5 = (x - 1)(x^3 + x^2 + 5x + 5)$$
$$= (x - 1)(x + 1)(x^2 + 5) = 0$$

Clearly, x = 1 and x = -1 are two real number solutions. The expression $(x^2 + 5)$ cannot be broken down to linear factors in the real numbers.

$(x^2 + 5) = 0$, or $x^2 = -5$, or

$$x = \pm\sqrt{-5} = \pm(\sqrt{5})(\sqrt{-1}) = \pm\sqrt{5}i .$$

So, the solutions are $\{1, -1, -\sqrt{5}i, \sqrt{5}i\}$. The corresponding polynomial function crosses the x-axis at x = 1 and x = -1.

Example (7):
What are the solutions of the polynomial equation
$x^4 + 15x^2 + 54 = 0$?

Factoring, we get $x^4 + 15x^2 + 54$
$$= (x^2 + 6)(x^2 + 9) = 0.$$

$(x^2 + 6) = 0$ when $x = \pm\sqrt{6}i$,

$(x^2 + 9) = 0$ when $x = \pm 3i$.

So, the solutions are $\{-\sqrt{6}i, \sqrt{6}i, -3i, 3i\}$. The corresponding polynomial function does not cross the x-axis.

################## Exercises ##################

(1) For the function $y = x^3 - 5x^2 + 13x - 65$:
    (a) Where are the x-intercepts?
       (Hint: It factors to $(x - 5)(x^2 + 13)$)
    (b) What is the end behavior for this function?

(2) For the function $y = x^4 + x^3 - 6x^2$:
    (a) Where are the x-intercepts?
    (b) What is the end behavior for this function?

(3) For the function $y = -x^4 - 6x^2 + 135$:
    (a) Where are the x-intercepts?
       (Hint: It factors to $-(x^2 - 9)(x^2 + 15)$)
    (b) What is the end behavior for this function?

(4) For the function $y = -x^3 + 15x^2 - 66x + 80$:
    (a) Where are the x-intercepts?
       (Hint: It factors to $-(x^2 - 10x + 16)(x - 5)$)
    (b) What is the end behavior for this function?

(3.9) Exponential and Logarithmic Equations

Exponential and logarithmic expressions, equations, and functions occur very frequently in mathematics, so we will consider a few examples in this chapter.

Example (1):

Solve for x in the equation: $2^x(8) = 2^{13x}$.

Note that $8 = 2^3$, so that $2^x(2^3) = 2^{13x}$, or $2^{x+3} = 2^{13x}$.

This means that $x + 3 = 13x$ when we equate the powers on the common base. Then $x = \frac{1}{4}$ is the solution. The reader should note that there is another illustrative way that the previous exponential equation $2^{x+3} = 2^{13x}$ could be solved. We should notice that: $\log_2(2^{x+3}) = \log_2(2^{13x})$. Then, using law (4) from the laws of logarithms in section (3.2), we have $x + 3 = 13x$, which says that $x = \frac{1}{4}$.

Example (2):

Solve for x in the equation: $2x = 10 \cdot \log_{10}\left(\frac{7 \times 10^{x+2}}{2 \times 10^{3x-5}}\right)$.

Dividing both sides by 2 and expanding the logarithm:
$x = 5 \cdot [\log_{10}(7 \times 10^{x+2}) - \log_{10}(2 \times 10^{3x-5})]$

Simplifying using the laws of logarithms:
$x = 5 \cdot [\log_{10}\left(\frac{7}{2}\right) + (x + 2) - (3x - 5)]$
$x = 5 \cdot [\log_{10} 3.5 - 2x + 7]$

$$x = 5 \cdot [(0.544) + 7 - 2x] = (37.720) - 10x$$

Then, $11x = 37.720$. So, $x = (3.429)$ is the solution, after rounding calculations to 3 decimal places.

Example (3):
Solve for x in the equation: $2^{2x} + 2^x - 8 = 0$.

Re-write as: $(2^x)^2 + 2^x - 8 = 0$. This is quadratic in (2^x).
Let m $=(2^x)$, and solve the quadratic equation:
$m^2 + m - 8 = 0$.

$$m = \frac{-1 \pm \sqrt{(1)^2 - (4)(1)(-8)}}{2} = \frac{-1 \pm \sqrt{33}}{2} = -3.372 \text{ or } 2.372.$$
Since m must be positive, we discard the negative solution, and then we have: $m = 2^x = 2.372$. Then using Law 4 of Logarithms from section (3.2): $\log_2(2^x) = \log_2(2.372)$.

So, x $= \log_2(2.372) = \frac{\ln(2.372)}{\ln(2)}$, when using Logarithm law (5) in section (3.2). We have x = 1.246 as the solution (after rounding all calculations to three decimal places).

Example (4):
Solve for x in the equation:
$\ln(x - 7) - \ln(2x + 13) + \ln(x) = \ln(3)$.

Using the laws of logarithms, this simplifies to:
$\ln\left(\frac{x(x - 7)}{2x + 13}\right) = \ln(3)$.

116

Making each side a power of e yields, $\left(\dfrac{x^2 - 7x}{2x + 13}\right) = 3$.

Then, we arrive at $x^2 - 13x - 39 = 0$.

Using the quadratic formula, $x = \dfrac{13 - 5\sqrt{13}}{2}$, or $\dfrac{13 + 5\sqrt{13}}{2}$,

that is, x is (-2.514) or (15.514). We must discard the negative solution because it leads to a problem in the original equation. So, the solution is x = 15.514.

Example (5):
Solve for x in the equation $2.4e^{-35x} = (1.6)$.

Firstly, we divide both sides by 2.4 to get:
$e^{-35x} = 0.6667$ (rounded to 4 decimal places).

Then take the natural logarithm of both sides:
$\ln(e^{-35x}) = \ln(0.6667) = -0.405465$
(rounded to 6 decimal places).

So we have $-35x = -0.405465$, or
$x = 0.011585$, rounded to 6 decimal places.

Figure (3.11) on the next page shows some exponential and logarithmic curves. We show the curves for f(x) = e^x and f(x) = e^{-x}, which correspond to curves (as x increases) of exponential growth and decay. Then we show the curve for the logarithmic curve f(x) = ln(x).

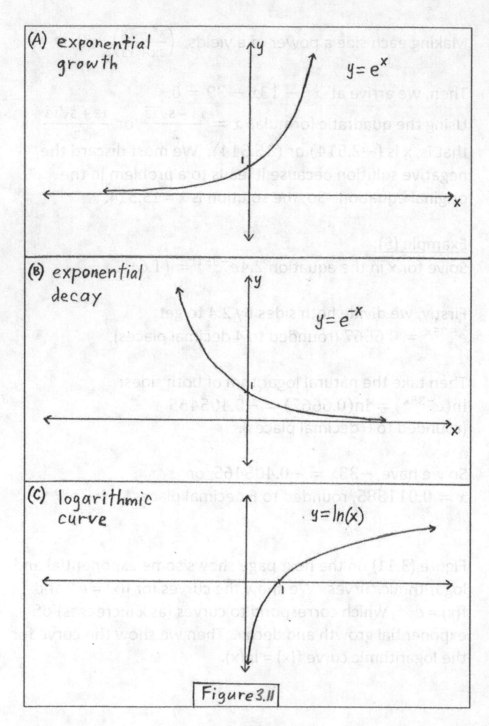

(A) exponential growth

$y = e^x$

(B) exponential decay

$y = e^{-x}$

(C) logarithmic curve

$y = \ln(x)$

Figure 3.11

118

There are many different models of the growth of biological populations. Many different factors must be considered such as the carrying capacity of the environment, and fighting and competitive behavior between individuals in the population as resources get strained. There are many factors that are hard to predict. However, here we will present the most basic model, which is a more simplistic type of model based on the fact that members of many populations can reproduce at a greater rate than the rate at which they die off. This is common with biological organisms, and if there is nothing to counteract the population growth they will increase in number greatly.

If we have a population of some organisms, which is growing in number at an increasing rate, then exponential models are a natural way of describing the growth. The most basic model is:

$$A(t) = (A_0)e^{kt},$$

where t is time, A(t) is the number of organisms at time t, A_0 is the number of organisms at time t = 0, and k is the rate of growth (k is expressed as a decimal number in the model, which we can interpret as a percentage). We will now apply this model to human population growth.

Example (5):
Let time t = 0 correspond to the beginning of the year 2018. If we assume that at the beginning of 2018, the human

population is 7.4 billion, and that the rate of population growth per year is 2.0% (k = .02), then our model is:

$A(t) = (7.4 \times 10^9)e^{(.02)t}$, where time t is in years.

This is probably a good model for the next few decades assuming no major calamities. Our model will most likely not be valid if we try to extrapolate too far into the future, because we cannot expect in general for current conditions and trends to stay as they are.

(A) Using this model, we can say that at the beginning of 2030 (t = 12), the human population will be:

$A(12) = (7.4 \times 10^9)e^{(.02)(12)} \approx 9.534$ billion people.

(B) Assuming that current trends continue, to figure when the human population will be 10 billion people, we can set $A(t) = 10,000,000,000 = (1.0 \times 10^{10}) = (10 \times 10^9)$ and solve for t in our model.

So, $(10 \times 10^9) = (7.4 \times 10^9)e^{(.02)t}$, or

$\left(\dfrac{10 \times 10^9}{7.4 \times 10^9}\right) = e^{(.02)t}$, or $1.351 = e^{(.02)t}$

Taking the natural logarithm of both sides, we get,
$\ln(1.351) = \ln\left(e^{(.02)t}\right) = (.02)t$, or
$0.300845 = (.02)t$.

Then, solving for t, we get t = 15.04 years, which would be about January of 2033.

Example (6):
The basic structure of the exponential growth model can be used for exponential decay. The only difference is that the growth parameter k is negative.

(A) Assume that at time t = 0 (t in years), a certain sample contains 6.88×10^{22} atoms of a certain radioactive element which has a half-life of 10,400 years. Then we can use the same type of exponential model (but with k < 0) to model exponential decay:

$$A(t) = (A_0)e^{kt}, \text{ where t is in years } (t \geq 0).$$

Since we know that the half-life for this element is 10,400 years, then we know that the number of atoms which have decayed after 10,400 years is (3.44×10^{22}). Plugging things into our model, we can figure k.

$$(3.44 \times 10^{22}) = (6.88 \times 10^{22})e^{k(10,400)}, \text{ or } \frac{1}{2} = e^{k(10,400)}.$$

Taking the natural logarithm of both sides,
$$\ln\left(\frac{1}{2}\right) = \ln\left(e^{k(10,400)}\right), \quad \text{or} \quad -(.69315) = k(10,400).$$
This gives us k = $\left(\frac{-(.69315)}{10,400}\right) = -(.00007) = -(.007)\%$.
This is the yearly rate of decay. So, our complete model is:

$$A(t) = (6.88 \times 10^{22})e^{-(.00007)t} .$$

(B) To figure how many years until the number of atoms is (1.5×10^{22}), we set up the equation:

$$(1.5 \times 10^{22}) = (6.88 \times 10^{22})e^{-(.00007)t}, \text{ or}$$

$$\left(\frac{1.5 \times 10^{22}}{6.88 \times 10^{22}}\right) = e^{-(.00007)t}.$$

Taking the logarithm of both sides gives us:

$$\ln\left(\frac{1.5 \times 10^{22}}{6.88 \times 10^{22}}\right) = \ln\left(e^{-(.00007)t}\right), \text{ or}$$

$$(-1.52315) = -(.00007)t. \quad \text{So, } t \approx 21,759 \text{ years.}$$

(C) How many years will pass before there are
$A(t) = 8.0 \times 10^{21}$ atoms remaining?

Setting up the equation from our model:

$$(8.0 \times 10^{21}) = (6.88 \times 10^{22})e^{-(.00007)t}, \quad \text{or}$$
$$\left(\frac{8.0 \times 10^{21}}{6.88 \times 10^{22}}\right) = e^{-(.00007)t}, \quad \text{or} \quad (0.11628) = e^{-(.00007)t}.$$

Taking the logarithm of both sides:

$$\ln(0.11628) = \ln\left(e^{-(.00007)t}\right) = -(.00007)t, \quad \text{or}$$
$$-(2.15176) = -(.00007)t.$$

Solving for t we get, $t \approx 30,739$ years.

Example (7):
The level of sound in decibels (L) for humans is modeled in the following way:

$$L = 10 \cdot \log_{10}\left(\frac{I}{I_0}\right) \text{ decibels.}$$

Where I is the sound intensity in $\left(\frac{\text{Watts}}{(\text{meter})^2}\right)$ for a given sound, and $I_0 = 1.0 \times 10^{-12}\left(\frac{\text{Watts}}{(\text{meter})^2}\right)$ is the threshold of human hearing.

How many decibels is the sound of a fire alarm that produces a sound intensity of $I = 9.8 \times 10^{-10}\left(\frac{\text{Watts}}{(\text{meter})^2}\right)$?

$$L = 10 \cdot \log_{10}\left(\frac{9.8 \times 10^{-10}}{1.0 \times 10^{-12}}\right) = 29.9 \text{ decibels.}$$

Example (8):
A geologist has developed a new method of measuring the energy released in an earthquake using a logarithmic scale. Using $E_0 = 2.6 \times 10^{10}$ Joules as the energy released in a quake of magnitude 0, measure the magnitude M with:

$M = 100 \cdot \log_{10}\left(\frac{E}{E_0}\right)$ Eigens. (M has magnitude in Eigens).

If an earthquake releases $E = 9.7 \times 10^{17}$ joules of energy, what is the magnitude on this new scale? The magnitude is:

$M = 100 \cdot \left(\log_{10}\left(\frac{9.7 \times 10^{17}}{2.6 \times 10^{10}}\right)\right) = 757.2$ Eigens.

################# Exercises #################

Solve the equations for x:

(1) $2^2 \cdot 2^{x+3} = 8^x \cdot 16^x$

(2) $10^x + 20(10)^x = 2100$

(3) $\log_2(x) + \log_2(2x) - \log_2(6x) = 1$

(4) $e^{4x-3} = e^{50x+12}$

(5) $\log_{10}(x) - \log_{10}(x-3) = \log_{10}(x-9)$

(6) $(e^x)^2 \cdot e^5 = e^x$

(3.10) Inequalities

We have four basic forms for an inequality:
$a < b, \ a > b, \ a \le b, \ a \ge b.$

Basic Rules
Without any loss of generality, we will consider the case
of $a < b$.

(a) When we have an inequality of the form a < b, then we
can add any real number c (positive or negative) to both

124

sides and the sense of the inequality (its direction) will not change.

So, if a < b, then (a + c) < (b + c) .

(b) When we have a < b, and multiply or divide by a positive number c, then the sense of the inequality does not change.

So, if a < b, and c is positive,

Then ac < bc, or $\frac{a}{c} < \frac{b}{c}$.

(c) However, when we have a < b, and multiply or divide by a negative number c, then the sense of the inequality does change.

So, if a < b, and c is negative,

Then ac > bc, or $\frac{a}{c} > \frac{b}{c}$.

(d) When we have a < b, and a and b are both positive or both negative, then if we take the reciprocal of both sides, then the sense of the inequality changes. That is,

If a < b, then $\frac{1}{a} > \frac{1}{b}$. (a and b are of the same sign).

It may surprise some that when we have a < b, where a is negative and b positive, then here it is not true that $\frac{1}{a} > \frac{1}{b}$.

Linear Inequalities

What we will call a linear inequality is a statement, in one of our four forms, that involves one variable to the first power. Without loss of generality, we'll consider $ax < b$.

If a is positive, then $ax < b$ implies $x < \dfrac{b}{a}$. The sense of the inequality does not change.

If a is negative, then $ax < b$ implies $x > \dfrac{b}{a}$. The sense of the inequality does change.

The solution set for these inequalities is an interval subset of the real number line. For example, $x < \dfrac{b}{a}$ corresponds to the interval $\left(-\infty, \dfrac{b}{a}\right)$, and $x > \dfrac{b}{a}$ corresponds to $\left(\dfrac{b}{a}, \infty\right)$.

Example (1):

(a) If $\dfrac{3}{2}x \geq -3$, then dividing both sides by $\dfrac{3}{2}$, which is of course equivalent to multiplying both sides by $\dfrac{2}{3}$, leads to

$x \geq (-3)\left(\dfrac{2}{3}\right)$ or $x \geq -2$. So, the solution set is $[-2, \infty)$.

(b) If $9x \leq 14$, then $x \leq \dfrac{14}{9}$. The solution set is $\left(-\infty, \dfrac{14}{9}\right]$.

(c) If $-4x > -9$, then $x < \dfrac{9}{4}$. The solution set is $\left(-\infty, \dfrac{9}{4}\right)$.

(d) If $-2x < 10$, then $x > -5$. The solution set is $(-5, \infty)$.

Example (2):
(a) If we have the three part inequality $a < cx < b$,
Then dividing by c, we get:

$\frac{a}{c} < x < \frac{b}{c}$, when c is positive, and

$\frac{b}{c} < x < \frac{a}{c}$, when c is negative. The sense of the inequality

changes when we divide by a negative number.

(b) If $-3 < -7x < -1$, then dividing by -7 leads to:
$\frac{1}{7} < x < \frac{3}{7}$. The solution set is $\left(\frac{1}{7}, \frac{3}{7}\right)$.

(c) If $2 \leq 5x \leq 30$, then dividing by 5 leads to:
$\frac{2}{5} \leq x \leq 6$. The solution set is $\left[\frac{2}{5}, 6\right]$.

Quadratic Inequalities

Example (3):
(a) For a > 0, If $x^2 < a$, then $-\sqrt{a} < x < \sqrt{a}$.
The solution set is the interval $(-\sqrt{a}, \sqrt{a})$.

(b) For a > 0, If $x^2 > a$, then $x < -\sqrt{a}$ or $x > \sqrt{a}$.
The solution set is $(-\infty, -\sqrt{a}) \cup (\sqrt{a}, \infty)$.

(c) If $x^2 < 9$, then $-3 < x < 3$. The solution set is $(-3,3)$.

(d) If $(x - 2)^2 \leq 16$, then $-4 \leq (x - 2) \leq 4$. Then add 2 to each part of this inequality, which leads to $-2 \leq x \leq 6$. The solution set is $[-2,6]$.

################## Exercises ##################

Find the solution set for each of the following inequalities.

(1) $\frac{1}{4}x \geq \frac{1}{16}$ (2) $(3x + 3 - 5x) < (1 + 2x)$

(3) $(8x) > (12 - x)$ (4) $\frac{3}{7}x \leq \frac{5}{13}$

(5) $-3x > -29$ (6) $|x| < 5$ (7) $|x| \geq 2$

(8) $|x - 4| < \frac{1}{2}$ (9) $-1 \leq 3x \leq 1$

(10) $-26 < 13x < 39$ (11) $-1 < 23x < 12$

(12) $\frac{1}{2} < 2x + 4 < 3$ (13) $x^2 \leq 14$

(14) $1 < x^2 < 9$

(3.11) Systems of Linear Equations

When we have the system of two linear equations:

$$\begin{cases} ax + by = c \\ dx + ey = f \end{cases}, \text{ where } \{a, b, c, d, e, f\} \text{ are numbers.}$$

what we have is a set consisting of two lines in the x-y plane, and what we want to know are the points (x,y) that satisfy both equations simultaneously.

Case (1): Consider the system $\begin{cases} 2x - 3y = 8 \\ -2x + 4y = 2 \end{cases}$.

Adding these two equations together leads to y = 10. Then going back to either equation and substituting 10 for y allows us to find x. Using the first equation, we have that

$$2x - 3(10) = 8 \quad \rightarrow \quad 2x = 38 \quad \rightarrow \quad x = 19.$$

So, the solution is the single point $(x, y) = (19, 10)$. The two lines intersect at this one point.

Case (2): Consider the system $\begin{cases} \frac{1}{2}x - 2y = -1 \\ \frac{3}{2}x - 6y = -3 \end{cases}$.

Since the first equation is a multiple of the second (or vice versa), these two lines are one and the same line that are represented differently (multiply the first equation by 3). Therefore, there are an infinite number of solutions to the system. They are all the points on the one line.

Case (3): Consider the system $\begin{cases} 2x + 3y = 4 \\ 2x + 3y = 6 \end{cases}$.

It is impossible for these two lines to intersect because they are parallel. We can write them both in slope-intercept form:

$$y = -\frac{2}{3}x + \frac{4}{3}$$
$$y = -\frac{2}{3}x + 2$$

These two lines have the same slope but they have different y-intercepts, so they are parallel. So, there is no point (x,y) in the plane that satisfies both equations simultaneously. In other words, there is no solution to this system.

So, we can see that a system of linear equations can have one solution, an infinite number of solutions, or no solution.

Example (1):

Solve the system $\begin{cases} x - 9y = 1 \\ x + 2y = 4 \end{cases}$.

Firstly, we want to eliminate one variable. Multiply the first equation by -1, which leads to:

$$\begin{cases} -x + 9y = -1 \\ x + 2y = 4 \end{cases}$$

Then add the two equations together to eliminate x. We get: $11y = 3$. So, $y = \frac{3}{11}$. We can plug this value for y into either equation, we'll choose the second. This leads to:

$x + 2\left(\frac{3}{11}\right) = 4$. This leads to: $x = \frac{38}{11}$.

So, the single point $(x, y) = \left(\frac{38}{11}, \frac{3}{11}\right)$ is the solution to this system.

Example (2):

Solve the system $\begin{Bmatrix} 3x - 2y = 2 \\ x + y = 6 \end{Bmatrix}$.

Firstly, we want to eliminate one variable. Multiply the second equation by -3, which leads to:

$$\begin{Bmatrix} 3x - 2y = 2 \\ -3x - 3y = -18 \end{Bmatrix}.$$

Then add the two equations together to eliminate x. We get: $-5y = -16$. So, $y = \frac{16}{5}$. We can plug this value for y into either equation, we'll choose the first. This leads to:

$3x - 2\left(\frac{16}{5}\right) = 2$. This leads to: $x = \frac{14}{5}$.

So, the single point $(x, y) = \left(\frac{14}{5}, \frac{16}{5}\right)$ is the solution to this system.

################ Exercises ##################

Solve the following systems of equations.

(1) $\begin{cases} x - y = 1 \\ 3x + y = 5 \end{cases}$ (2) $\begin{cases} 2x - 7y = 12 \\ 8x - 28y = 48 \end{cases}$

(3) $\begin{cases} 2x + 3y = -1 \\ -2x - 4y = -7 \end{cases}$ (4) $\begin{cases} x + 3y = 4 \\ 2x - y = 1 \end{cases}$

(3.12) Summation Algebra

Subscript Notation
We often deal with a set of several variables such as:

$$\{X_1, X_2, \ldots, X_n\} \text{ or } \{x_1, x_2, \ldots, x_m\},$$

which for example could be a set of n random variables, or a collection of m observations from some experiment.

When we have several variables like this, we could use different letters for each of them, but that would be cumbersome and there are only 26 letters in the English alphabet. Sometimes the number of variables could be very large, so we have to use numeric subscripts to distinguish between each of the variables in the list and to work with them algebraically. The subscripts are of course the symbols to the lower right of the variables, and identify an individual variable. This is much more flexible, and can be applied to many different types of situations.

Sometimes we have variables that are part of a two-dimensional classification scheme, where we must have variables with two subscripts, such as X_{ij} , where i and j may range over two different index sets (each with a finite number or even infinite number of members). We could have a three-dimensional classification scheme, so we must have variables with three subscripts, such as X_{ijk} , where i, j, and k may range over three different index sets (each with a finite number or even infinite number of members). This type of thing can be extended to even more than three dimensions.

Summations
Summation algebra is very common in mathematics, and especially in probability and statistics, because we frequently encounter sums of many different types, and they can contain very many terms. For example, the sum of the first 35 positive integers, call it S, can be written:

$$S = (1 + 2 + 3 + 4 + 5 + 6 + 7 + 8 + 9 + 10 + 11 + 12 + 13 + 14$$
$$+ 15 + 16 + 17 + 18 + 19 + 20 + 21 + 22 + 23 + 24 + 25$$
$$+ 26 + 27 + 28 + 29 + 30 + 31 + 32 + 33 + 34 + 35).$$

Fortunately, mathematicians have created a shorthand notation for this. We use the Greek capital letter sigma Σ to denote a sum. We can use the notation $\sum_{i=1}^{35}(i)$ as a shorthand for the sum S above.

We can write $\sum_{i=1}^{1000}(i)$ for the sum of the first one thousand positive integers. The $\sum_{i=1}^{N}(i)$ means add up the numbers i, where i begins with 1, then i increases to 2, then i increases to 3, and so on, until the process ends when $i = 1000$. So, $\sum_{i=1}^{N}(i)$ is a shorthand for $(1 + 2 + 3 + \cdots + N)$, no matter how large N may be.

The sum $\sum_{i=1}^{N}(i^2)$ is an easy way to write:
$(1^2 + 2^2 + 3^2 + \cdots + N^2)$.

The sum $\sum_{i=1}^{N}(i^3)$ is an easy way to write:
$(1^3 + 2^3 + 3^3 + \cdots + N^3)$.

We can see how much energy we can save with this type of notation, especially if we wanted to represent the sum of the set of quantities (where we have 10,000 of them):

$$\left\{\left(\frac{X_i - 23.7}{5.6}\right), i = 1, 2, 3, \ldots, 10{,}000\right\},$$
where we have the 10,000 variables $\{X_1, X_2, \ldots, X_{10,000}\}$.

We simply use the shorthand: $\sum_{i=1}^{10,000}\left(\frac{X_i - 23.7}{5.6}\right)$

to mean the sum $\left\{\left(\frac{X_1 - 23.7}{5.6}\right) + \cdots + \left(\frac{X_{10,000} - 23.7}{5.6}\right)\right\}$.

The letter (i) is called the index of the summation. We could use any suitable letter. The three sums:

$$\sum_{i=1}^{N}(i), \quad \sum_{k=1}^{N}(k), \quad \sum_{x=1}^{N}(x)$$

all represent the sum of the first N positive integers. In many cases, the context dictates certain commonly used notations and symbols.

The reader may be interested in three formulas, which are used to represent certain sums:

$$\sum_{i=1}^{n}(i) = \frac{(n)(n+1)}{2}$$

$$\sum_{i=1}^{n}(i^2) = \frac{(n)(n+1)(2n+1)}{6}$$

$$\sum_{i=1}^{n}(i^3) = \left(\frac{(n)(n+1)}{2}\right)^2$$

As an example, determine the sum $\sum_{i=1}^{5}(2i - 7)$:

The $\sum_{i=1}^{5}(2i - 7)$

$$= \sum_{i=1}^{5}(2i) - \sum_{i=1}^{5}(7)$$

$$= (2(1) + 2(2) + 2(3) + 2(4) + 2(5)) - (7 + 7 + 7 + 7 + 7)$$

$$= (2 + 4 + 6 + 8 + 10) - (35)$$

$$= (-5).$$

Another way that this can be figured is to use the first of the summation formulas presented above:

$$\sum_{i=1}^{5}(2i - 7) = \sum_{i=1}^{5}(2i) - \sum_{i=1}^{5}(7)$$

$$= 2\sum_{i=1}^{5}(i) - (5)(7)$$

$$= 2\left(\frac{(5)(5 + 1)}{2}\right) - 35$$

$$= 2(15) - 35$$

$$= (-5).$$

Double Summations

Double summations occur quite frequently in probability and statistics. The relationship between two variables is often of interest in mathematics because it then leads us to a definition of the relationship between a "collection of several variables" with itself, by considering the relationship between all the possible pairs of two variables.

For example, a double summation that will occur later in this book is:

$$\sum_{i=1}^{N}\sum_{j=1}^{M} p_{XY}(x_i, y_j).$$

If $p_{XY}(x_i, y_j)$ can be factored into the form $p_X(x_i)p_Y(y_j)$, then we can rewrite this double sum as:

$$\sum_{i=1}^{N}\sum_{j=1}^{M} p_X(x_i)p_Y(y_j) = \sum_{i=1}^{N} p_X(x_i) \cdot \sum_{j=1}^{M} p_Y(y_j).$$

This algebraic manipulation is possible because all factors to the right of the double sum notation that don't involve the index j can be brought across the summation symbol since they involve i in this case, but not j.

As another example of a double sum used later, the result

$$\sum_{i=1}^{N}\sum_{j=1}^{M}(i+j) = \sum_{j=1}^{M}\sum_{i=1}^{N}(i+j)$$

is true since we have finite sums. The reader will be asked to verify that this result is true for finite numbers N and M in the exercises.

################# Exercises #################

(1) Find: $\sum_{i=4}^{10} 3i$

(2) Show that for the numbers {1,3,5,7,9},
    where $x_1 = 1$, $x_2 = 3$, $x_3 = 5$, $x_4 = 7$, and $x_5 = 9$,
    and with average $\bar{x} = \frac{\sum_{1}^{5} x_i}{5}$ that:
    $\sum_{i=1}^{5}(x_i - \bar{x}) = 0$.

(3) Show that $\sum_{m=1}^{4}\sum_{n=1}^{3}(m+n) = \sum_{n=1}^{3}\sum_{m=1}^{4}(m+n)$.

(4) Determine each of the sums:

  (a) $\sum_{i=1}^{20}(i^2 - 6)$ (b) $\sum_{i=1}^{100}(i + 2)$ (c) $\sum_{i=1}^{5} i^3$

***** GEOMETRY *****

(3.13) Basic Plane and Solid Geometry

Geometry is a familiar subject where algebraic symbolism and equations are abundant. All students should be familiar with the basic results in plane geometry, and with a few formulas in solid geometry.

BASIC PLANE GEOMETRY

For the sections on Circles, Segments and Lines, Rays, Angles, Radian measure for an angle, Arc length, and Sectors, refer to Figure (4.3) on the next page.

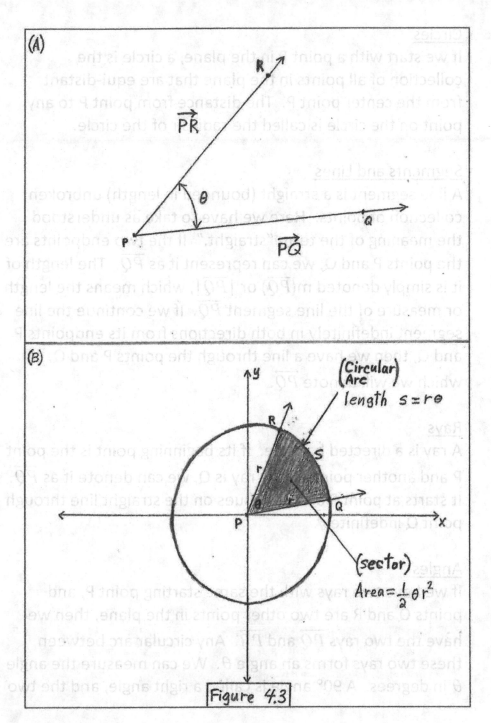

Figure 4.3

Circles

If we start with a point P in the plane, a circle is the collection of all points in the plane that are equi-distant from the center point P. The distance from point P to any point on the circle is called the radius r of the circle.

Segments and Lines

A line segment is a straight (bounded in length) unbroken collection of points. Here we have to take as understood the meaning of the term "straight." If the two endpoints are the points P and Q, we can represent it as $\overline{PQ}$. The length of it is simply denoted $m(\overline{PQ})$ or $|\overline{PQ}|$, which means the length or measure of the line segment $\overline{PQ}$. If we continue the line segment indefinitely in both directions from its endpoints P and Q, then we have a line through the points P and Q, which we will denote $\overleftrightarrow{PQ}$.

Rays

A ray is a directed half line. If its beginning point is the point P and another point on the ray is Q, we can denote it as $\overrightarrow{PQ}$. It starts at point P and continues on the straight line through point Q indefinitely.

Angles

If we have two rays with the same starting point P, and points Q and R are two other points in the plane, then we have the two rays $\overrightarrow{PQ}$ and $\overrightarrow{PR}$. Any circular arc between these two rays forms an angle θ. We can measure the angle θ in degrees. A 90° angle is called a right angle, and the two rays $\overrightarrow{PQ}$ and $\overrightarrow{PR}$ would be perpendicular. A 180° angle would be one half of a complete rotation around any circle centered at P. In this case the rays $\overrightarrow{PQ}$ and $\overrightarrow{PR}$ would be directed in opposite directions and would form the line $\overleftrightarrow{QR}$ through the three points P, Q, and R. The point P would lie on the line somewhere between points Q and R. A 270° angle would be $\left(\frac{3}{4}\right)$ths of a circular rotation about the point P, and the rays $\overrightarrow{PQ}$ and $\overrightarrow{PR}$ would again be mutually perpendicular to each other, but in other directions. A 360° angle would be one full circular rotation about the point P. So, a circle is divided into 360 angles each of measure 1°. A degree is divided into 60 minutes, and a minute is divided into 60 seconds. This system of angular measure is handed down to us from middle eastern peoples.

Radian Measure of an Angle
Suppose we have a circle with center point P and radius r, where r > 0. Between two rays $\overrightarrow{PQ}$ and $\overrightarrow{PR}$, where the two points Q and R are on the circle, we have the angle θ. We can describe the measure of the angle θ in so-called radians instead of degrees. Let the arc of the circle from point Q to point R be of length s. Then the radian measure of angle θ is the ratio $\left(\frac{s}{r}\right)$. This says that the radian measure of an angle θ is the multiple of radius lengths for the arc length s that the angle θ sweeps out. It has been known since antiquity that the ratio of the circumference of any circle to its diameter (the diameter is twice the radius) is that mysterious irrational number π, the value of which has been calculated to many thousands of decimal places. We will

approximate π as 3.141592654. Therefore, the ratio of the circumference to the radius is 2π. That is, calling the circumference C, the ratio $\frac{C}{r} = 2\pi$. For this reason we can say that 360° is 2π radians. So a 90° angle would be of measure $\frac{2\pi}{4} = \frac{\pi}{2}$ radians, a 180° angle would be of measure $\frac{2\pi}{2} = \pi$ radians, and a 270° angle would be of measure $\frac{3}{4}(2\pi) = \frac{3}{2}\pi$ radians. We can go back and forth between degrees and radians by use of conversion factors. For an angle θ in degrees, the radian measure of θ is:

$$\theta \text{ (in radians)} = \left(\frac{\pi \text{ radians}}{180 \text{ degrees}}\right) \cdot (\theta \text{ degrees}).$$

For an angle θ in radians, the degree measure of θ is:

$$\theta \text{ (in degrees)} = \left(\frac{180 \text{ degrees}}{\pi \text{ radians}}\right) \cdot (\theta \text{ radians}).$$

As an example, 1 radian $= \left(\frac{180 \text{ degrees}}{\pi \text{ radians}}\right) \cdot (1 \text{ radian})$

$$= \left(\frac{180}{\pi}\right) \text{ degrees} \approx 57.296°$$

The criterion for determining which conversion factor to use is by choosing the one where either the units (radians) or the units (degrees) cancels out. As another example,

$$2.35 \text{ radians} = \left(\frac{180 \text{ degrees}}{\pi \text{ radians}}\right) \cdot (2.35 \text{ radians})$$

$$= \left(\frac{(180) \cdot (2.35)}{\pi}\right) \text{ degrees} \approx 134.645°$$

Arc Length and Sectors

If angle θ is measured in radians, then we have the relation $s = r\theta$, where s is the arc length. So, if the angle θ has measure 2π radians, then we can see that the distance around the circle, the circumference of the circle, is $2\pi r$. For a circle centered at point P, with two different rays $\overrightarrow{PQ}$ and $\overrightarrow{PR}$ (where points Q and R are on the circle) that form the angle θ between the rays $\overrightarrow{PQ}$ and $\overrightarrow{PR}$, then the region between rays $\overrightarrow{PQ}$ and $\overrightarrow{PR}$ and the circle is called a sector of the circle. It turns out that the area of a sector is $\frac{1}{2}\theta r^2$, for $0 \le \theta \le 2\pi$. So, if the angle θ has measure 2π radians, we can see that the area of the region enclosed by the circle is $\frac{1}{2}(2\pi)r^2 = \pi r^2$.

For the following sections on polygons, refer to Figure (4.5) on the next page.

Polygons

A polygon is an n-sided convex geometric figure in the plane. We will only consider polygons of n = 3 or 4 sides. The sum of the interior angles for a polygon of n sides is $(n-2) \cdot 180°$. When n = 3, we have a triangle. When n = 4, we have some kind of quadrilateral, and the ones that we will consider here are the square, rectangle, parallelogram, and the trapezoid.

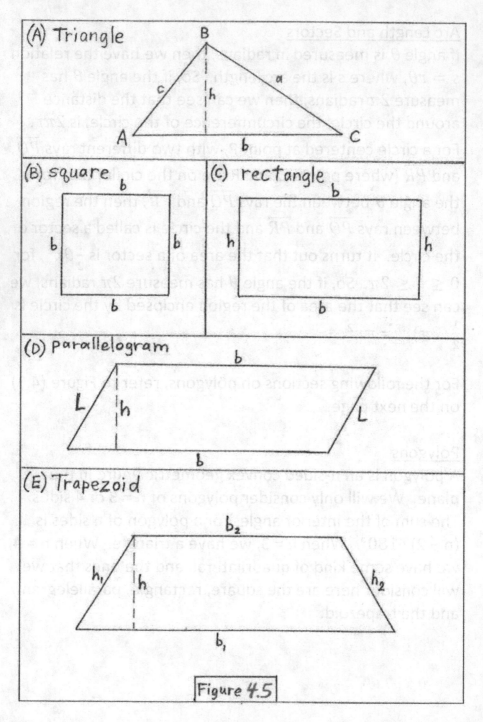

Figure 4.5

(A) For a <u>triangle</u>, which is a three sided polygon, the sum of the interior angles is 180° and the area is $A = \frac{1}{2}(b)(h)$, where b is the length of the base and h is the height. The sum of the lengths of the sides of a polygon is the perimeter P. For a triangle, with sides of lengths a, b, and c, the perimeter is $P = (a + b + c)$.

A <u>right triangle</u> is one where one of the interior angles is 90° (the right angle), and the other two interior angles are each less than 90° (these other two interior angles add to 90°). The side opposite the 90° angle is the longest side and is called the hypotenuse. The other two sides are the legs of the right triangle and are each of length less than the length of the hypotenuse. For a right triangle in a plane with legs of lengths a and b, and the hypotenuse is of length c, the Pythagorean Theorem says that: $a^2 + b^2 = c^2$. This Pythagorean Theorem is one of the most important results in all of mathematics. It has many applications.

(B) A <u>square</u> is a four sided polygon where the four interior angles are right angles and the four sides are equal in length. If each side of the square is of length b, then the area of the square is $A = b^2$ and the perimeter is $P = 4b$.

(C) A <u>rectangle</u> is a four sided polygon where the four interior angles are right angles, but the top and bottom are of length b and the two sides are of length h, where b is not necessarily equal to h. The area is A = bh, and the perimeter is simply P = 2b + 2h.

(D) For a <u>parallelogram</u>, the two pairs of opposite sides are parallel, but the interior angles are not necessarily right angles. If the base length is b and the height is h, then the area is A = bh. The perimeter can be determined if we know the slant length L (where L ≥ h) and it would be P = 2b + 2L.

(E) For a <u>trapezoid</u>, where the bottom is of length b_1, the top is of length b_2, $b_1 \neq b_2$, and the slanted sides are of lengths h_1 and h_2, the perimeter is P = $(h_1 + h_2 + b_1 + b_2)$. If the height of the trapezoid is h, then the area of it is $A = \frac{1}{2}(b_1 + b_2) \cdot (h)$.

BASIC SOLID GEOMETRY

For the following sections on solid geometry, refer to Figure (4.7) and Figure (4.8) on the next two pages.

Cubes
A cube is an object in three dimensions analogous to the square that we have in two dimensions. In fact, it consists of a top, bottom, and four sides that are all squares. If the length of all sides is b, then the surface area is $S = 6b^2$, and the volume is simply $V = b^3$ (length × width × height).

Rectangular Solid
A rectangular solid is a three dimensional object analogous to the rectangle in two dimensions. If the length is l, the width is w, and the height is h, then its volume is simply $V = (l \cdot w \cdot h)$. Its surface area is S = (2lw + 2lh + 2wh).

146

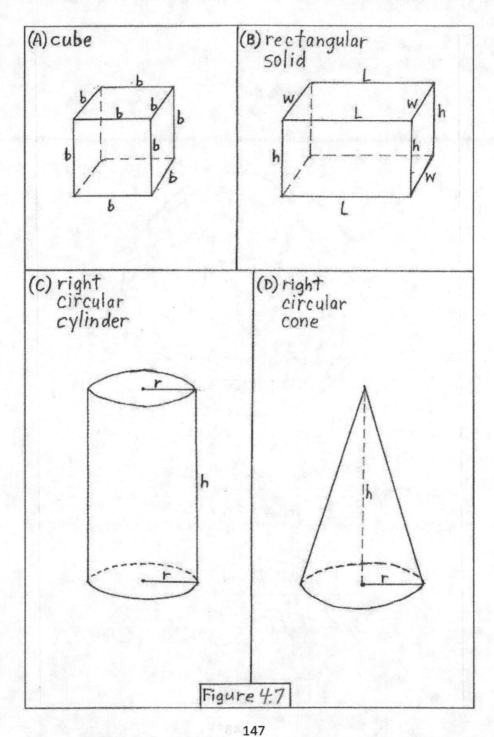

(A) cube

(B) rectangular solid

(C) right circular cylinder

(D) right circular cone

Figure 4.7

147

Sphere

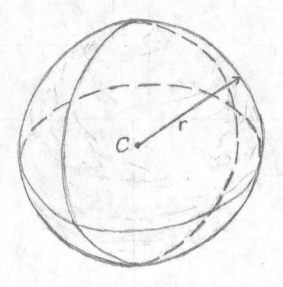

Volume

$$V = \frac{4}{3}\pi r^3$$

Surface Area

$$S = 4\pi r^2$$

Figure 4.8

Right Circular Cylinder

If we have an upright cylinder of height h with a circular base of radius r, we call this a right circular cylinder because the angle between the base and its vertical direction is a right angle. This kind of object is like a soup can of sorts. The surface area is $S = (2(\pi r^2) + (2\pi r)h)$. The first term in this formula for S is the sum of the areas of the circular base and the circular top, and the second term is the area of the side, which is the circumference of the base multiplied by the height. The volume is $V = (\pi r^2) \cdot (h)$, the area of the base multiplied by the height.

Right Circular Cone

If we have an upright circular cone, that is, a cone where the angle between the circular base of radius r and the vertical axis of the cone of length h is a right angle, then the volume turns out to be $V = \frac{1}{3}\pi r^2 h$, and the surface area is calculated using methods of calculus to be $S = \pi r h \sqrt{r^2 + h^2}$.

Sphere

A sphere is the set of points in 3-dimensional space that are equidistant from a single point C. By methods of calculus we can determine the enclosed volume to be $V = \frac{4}{3}\pi r^3$ and the area of the spherical surface is $S = 4\pi r^2$, where r is the radius of the sphere.

(4) <u>FUNDAMENTALS OF TRIGONOMETRY AND THE CALCULUS</u>

***** TRIGONOMETRY *****

(4.1) <u>The Trigonometric Functions</u>

For the right triangle in Figure (5.2) on the next page, with sides labeled x, y, and r, we define the six trigonometric (trig) functions for the angle θ.

$$sine(\theta) = sin(\theta) = \frac{y}{r} \qquad cosine(\theta) = cos(\theta) = \frac{x}{r}$$

$$tangent(\theta) = tan(\theta) = \frac{y}{x} \qquad cotangent(\theta) = cot(\theta) = \frac{x}{y}$$

$$secant(\theta) = sec(\theta) = \frac{r}{x} \qquad cosecant(\theta) = csc(\theta) = \frac{r}{y}$$

$$tan(\theta) = \frac{y}{x} = \frac{\left(\frac{y}{r}\right)}{\left(\frac{x}{r}\right)} = \frac{\sin(\theta)}{\cos(\theta)} = \frac{1}{\cot(\theta)}$$

$$cot(\theta) = \frac{x}{y} = \frac{\left(\frac{x}{r}\right)}{\left(\frac{y}{r}\right)} = \frac{\cos(\theta)}{\sin(\theta)} = \frac{1}{\tan(\theta)}$$

$$sec(\theta) = \frac{r}{x} = \frac{1}{\left(\frac{x}{r}\right)} = \frac{1}{\cos(\theta)}$$

$$csc(\theta) = \frac{r}{y} = \frac{1}{\left(\frac{y}{r}\right)} = \frac{1}{\sin(\theta)}$$

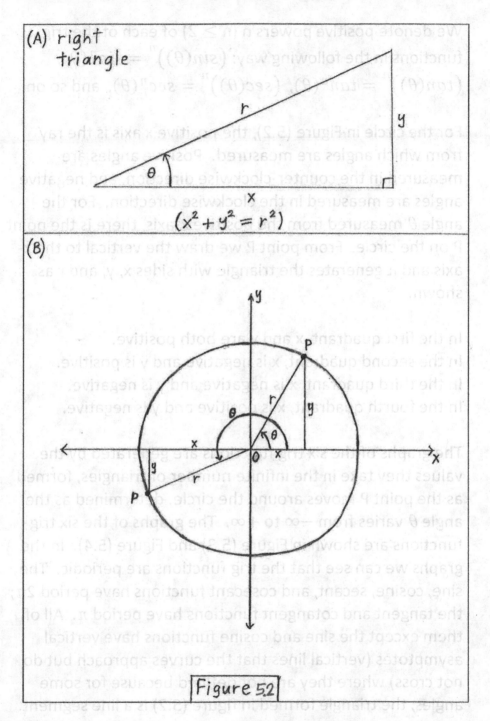

(A) right triangle

$(x^2 + y^2 = r^2)$

(B)

Figure 52

151

We denote positive powers n (n $\geq$ 2) of each of the trig functions in the following way: $(sin(\theta))^n = sin^n(\theta)$, $(tan(\theta))^n = tan^n(\theta)$, $(sec(\theta))^n = sec^n(\theta)$, and so on.

For the circle in Figure (5.2), the positive x axis is the ray from which angles are measured. Positive angles are measured in the counter-clockwise direction, and negative angles are measured in the clockwise direction. For the angle θ measured from the positive x-axis, there is the point P on the circle. From point P we draw the vertical to the x-axis and it generates the triangle with sides x, y, and r as shown.

In the first quadrant, x and y are both positive.
In the second quadrant, x is negative and y is positive.
In the third quadrant, x is negative and y is negative.
In the fourth quadrant, x is positive and y is negative.

The graphs of the six trig functions are generated by the values they take in the infinite number of triangles, formed as the point P moves around the circle, determined as the angle θ varies from $-\infty$ to $+\infty$. The graphs of the six trig functions are shown in Figure (5.3) and Figure (5.4). In the graphs we can see that the trig functions are periodic. The sine, cosine, secant, and cosecant functions have period 2π; the tangent and cotangent functions have period π. All of them except the sine and cosine functions have vertical asymptotes (vertical lines that the curves approach but do not cross) where they are not defined because for some angles, the triangle formed in figure (5.2) is a line segment.

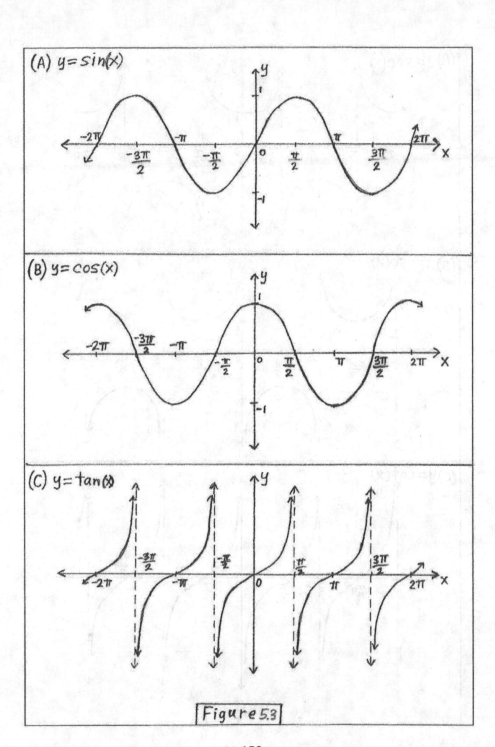

(A) $y = \sin(x)$

(B) $y = \cos(x)$

(C) $y = \tan(x)$

Figure 5.3

153

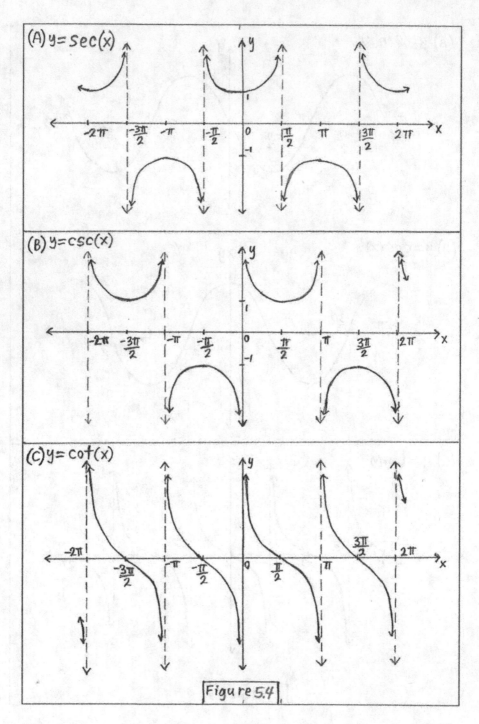

(A) $y = \sec(x)$

(B) $y = \csc(x)$

(C) $y = \cot(x)$

Figure 5.4

154

For example, the tan(90°) is not defined. In the generated triangle, there would be a problem with division by zero. The trig functions at some common angles θ are:

θ	$\sin\theta$	$\cos\theta$	$\tan\theta$	$\csc\theta$	$\sec\theta$	$\cot\theta$
0°	0	1	0	--	1	--
30°	$1/2$	$\sqrt{3}/2$	$1/\sqrt{3}$	2	$2/\sqrt{3}$	$\sqrt{3}$
45°	$1/\sqrt{2}$	$1/\sqrt{2}$	1	$\sqrt{2}$	$\sqrt{2}$	1
60°	$\sqrt{3}/2$	$1/2$	$\sqrt{3}$	$2/\sqrt{3}$	2	$1/\sqrt{3}$
90°	1	0	--	1	--	0
120°	$\sqrt{3}/2$	$-1/2$	$-\sqrt{3}$	$2/\sqrt{3}$	-2	$-1/\sqrt{3}$
135°	$1/\sqrt{2}$	$-1/\sqrt{2}$	-1	$\sqrt{2}$	$-\sqrt{2}$	-1
150°	$1/2$	$-\sqrt{3}/2$	$-1/\sqrt{3}$	2	$-2/\sqrt{3}$	$-\sqrt{3}$
180°	0	-1	0	--	-1	--
270°	-1	0	--	-1	--	0

(4.2) Trigonometric Identities

The reader may find this example of the use of complex numbers to be quite interesting. We will derive some of the more important trig identities that students should know.

The reader should be familiar with the important results relating the complex numbers and trigonometry, known as the Euler identities. The derivation of these identities will involve calculus, and is done in section 4.23, when we discuss infinite series:

$$e^{i\theta} = (cos\theta) + i(sin\theta)$$
$$e^{-i\theta} = (cos\theta) - i(sin\theta)$$

We will derive some important trig identities with the use of the complex number system. The algebraic form of a complex number z is z = a + bi. The magnitude of z is: $|z| = \sqrt{a^2 + b^2}$. We can call this r. The polar form for z involves writing z = a + bi = $re^{i\theta}$. θ is the angle from the positive x-axis to the radius vector r (from the origin of the complex plane to the point a + bi), and we can call it the angle of z. The quantities r and θ can be used to locate any point in the plane, just as we can use x and y. We call the x-y coordinate system a rectangular coordinate system, and the use of r and θ we call polar coordinates.

The magnitude of $e^{i\theta}$ is the magnitude of $(cos\theta + isin\theta)$, which is:

$$\left|e^{i\theta}\right| = \sqrt{(\cos\theta + i\sin\theta) \cdot (\cos\theta - i\sin\theta)}$$
$$= \sqrt{\cos^2(\theta) + \sin^2(\theta)}.$$

This comes from section (2.8) where we learned that the magnitude of a complex number z is the square root of the product of the number z and its conjugate $\bar{z}$.

From previously, we know that $\cos(\theta) = \frac{x}{r}$ and $\sin(\theta) = \frac{y}{r}$, so $\cos^2(\theta) + \sin^2(\theta) = \frac{x^2}{r^2} + \frac{y^2}{r^2} = \frac{(x^2 + y^2)}{r^2} = \frac{r^2}{r^2} = 1$, which makes use of the Pythagorean theorem. So, we have our first important trig identity:

$$\cos^2(\theta) + \sin^2(\theta) = 1. \qquad\qquad (4.2.1)$$

(Note: This shows that $\left|e^{i\theta}\right| = 1$).

Writing: $1 + \frac{y^2}{x^2} = 1 + \tan^2(\theta) = \frac{(x^2 + y^2)}{x^2} = \frac{r^2}{x^2} = \sec^2(\theta)$, we have our second important trig identity:

$$1 + \tan^2(\theta) = \sec^2(\theta). \qquad\qquad (4.2.2)$$

Writing: $1 + \frac{x^2}{y^2} = 1 + \cot^2(\theta) = \frac{(x^2 + y^2)}{y^2} = \frac{r^2}{y^2} = \csc^2(\theta)$, we have our third important trig identity:

$$1 + \cot^2(\theta) = \csc^2(\theta). \qquad\qquad (4.2.3)$$

##########

Now, for two complex numbers $z_1 = (a_1 + b_1 i)$ and $z_2 = (a_2 + b_2 i)$, with polar forms $z_1 = r_1 e^{i\theta_1}$ and $z_2 = r_2 e^{i\theta_2}$ respectively, the product:

$z_1 \cdot z_2 = (r_1 e^{i\theta_1})(r_2 e^{i\theta_2}) = (r_1 r_2)e^{i(\theta_1 + \theta_2)}$. The magnitudes are multiplied and the angles of the two complex numbers are added. The quotient of z_1 and z_2 is $\dfrac{z_1}{z_2} = \dfrac{r_1 e^{i\theta_1}}{r_2 e^{i\theta_2}} = \left(\dfrac{r_1}{r_2}\right) e^{i(\theta_1 - \theta_2)}$. The magnitudes are divided and the denominator complex number angle is subtracted from the numerator complex number angle.

$$e^{i(\theta_1 + \theta_2)} = \cos(\theta_1 + \theta_2) + i\sin(\theta_1 + \theta_2)$$
$$e^{i\theta_1} \cdot e^{i\theta_2} = (\cos\theta_1 + i\sin\theta_1)(\cos\theta_2 + i\sin\theta_2)$$
$$= (\cos\theta_1 \cos\theta_2 - \sin\theta_1 \sin\theta_2)$$
$$+ i(\sin\theta_1 \cos\theta_2 + \cos\theta_1 \sin\theta_2)$$

$$e^{i(\theta_1 - \theta_2)} = \cos(\theta_1 - \theta_2) + i\sin(\theta_1 - \theta_2)$$
$$e^{i\theta_1} \cdot e^{-i\theta_2} = (\cos\theta_1 + i\sin\theta_1)(\cos\theta_2 - i\sin\theta_2)$$
$$= (\cos\theta_1 \cos\theta_2 + \sin\theta_1 \sin\theta_2)$$
$$+ i(\sin\theta_1 \cos\theta_2 - \cos\theta_1 \sin\theta_2)$$

From this we have the sum and difference identities:

$$\cos(\theta_1 \pm \theta_2) = \cos\theta_1 \cos\theta_2 \mp \sin\theta_1 \sin\theta_2 \qquad (4.2.4)$$
$$\sin(\theta_1 \pm \theta_2) = \sin\theta_1 \cos\theta_2 \pm \cos\theta_1 \sin\theta_2 \qquad (4.2.5)$$

##########

Now we can derive two double angle formulas.

From identity (4.2.5), if $\theta_1 = \theta_2$, then we have:

$$sin(2\theta) = 2sin\theta cos\theta \qquad (4.2.6)$$

From identity (4.2.4), if $\theta_1 = \theta_2$, then we have:

$$cos(2\theta) = cos^2(\theta) - sin^2(\theta) \qquad (4.2.7)$$
$$= 2cos^2(\theta) - 1$$
$$= 1 - 2sin^2(\theta)$$

##########

Now we can derive two half-angle formulas:
From the last two formulations of identity (4.2.7):

$$cos^2(\theta) = \frac{1+ \cos(2\theta)}{2} \qquad (4.2.8)$$
$$sin^2(\theta) = \frac{1- \cos(2\theta)}{2} \qquad (4.2.9)$$

(4.3) Inverse Trigonometric Functions

When we restrict the domain of the trigonometric functions, we can define six inverse trig functions. We will call them the arcsin(x), arccos(x), arctan(x), arcsec(x), arccsc(x), and the arccot(x) functions.

Another notation which is commonly used for these functions is the $sin^{-1}(x)$, $cos^{-1}(x)$, $tan^{-1}(x)$, $sec^{-1}(x)$, $csc^{-1}(x)$, and $cot^{-1}(x)$. We will only concern ourselves with the first three of these inverse trig functions. The remaining ones are completely analogous in nature.

The trig functions assign to an angle a number. So, the inverse trig functions map a number back to an angle. We must restrict the domains so that we can have the inverse trig functions. The restricted domains are called principle value ranges. For the arcsin and arctan functions, the principle value range is $\left[-\frac{\pi}{2}, \frac{\pi}{2}\right]$. For the arccos function, the principle value range is $[0, \pi]$. What this means is this:

The arcsin(x) returns an angle θ in the interval $\left[-\frac{\pi}{2}, \frac{\pi}{2}\right]$ such that the $\sin(\theta) = x$.

The arctan(x) returns an angle θ in the interval $\left[-\frac{\pi}{2}, \frac{\pi}{2}\right]$ such that the $\tan(\theta) = x$.

The arccos(x) returns an angle θ in the interval $[0, \pi]$ such that the $\cos(\theta) = x$.

(4.4) Trigonometry Applications

Example (1):

The $\sin^{-1}\left(\frac{1}{2}\right) = \frac{\pi}{6}$, or 30°, since the $\sin(30°) = \sin\left(\frac{\pi}{6}\right) = \frac{1}{2}$.

The $\tan^{-1}(-1) = -\frac{\pi}{4}$, or -45°, since the $\tan(-45°) = \tan\left(-\frac{\pi}{4}\right) = -1$.

The $\cos^{-1}(0) = \frac{\pi}{2}$, or 90°, since the $\cos(90°) = \cos\left(\frac{\pi}{2}\right) = 0$.

Example (2):
The arctan(3) is an angle θ such that the $\tan(\theta) = 3$. Using the inverse tangent button on my calculator tells me that $\theta \approx 71.57°$.

Example (3):
In part E of Figure (4.5), suppose that we wanted to find the area of the trapezoid. We would have to know the height h that is indicated. If we knew that the angle between sides h_1 and b_1 was 60°, and that $b_1 = 10$, $b_2 = 8$, and $h_1 = 3$, then since the $\sin(60°) = \frac{h}{h_1} = \frac{\sqrt{3}}{2}$, we could determine that

$h = \left(\frac{\sqrt{3}}{2}\right) h_1 = \left(\frac{\sqrt{3}}{2}\right)(3)$. Then the area of the trapezoid is

$A = \left(\frac{1}{2}\right)(b_1 + b_2)h = \left(\frac{1}{2}\right)(18)\left(\frac{\sqrt{3}}{2}\right)(3)$. This is approximately 23.38 .

Example (4):
In part A of Figure (4.5), suppose that we wanted to find the area of the triangle. We would have to know the height h that is indicated. If we knew that the angle between sides $\overline{AB}$ and $\overline{AC}$ was 53°, and that b = 20 and $|\overline{AB}| = 8.5$, then since the $\sin(53°) \approx 0.7986 = \frac{h}{|\overline{AB}|} = \frac{h}{8.5}$, we could determine that $h \approx (0.7986)(8.5) \approx (6.79)$.

So, the area of the triangle is approximately $A = \left(\frac{1}{2}\right) bh$

$= \left(\frac{1}{2}\right)(20)(6.79) = (67.9)$.

Example (5):

If we have a force vector $\vec{F}$, a quantity with a magnitude and direction, it can be written in terms of its x and y direction components $\vec{F}_x$ and $\vec{F}_y$.

(For the readers familiar with vector algebra, we know that $\vec{F} = \vec{F}_x + \vec{F}_y$).

If θ is the angle between the horizontal x-component $\vec{F}_x$ and $\vec{F}$, then: $cos(\theta) = \frac{|\vec{F}_x|}{|\vec{F}|}$, and $sin(\theta) = \frac{|\vec{F}_y|}{|\vec{F}|}$.

So, we have:
$|\vec{F}_x| = |\vec{F}|(\cos(\theta))$, and $|\vec{F}_y| = |\vec{F}|(\sin(\theta))$.

For example, if $|\vec{F}| = 34.7$, and $\theta = 40°$, then
$|\vec{F}_x| = (34.7)(\cos(40°)) = (34.7)(0.766) = 26.58$
$|\vec{F}_y| = (34.7)(\sin(40°)) = (34.7)(0.643) = 22.31$

Within rounding errors, the reader can verify that because the x and y directions are perpendicular to each other, the Pythagorean theorem says that:

$$(|\vec{F}_x|)^2 + (|\vec{F}_y|)^2 = (|\vec{F}|)^2, \text{ or}$$

$$(26.58)^2 + (22.31)^2 = (34.7)^2$$

Example (6):
Suppose we have a right triangle with a hypotenuse of length r and legs of lengths x and y, in the x and y directions respectively. If x = 15 and y = 26, at what angle θ is the hypotenuse inclined to the horizontal.

Since the $\tan(\theta) = \frac{y}{x} = \frac{26}{15} = 1.7333$,

$\theta = tan^{-1}(\tan(\theta)) = tan^{-1}(1.7333) \approx 60.02°$

Example (7):
A cannon on level ground fires an artillery round with a velocity $v_0 = 315 \left(\frac{meters}{second}\right)$, at an angle $\theta = 51°$ to the horizontal.

Neglecting air resistance, the shell will travel in a parabolic path, where the distance in the x and y directions is given by the equations (t is in seconds):

$x = v_0(cos\theta)t$ (in meters).
$y = v_0(sin\theta)t - (4.9)t^2$ (in meters).

(A) How much time will the projectile be in the air?
The projectile starts at y = 0 and will stay in the air until y = 0 again. Determining the constants in the equation for the y position:

$y = (315)(\sin(51°))t - (4.9)t^2 = (244.8)t - (4.9)t^2$.
$y = t(244.8 - 4.9t)$.

163

This equation is solved when t = 0, but we are interested in the other solution for t. So, we solve $(244.8 - 4.9t) = 0$, which yields t ≈ 49.96 seconds.

(B) How far down range does the projectile travel before it hits the ground?
Figuring the constants in the equation for the x position:

$$x = (315)(\cos(51°))\, t.$$
$$x = (198.2)t.$$

Since the projectile stays in the air t = 49.96 seconds, it will strike the ground $(198.2)(49.96) \approx (9902.1)$ meters down range (9.9021 kilometers down range).

****

Most applications of trigonometry are in the physical sciences and engineering. The reader that may be in such a program will undoubtedly encounter many such applications in their studies. So far here in chapter 4, we have explained some of the basic concepts, notation, and formulas that we have in trigonometry, along with just a few examples of its many applications. With that said, this concludes our look at Trigonometry. After some exercises, we will move on to an introductory look at the single variable calculus.

(1) (a) If we have a right triangle oriented as in Part (A) of Figure (5.2), with x = 4, y = 3, and r = 5. What are the values of the six trig functions for the angle θ?
(b) What is the angle measure of θ in degrees and radians? (round to two decimal places)
(c) The angle between the sides of lengths x and y is of course 90°. What is the measure of the angle that is between the sides of lengths y and r, in degrees?

(2) If we have a right triangle oriented as in Part (A) of figure (5.2), with x = 7.3 and y = 8.9, what is the length r of the hypotenuse and what is the measure of the angle θ in degrees? (round to two decimal places)

(3) If we have a force vector $\vec{F}$ in the x-y plane with initial point at coordinates (2,3) and terminal point with coordinates (10,-4):
(a) What is the magnitude of $\vec{F}$? (Hint: use the pythagorean theorem with the change in x and the change in y)
(b) What is the magnitude of the components of $\vec{F}$ in the x and y directions?
(c) What is the inclination of $\vec{F}$ to the horizontal direction? (round to two decimal places)

(4) Show that for an angle θ,
$$sec(\theta)sin^2(\theta) + cos^2(\theta)sec(\theta) = \frac{sec^2(\theta)}{cos(\theta)} - \frac{tan^2(\theta)}{cos(\theta)}$$

(5) Use: $\sin(\theta_1 - \theta_2) = \sin(\theta_1)\cos(\theta_2) - \cos(\theta_1)\sin(\theta_2)$
to show that the $\sin(15°) = \left(\frac{\sqrt{3}-1}{2\sqrt{2}}\right)$, using only the table
of trig function values for common angles in section 4.1.
(Hint: choose two common angles whose difference is
15°).

(6) If we have a triangle with interior angles A, B, and C,
with opposite sides a, b, and c respectively, then the
Law of Sines says:
$$\frac{\sin(A)}{a} = \frac{\sin(B)}{b} = \frac{\sin(C)}{c}.$$

If A = 36°, a = 5, and b = 7:

(a) What is the measure of angle B (in degrees)?
(b) What is the measure of angle C (in degrees)?
(c) What is the length of side c?

(7) If we have a triangle with interior angles A, B, and C,
with opposite sides a, b, and c respectively, then the
Law of Cosines says:
$$c^2 = a^2 + b^2 - 2abcos(C).$$

If a = 6, and b = 8, then,

(a) What is c if C = 20°?
(b) What is c if C = 90°?
(c) What is c if C = 150°?
(d) What is the form of the Law of cosines if C = 90°?

***** THE CALCULUS *****

Calculus is a branch of math that is very useful when we have to deal with continuously varying quantities. For example, it was first developed to deal with problems in physics, where things change in time such as a planet's position, velocity, and acceleration. It has since been applied to a large variety of problems in other sciences as well, because they have become increasingly sophisticated with regard to mathematics. It helps us deal with rates of change, and is of use in optimization, and also helps us in solving many different kinds of problems in geometry, to name just a few of its applications. Calculus is also of great importance in the theory underlying much of statistical methodology. We cannot possibly do justice to the infinite array of problems which can be solved with the methods of calculus. The reader should note that the treatment of probability and statistics later in this book is algebraic in nature and does not involve calculus.

Calculus involves two main tools called the derivative and the integral, and because of this the subject is broken into two main parts called the differential and the integral calculus. These two tools involve an extensive collection of formulas and methods which have to be learned in order to use this branch of mathematics for making calculations. The main concepts of calculus involve limiting processes which are sequential in nature, so we'll start with sequences.

(4.5) Sequences

A sequence is a function whose domain is the set of positive integers. We can denote it by:

$$\{a_n\} = \{a_n\}_{n=1}^{\infty} = \{a_1, a_2, a_3, \ldots\}, \text{ where we have the}$$

function f(n) = a_n, for (n = 1, 2, 3, . . .). We call a_n the nth term of the sequence. Sometimes the index n may start at some integer other than 1 (0 is common), but this should cause no confusion.

The notation above is exactly like the set notation that we learned in chapter 2. We said that for a set the order that the members are listed is unimportant, as long as there is no confusion as to what is in the set. However, we should think of a sequence a bit differently. A sequence is an ordered set where each of the members has a definite position. Some examples are:

(I) $\{a_n\} = \{n\} = \{1, 2, 3, \ldots\}$

(II) $\{a_n\} = \left\{\frac{1}{n}\right\} = \left\{1, \frac{1}{2}, \frac{1}{3}, \frac{1}{4}, \ldots\right\}$

(III) $\{a_n\} = \left\{\frac{n}{n+1}\right\} = \left\{\frac{1}{2}, \frac{2}{3}, \frac{3}{4}, \frac{4}{5}, \ldots\right\}$

168

(IV) $\{a_n\} = \{(-1)^{n+1}\} = \{1, -1, 1, -1, ...\}$

Convergence and Divergence

Generally, for a sequence our main concern is whether it converges or diverges. This means does the sequence approach a single number arbitrarily closely after a certain point, or not. If it does, then it is said to be a convergent sequence and it converges to that specific number. If it does not, then it is said to be a divergent sequence.

(A) In the four examples above, the first sequence $\{n\}$ is divergent because the numbers in the sequence simply approach $+\infty$ (not a single number).

(B) In the second example, the sequence $\left\{\frac{1}{n}\right\}$ approaches the single number 0 arbitrarily close, so we would say that the sequence converges to 0. It never reaches 0, but it gets as close as we would like (think of an arbitrarily small positive number ε), for all terms after a certain point. We could express this fact with a certain notation called limit notation (abbreviated lim):

$$\lim_{n \to \infty} \left(\frac{1}{n}\right) = 0.$$

(C) In the third example, the sequence $\left\{\frac{n}{n+1}\right\}$ approaches the single number 1 arbitrarily closely, so we would say that the sequence converges to 1. It never reaches 1, but it gets as close as we would like for all terms after a certain point. Then we can say:

169

$$\lim_{n \to \infty} \left(\frac{n}{n+1} \right) = 1 \, .$$

(D) In the fourth example, the terms continually oscillate between 1 and -1. So, the terms of the sequence do not approach a single number arbitrarily closely and we would say that the sequence diverges.

At this point, we will give the technical definition of what it means for a sequence to be convergent:

<u>Definition of Convergence</u>
Suppose we have a sequence $\{a_n\}$. Choose an arbitrarily small positive number ε. Then $\{a_n\}$ converges to a real number L if there exists a positive integer N such that $|a_n - L| < \varepsilon$ for all n > N.

(A) In the second example above, for the sequence $\left\{ \frac{1}{n} \right\}$, we would prove that it converges to L = 0 in the following way. Choose an arbitrarily small positive number ε. As an example, let $\varepsilon = \frac{1}{1000}$. If we choose N = 10,000, then:
$$|a_n - L| = \left| \frac{1}{n} - 0 \right| = \left| \frac{1}{n} \right| < \frac{1}{1000} \, , \text{ for all n > 10,000.}$$
That is, we have found a large enough integer N such that $|a_n - L| < \varepsilon$, when n > N. Since ε can be chosen as small as we like and we can find our N, this proves that the sequence $\left\{ \frac{1}{n} \right\}$ converges to 0.

(B) In the third example above, for the sequence $\left\{\dfrac{n}{n+1}\right\}$, we would prove that it converges to L = 1 in the following way. Choose an arbitrarily small positive number ε. For example, let $\varepsilon = \dfrac{1}{1,000,000}$. If we choose N = 10,000,000, then:

$$|a_n - L| = \left|\frac{n}{n+1} - 1\right| = \left|\left(1 - \frac{1}{n+1}\right) - 1\right| = \left|\frac{1}{n+1}\right| < \frac{1}{1,000,000}$$

for all n > 10,000,000. So, we have found a large enough integer N such that $|a_n - L| < \varepsilon$, when n > N. Since ε can be chosen as small as we like and we could find our N, this proves that the sequence $\left\{\dfrac{n}{n+1}\right\}$ converges to 1.

(C) Now consider the sequence $\{2 + e^{-n}\}$. We want to prove that it converges to L = 2. Choose $\varepsilon = e^{-100}$. Then we need to find an integer N large enough such that $|a_n - L| < \varepsilon$, when n > N. We can choose N = 500. Then since $|e^{-n}| < e^{-100}$ for all n > 500, it follows that $|a_n - L| = |(2 + e^{-n}) - 2| = |e^{-n}| < \varepsilon$ for all n > N. Since ε can be chosen as small as we like and we can find our N, this proves that the sequence $\{2 + e^{-n}\}$ converges to 2.

More About Sequences
The fourth sequence from above was defined to be:

$$\{a_n\} = \{(-1)^{n+1}\} = \{1, -1, 1, -1, \ldots\}.$$

Note that the following three subsequences can be chosen from it:

$$\{1, 1, 1, \ldots\}, \{-1, -1, -1, \ldots\}, \{1, 1, 1, -1, 1, 1, 1, -1,\}$$

The first two of these are convergent sequences, converging to 1 and -1 respectively, but the third one diverges. We had stated that the sequence from which they come diverges. We will now present an important theorem without proof:

Theorem 1: A sequence has infinitely many subsequences, and the sequence is convergent to a real number L iff all of its subsequences are convergent to the same number L.

We will now present two additional theorems that should be clear to the reader without proof:

Theorem 2: If we have a sequence $\{a_n\}$ that is monotone increasing (meaning that $a_{n+1} > a_n$ for every two adjacent terms in the sequence), and if the collection of all terms is bounded above (meaning that there is a number M such that every term of the sequence is less than M), then the sequence converges to some real number L.

Theorem 3: If we have a sequence $\{a_n\}$ that is monotone decreasing (meaning that $a_{n+1} < a_n$ for every two adjacent terms in the sequence), and if the collection of all terms is bounded below (meaning that there is a number M such that every term of the sequence is greater than M), then the sequence converges to some real number L.

Uniform Convergence

Concerning sequences, up to this point we have dealt with the "pointwise convergence" of a sequence of numbers. That is, a sequence of numbers converging to some single number. We will now briefly illustrate the idea of "uniform convergence" of a sequence of functions on an interval (a,b), to a single function. This is actually a topic which is considered at an advanced level, but we can convey the basic ideas at this point without difficulty. We discuss this topic here because we will have to note its importance later in our introduction to some of the basic ideas of single variable calculus. Refer to Figure (5.45) in this discussion of uniform convergence. Note that we will use the intervals (0,1) in part(A) and [0,1] in part(B).

In Figure (5.45)-part (A), consider the sequence of functions (Note: not a sequence of numbers, but functions):

$$\{y_n\} = \{f_n(x)\}_{n=2}^{\infty} = \left\{\frac{x^n}{(n-1)}\right\}_{n=2}^{\infty} \text{ , defined on } (0,1).$$

Consider a single number in (0,1) such as $x = \frac{2}{3}$. Then we have the sequence of numbers:

$$\left\{f_n\left(\frac{2}{3}\right)\right\}_{n=2}^{\infty} = \left\{\left(\frac{2}{3}\right)^2, \frac{\left(\frac{2}{3}\right)^3}{2}, \frac{\left(\frac{2}{3}\right)^4}{3}, \dots\right\}.$$

We have a similar sequence for every element of (0,1), and we can see that each such sequence would converge to 0.

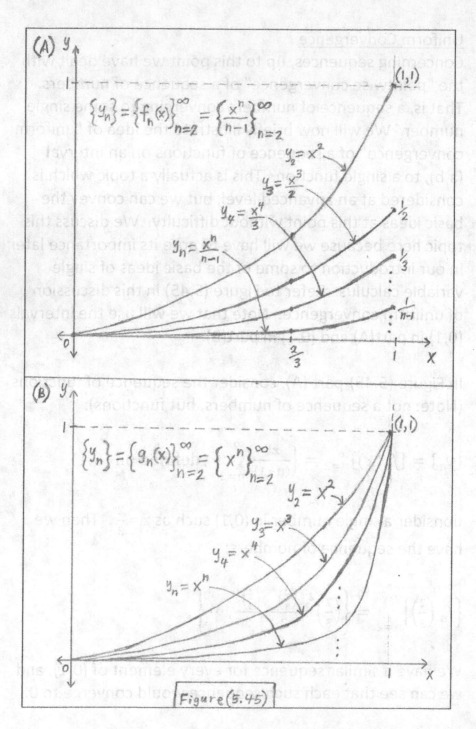

$$\{y_n\} = \{f_n(x)\}_{n=2}^{\infty} = \left\{\frac{x^n}{n-1}\right\}_{n=2}^{\infty}$$

$$y_2 = x^2$$

$$y_3 = \frac{x^3}{2}$$

$$y_4 = \frac{x^4}{3}$$

$$y_n = \frac{x^n}{n-1}$$

$$\{y_n\} = \{g_n(x)\}_{n=2}^{\infty} = \{x^n\}_{n=2}^{\infty}$$

$$y_2 = x^2$$

$$y_3 = x^3$$

$$y_4 = x^4$$

$$y_n = x^n$$

Figure (5.45)

174

Suppose we have an arbitrarily small positive number ε, $0 < \varepsilon < 1$, and any x in the interval (0,1). For any given n, (n = 2,3,4, ...), $f_n(x)$ is bounded above on (0,1) by $\left(\frac{1}{n-1}\right)$. All we need to do is note that we can find a large enough integer N such that $\frac{1}{(n-1)} < \varepsilon$ whenever $n > N$. How about if we let N = the first integer greater than the quantity $\left(\frac{2}{\varepsilon} + 1\right)$. This will work because whenever $n > N$, $\frac{1}{(n-1)} < \frac{1}{\left(\frac{2}{\varepsilon}+1\right)-1} = \frac{\varepsilon}{2} < \varepsilon$. So, we have established that for any arbitrarily small positive number ε, and for every x in (0,1), we can find a large enough number N so that we can say that:

$$|f_n(x) - 0| < \varepsilon \text{ ,whenever n } > N, \text{ and x is in (0,1).}$$

This says that the sequence of functions $\{f_n(x)\}$ is converging uniformly to the single function $f(x) = 0$ on the entire interval (0,1). We can write $\{f_n(x)\} \to f(x)$ uniformly on (0,1), where $f(x)$ is the zero function on (0,1).

In Figure (5.45)-part(B), we show a sequence of functions $\{g_n(x)\} = \{x^n\}$ that do not converge uniformly on the closed interval [0,1] to the zero function $f(x) = 0$.

If we let $\varepsilon > 0$, then for each function $g_n(x)$ in the sequence there will be an interval (c,1], where $0 < c < 1$, such that $|g_n(x) - 0| > \varepsilon$,for all x in (c,1]. So, $\{g_n(x)\}$ does not converge uniformly to $f(x) = 0$ on [0,1].

(4.6) Limits and Continuity

One-Sided and Two-Sided Limits
Consider the function y = f(x) defined on the closed interval [a,b] and a real number c in the open interval (a,b), then we will consider the three limit situations:

(A) $\lim\limits_{x \to c^-} f(x)$ (B) $\lim\limits_{x \to c^+} f(x)$ (C) $\lim\limits_{x \to c} f(x)$

Limits (A) and (B) are one-sided limits, limit (C) is the corresponding two-sided limit (or just the limit). For purposes of illustration, let's assume that [a,b] is [0,10] and c = 3.

(A) When we consider the one-sided limit $\lim\limits_{x \to 3^-} f(x)$, what should come to mind is a sequence that converges to 3 from the left of 3, such as: {2.9, 2.99, 2.999, ...}. Then what the notation $\lim\limits_{x \to 3^-} f(x)$ means is the limit of the sequence of y-values: {f(2.9), f(2.99), f(2.999), ...}   (4.61)

(B) When we consider the one-sided limit $\lim\limits_{x \to 3^+} f(x)$, what should come to mind is a sequence that converges to 3 from the right of 3, such as: {3.1, 3.01, 3.001, ...}. Then what the notation $\lim\limits_{x \to 3^+} f(x)$ means is the limit of the sequence of y-values: {f(3.1), f(3.01), f(3.001), ...}   (4.62)

(C) When we consider the two-sided limit $\lim_{x \to 3} f(x)$, what should come to mind is a sequence that converges to 3 from both the left and right of 3, something such as the sequence {2.9, 3.1, 2.99, 3.01, 2.999, 3.001, ...}. Then what the notation $\lim_{x \to 3} f(x)$ means is the limit of the y-values in the sequence: {f(2.9), f(3.1), f(2.99), f(3.01), ...}   (4.63)

Clearly, for the two-sided limit $\lim_{x \to 3} f(x)$ to exist, we need the $\lim_{x \to 3^-} f(x) = \lim_{x \to 3^+} f(x)$, because the sequences (4.61) and (4.62) are subsequences of sequence (4.63). This is an important point: The limit exists iff the corresponding limits from the left and the right are both equal.

Properties of Limits
There are some properties of limits which are important to note. If we have two functions f(x) and g(x) and the real number c, and $\lim_{x \to c} f(x) = L$ and $\lim_{x \to c} g(x) = M$, where L and M are finite numbers, then the following are true:

(A) $\lim_{x \to c} cf(x) = c \lim_{x \to c} f(x) = cL$

(B) $\lim_{x \to c}(f(x) + g(x)) = \lim_{x \to c} f(x) + \lim_{x \to c} g(x) = L + M$

(C) $\lim_{x \to c} f(x) \cdot g(x) = \left(\lim_{x \to c} f(x)\right) \cdot \left(\lim_{x \to c} g(x)\right) = L \cdot M$

(D) $\lim_{x \to c} \dfrac{f(x)}{g(x)} = \dfrac{\left(\lim_{x \to c} f(x)\right)}{\left(\lim_{x \to c} g(x)\right)} = \dfrac{L}{M}$, provided that $M \neq 0$

(E) $\lim_{x \to c} \sqrt[n]{f(x)} = \sqrt[n]{\left(\lim_{x \to c} f(x)\right)} = \sqrt[n]{L}$

Continuity

For the function $y = f(x)$ to be continuous on the open interval (a,b), we need to have $\lim_{x \to c} f(x) = f(c)$ for every number c in the open interval (a,b). If in addition, we have the limit $\lim_{x \to a^+} f(x) = f(a)$ and the limit $\lim_{x \to b^-} f(x) = f(b)$, then we can say that f(x) is continuous on the closed interval [a,b].

When a function f(x) is continuous on [a,b], then we can draw the graph of f(x) on [a,b] without having to lift the pencil from the paper. The continuous function f(x) need not be entirely smooth on [a,b] (it can have sharp points so to speak), but the graph of f(x) must be connected on [a,b].

There are four ways that f(x) could fail to be continuous on the closed interval [a,b] at the point where x = c, and these four ways are illustrated in Figure (5.51) on the next page.

(A) The function could be defined at x = c, but $\lim_{x \to c} f(x)$ does not exist at x = c because f(x) is approaching two different y-values as x approaches c from the left and from the right.

(B) The $\lim_{x \to c} f(x)$ exists at x = c, but the function is not defined at x = c. In this case there would be a hole in the graph of f(x) at x = c.

(C) The $\lim_{x \to c} f(x)$ exists at x = c and the function is defined at x = c, but the two are not equal.

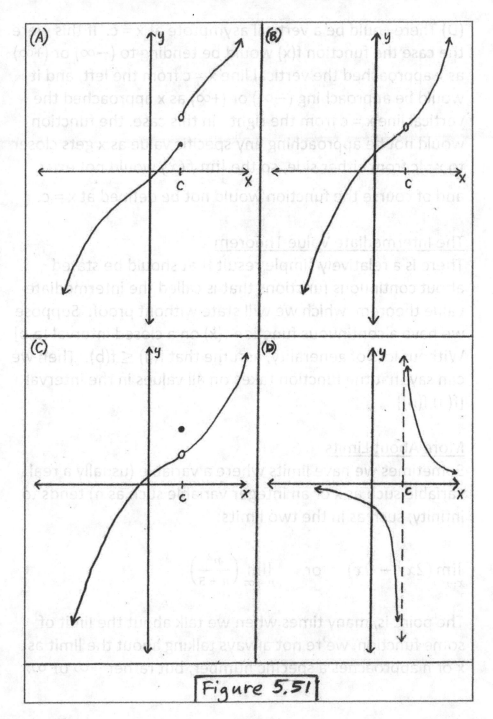

Figure 5.51

(D) There could be a vertical asymptote at x = c. If this were the case the function f(x) would be tending to $(-\infty)$ or $(+\infty)$ as x approached the vertical line x = c from the left, and it would be approaching $(-\infty)$ or $(+\infty)$ as x approached the vertical line x = c from the right. In this case, the function would not be approaching any specific value as x gets closer to x = c from either side, so the $\lim_{x \to c} f(x)$ would not exist, and of course the function would not be defined at x = c.

The Intermediate Value Theorem

There is a relatively simple result that should be stated about continuous functions that is called the intermediate value theorem, which we will state without proof. Suppose we have a continuous function f(x) on a closed interval [a,b]. Without loss of generality, assume that f(a) ≤ f(b). Then we can say that the function takes on all values in the interval [f(a),f(b)].

More About Limits

Sometimes we have limits where a variable (usually a real variable such as x or an integer variable such as n) tends to infinity, such as in the two limits:

$$\lim_{x \to \infty} (2x^3 - 9x) \quad \text{or} \quad \lim_{n \to \infty} \left(\frac{n^2}{n + 5} \right).$$

The point is, many times when we talk about the limit of some function, we're not always talking about the limit as x or n approaches a specific number, but rather $-\infty$ or ∞.

There are some properties of limits which are important to note. Suppose we have two functions f(x) and g(x).
In many situations, when we have two functions f(x) and g(x) and the number c, the limit $\lim\limits_{x\to\infty}\dfrac{f(x)}{g(x)}$ or $\lim\limits_{x\to c}\dfrac{f(x)}{g(x)}$ may approach $\dfrac{0}{0}$ or $\dfrac{\infty}{\infty}$. These are called indeterminate forms. There are several of them in calculus such as $(\infty - \infty)$, $(0 \cdot \infty)$, (0^0), (1^∞) and (∞^0). The limit in these cases may or may not exist (meaning that it is a real number or it tends to infinity), we simply do not know. In these cases, we may be able to determine the limit if it exists, or determine that the limit does not exist, by doing an algebraic manipulation and/or by using L'Hospitals's Rule, which we will learn about later. Here, we want to show how a limit involving one of the two indeterminate forms $\dfrac{0}{0}$ or $\dfrac{\infty}{\infty}$ may be found through an algebraic manipulation.

Example (1):
Before we get to these indeterminate forms, let's consider some rather straightforward examples of limits where we can evaluate the limit by substitution, or by some simple reasoning.

(A) The $\lim\limits_{x\to 10} (x^2 + 3x)$ can be found by simply substituting 10 for x. So, $\lim\limits_{x\to 10} (x^2 + 3x) = (10^2 + 3(10)) = 130$. The function is continuous for all real numbers, so there is nothing tricky here.

(B) The $\lim\limits_{x \to -27} (\sqrt[3]{x})$ can be found by simply substituting -27 for x. So, $\lim\limits_{x \to -27} (\sqrt[3]{x}) = (\sqrt[3]{-27}) = -3$. Once again, the function is continuous for all real numbers, and once again evaluating the limit presents no problems.

(C) The $\lim\limits_{x \to 0} (e^x)$ can be found by simply substituting 0 for x. So, the $\lim\limits_{x \to 0} (e^x) = (e^0) = 1$. The function $f(x) = e^x$ is continuous for all real numbers, so there is no difficulty here.

(D) The $\lim\limits_{x \to 0} \left(\frac{1}{x}\right)$ does not lead to an indeterminate form, but it does require a little bit of reasoning. As x approaches 0 from the left, the $\left|\frac{1}{x}\right|$ tends to $+\infty$. However, $\left(\frac{1}{x}\right)$ is always a negative number when x is to the left of 0. So, the limit $\lim\limits_{x \to 0^-} \left(\frac{1}{x}\right) = -\infty$. As x approaches 0 from the right, the $\left|\frac{1}{x}\right|$ tends to $+\infty$, and $\left(\frac{1}{x}\right)$ is always a positive number when x is to the right of 0. So, the limit $\lim\limits_{x \to 0^+} \left(\frac{1}{x}\right) = +\infty$. The function $f(x) = \left(\frac{1}{x}\right)$ is not defined at x = 0, in fact this function f(x) has a vertical asymptote at x = 0, and the curve for f(x) goes downward to $-\infty$ as x approaches 0 from the left and the curve for f(x) goes upward to $+\infty$ as x approaches 0 from the right. So the limit does not exist.

(E) The $\lim\limits_{x \to \infty} \left(\frac{1}{x}\right)$ and the $\lim\limits_{x \to -\infty} \left(\frac{1}{x}\right)$ once again do not lead to indeterminate forms, yet they do require a little bit of reasoning. As x tends to $+\infty$ or to $-\infty$, the $\left|\frac{1}{x}\right|$ approaches 0. Both of these limits are 0. However, with the first limit, the curve is always above the x-axis as it approaches 0 because the x-values would be positive. But with the second limit, the curve is always below the x-axis as it approaches 0 because the x-values would be negative.

Example (2):

Consider the function $f(x) = \frac{(x^2 - 4)}{(x + 2)}$. Find the following limits: $\lim\limits_{x \to 2} f(x)$ and $\lim\limits_{x \to -2} f(x)$.

(A) When x approaches 2 from both sides, the function is approaching $\frac{0}{4} = 0$. Since this is not an indeterminate form, we can conclude that $\lim\limits_{x \to 2} f(x) = 0 = f(2)$ and therefore, f(x) is continuous at x = 2.

(B) When x approaches x = -2 from both sides, the function is approaching $\frac{0}{0}$, which is an indeterminate form. So, we need to algebraically manipulate f(x) in order to determine if the $\lim\limits_{x \to -2} f(x)$ exists or not. We can do the following:

$$f(x) = \frac{(x^2 - 4)}{(x + 2)} = \frac{(x - 2)(x + 2)}{(x + 2)} = (x - 2), \text{ after cancelling terms.}$$

So, $\lim\limits_{x \to -2} f(x) = (-2 - 2) = -4$.

Since $\lim_{x \to -2} f(x) = -4$ from both sides of -2, rather than $(-\infty)$ or $(+\infty)$, we know that there is no vertical asymptote at $x = -2$. Also, we know that the y-values approach -4 as $x \to -2$, even though the function f(x) is not defined at $x = -2$. Therefore, $x = -2$ is not in the domain of f(x) and the graph of f(x) has a hole in it at the point (-2,-4). So, f(x) is not continuous at the point (-2,-4).

Example (3):

Consider the function $f(x) = \frac{(x^2 - 3x - 18)}{(x-6)}$. Find the following limits: $\lim_{x \to 6} f(x)$ and $\lim_{x \to -3} f(x)$.

(A) When x approaches -3 from both sides, the function is approaching $\frac{0}{-9} = 0$. Since this is not an indeterminate form, we can conclude that $\lim_{x \to -3} f(x) = 0 = f(-3)$ and therefore f(x) is continuous at $x = -3$.

(B) When x approaches $x = 6$ from both sides, the function is approaching $\frac{0}{0}$, which is an indeterminate form. So, we need to algebraically manipulate f(x) in order to determine if the $\lim_{x \to 6} f(x)$ exists or not. We can do the following:

$f(x) = \frac{(x^2 - 3x - 18)}{(x-6)} = \frac{(x-6)(x+3)}{(x-6)} = (x+3)$, after cancelling terms.

So, $\lim_{x \to 6} f(x) = (6+3) = 9$.

Since $\lim_{x \to 6} f(x) = 9$ from both sides of 6, rather than $(-\infty)$ or $(+\infty)$, we know that there is no vertical asymptote at x = 6. Also, we know that the y-values approach 9 as x → 6, even though the function f(x) is not defined at x = 6. Therefore, x = 6 is not in the domain of f(x) and the graph of f(x) has a hole in it at the point (6,9). So then, f(x) is not continuous at the point (6,9).

Example (4):

Evaluate the $\lim_{x \to \infty} \frac{3x^2-4x+7}{11x^2+5x-23}$. As x → ∞, the expression approaches $\frac{\infty}{\infty}$, which is an indeterminate form. We can't tell exactly what it is approaching, if any finite number at all. So, we must do some algebraic manipulation in order to see if it has a limit. Let's try this: multiply the numerator and denominator of the fraction by $\frac{1}{x^2}$.

$$\lim_{x \to \infty} \frac{3x^2-4x+7}{11x^2+5x-23} = \lim_{x \to \infty} \frac{(3x^2-4x+7)\cdot\left(\frac{1}{x^2}\right)}{(11x^2+5x-23)\cdot\left(\frac{1}{x^2}\right)} = \lim_{x \to \infty} \frac{3-\frac{4}{x}+\frac{7}{x^2}}{11+\frac{5}{x}-\frac{23}{x^2}}.$$

Some reflection should convince you that as x → ∞, each of the four fractions $\frac{4}{x}$, $\frac{7}{x^2}$, $\frac{5}{x}$, and $\frac{23}{x^2}$ approach 0. So, clearly the $\lim_{x \to \infty} \frac{3x^2-4x+7}{11x^2+5x-23} = \left(\frac{3-0+0}{11+0-0}\right) = \frac{3}{11}$.

Example (5):

Evaluate the $\lim_{x \to \infty} \frac{2x^2+6x-90}{3x+45}$. As x → ∞, the expression approaches $\frac{\infty}{\infty}$, which is an indeterminate form. Again, we

185

can't be sure of exactly what it is approaching. So, we must do some algebraic manipulation in order to see if it has a limit. Let's try multiplying the numerator and denominator of the fraction by $\frac{1}{x^2}$.

$$\lim_{x\to\infty} \frac{2x^2+6x-90}{3x+45} = \lim_{x\to\infty} \frac{(2x^2+6x-90)\cdot\left(\frac{1}{x^2}\right)}{(3x+45)\cdot\left(\frac{1}{x^2}\right)} = \lim_{x\to\infty} \frac{2+\frac{6}{x}-\frac{90}{x^2}}{\frac{3}{x}+\frac{45}{x^2}}.$$

Again, some reflection should convince you that as $x \to \infty$, each of the four fractions $\frac{6}{x}, \frac{90}{x^2}, \frac{3}{x}$, and $\frac{45}{x^2}$ approach 0. So, clearly the expression $\frac{2x^2+6x-90}{3x+45}$ is approaching $\left(\frac{2+0-0}{0+0}\right)$, which is approaching $(+\infty)$, as $x \to \infty$.

################# Exercises ##################

Find the following limits.

(1) $\lim_{x\to 3}\left(\frac{x^4-9}{x}\right)$ (2) $\lim_{x\to\infty}\left(\frac{1}{10}\right)^x$ (3) $\lim_{x\to 1^-}\left(\frac{1}{x-1}\right)$

(4) $\lim_{x\to 1^+}\left(\frac{1}{x-1}\right)$ (5) $\lim_{x\to 2^-}\left(\frac{x^2+2}{x-2}\right)$ (6) $\lim_{x\to\infty}\left(\frac{1}{\sqrt{x}}\right)$

(7) $\lim_{x\to 1}\left(\frac{2^x}{x}\right)$ (8) $\lim_{x\to 0^-}(|x|-x)$ (9) $\lim_{x\to 2^-}\left(\frac{x-2}{x+2}\right)$

(10) $\lim_{x\to 2^+}\left(\frac{x-2}{x+2}\right)$ (11) $\lim_{x\to 2^-}\left(\frac{x-2}{x}\right)$ (12) $\lim_{x\to 2^+}\left(\frac{x-2}{x}\right)$

(13) $\lim_{x\to 6^-}\left(\frac{x^2-5x-14}{x+2}\right)$ (14) $\lim_{x\to -2^-}\left(\frac{x^2-5x-14}{x+2}\right)$

186

(15) $\displaystyle\lim_{x\to 0^-}\left(\frac{|x|}{x}\right)$ (16) $\displaystyle\lim_{x\to 0^+}\left(\frac{|x|}{x}\right)$

(17) Is $f(x) = \dfrac{|x-1|}{x-1}$ continuous on [0,2] ?

(18) What value of A makes f(x) continuous at x = $-\dfrac{1}{2}$?

$$f(x) = \begin{cases} Ax + 1, & x > -\dfrac{1}{2} \\ 0 & , x \le -\dfrac{1}{2} \end{cases}.$$

(4.7) The Derivative

Average Rate of Change

If we have a function y = f(x) which is continuous on some closed interval [a,b], and x_1 and x_2 are two distinct points in this interval (where $x_1 < x_2$), then the average rate of change of f(x) on the sub-interval $[x_1, x_2]$ is defined to be:

$m = \left(\dfrac{f(x_2)-f(x_1)}{(x_2-x_1)}\right)$.   (See Figure (5.52) on the next page)

The change in f(x) can be called Δy and the change in x can be called Δx, so that m = $\left(\dfrac{\Delta y}{\Delta x}\right)$. "m" is the slope of the line through the two points with coordinates $(x_1, f(x_1))$ and $(x_2, f(x_2))$.

187

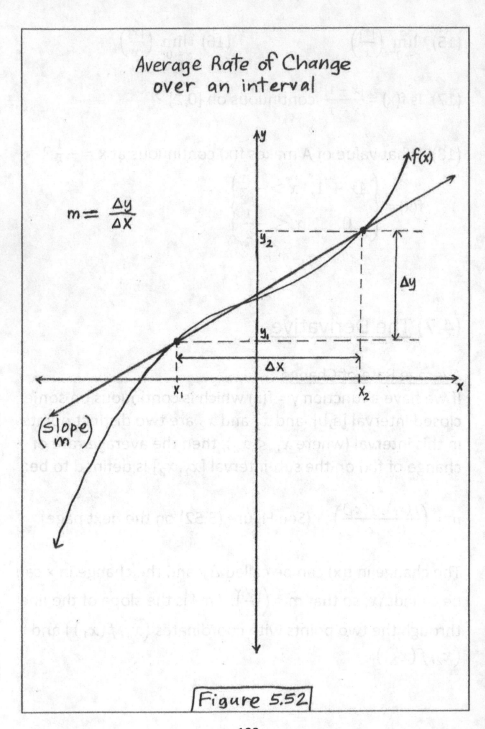

Figure 5.52

The Instantaneous Rate of Change

If the real number c is a member of the open set (a,b), then consider any sequence of points $\{x_i\} = \{x_1, x_2, x_3, \ldots\}$ where all the points in the sequence are in (a,b) and the sequence converges to c. Then the sequence:

$\{m_i\} = \left\{\frac{f(x_i)-f(c)}{(x_i-c)}\right\}$ is a sequence of average rates of change.

If this sequence converges to a real number L, then we call L the instantaneous rate of change of the function y = f(x) at the point (c,f(c)), and we write:

$$L = \lim_{i\to\infty}(m_i) = \lim_{i\to\infty}\left(\frac{(\Delta y)_i}{(\Delta x)_i}\right) = \lim_{i\to\infty}\left(\frac{f(x_i)-f(c)}{(x_i-c)}\right).$$

The Derivative

For the sequence $\{x_i\}$ above, let h = $(x_i - c)$. Since the terms of the sequence $\{x_i\}$ converge to c, h converges to 0. For the function f(x) on the closed interval [a,b], and the point c in the open interval (a,b), the number L above if it exists, can now be written $f'(c)$ and defined:

$$f'(c) = \lim_{h\to 0}\left(\frac{f(x+c)-f(c)}{h}\right)$$ (See Figure (5.53) on next page).

We call $f'(c)$ the value of the derivative for the function f(x) at the point $(c, f(c))$. It is the instantaneous rate of change of f(x) at x = c. That is how it is interpreted.

189

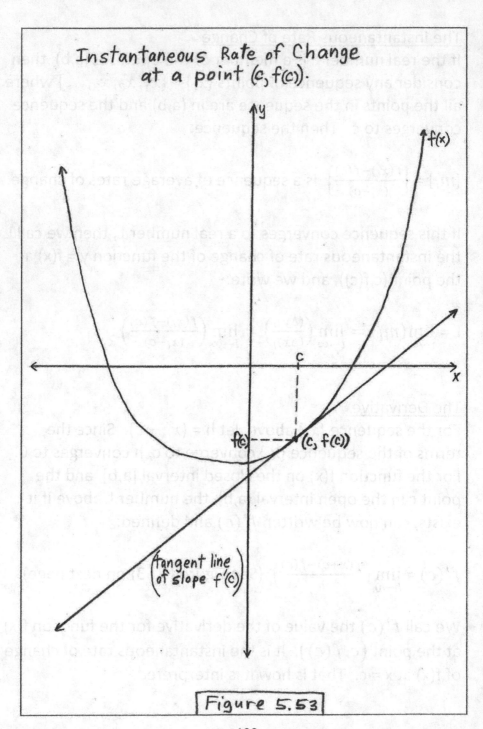

Figure 5.53

If $f'(c)$ exists for every x-value c in (a,b), then this defines a function $f'(x)$ for all x in the interval (a,b) and we call it the derivative function and we say that f(x) is differentiable on the open interval (a,b). We simply replace c with x and then say:

$$f'(x) = \lim_{h \to 0} \left(\frac{f(x+h)-f(x)}{h} \right)$$ (definition of the derivative).

$f'(c)$ is simply the value of the derivative function $f'(x)$ when x = c. If in addition, the following two limits exist:

$$f'(a) = \lim_{h \to 0^+} \left(\frac{f(a+h)-f(a)}{h} \right) \quad \text{and}$$

$$f'(b) = \lim_{h \to 0^-} \left(\frac{f(b+h)-f(b)}{h} \right),$$

then f(x) is differentiable on the closed interval [a,b].

Example (1):
Suppose $f(x) = 3x^2 + 2x$. Find the derivative of f(x).

$$f'(x) = \lim_{h \to 0} \left(\frac{f(x+h)-f(x)}{h} \right)$$

$$= \lim_{h \to 0} \left(\frac{3(x + h)^2 + 2(x + h) - (3x^2 + 2x)}{h} \right)$$

$$= \lim_{h \to 0} \left(\frac{3x^2 + 6xh + 3h^2 + 2x + 2h - 3x^2 - 2x}{h} \right)$$

$$= \lim_{h \to 0} \left(\frac{6xh + 2h + 3h^2}{h} \right)$$

$$= \lim_{h \to 0} (6x + 2 + 3h) = (6x + 2).$$

So, $f'(x) = (6x + 2)$ is the derivative of $f(x) = 3x^2 + 2x$.

Note that at every step of this derivation, except the last, if we substitute 0 for h, we get the indeterminate form $\frac{0}{0}$. So, we had to algebraically simplify the expression at each step, so that we could find $f'(x)$.

With the derivative $f'(x)$, we have found a function which gives us the instantaneous rate of change of f(x) at a value x. The derivative is the slope of the tangent line to the curve at a single point, and it is interpreted as a rate of change. For the function $f(x) = 3x^2 + 2x$ of Example (1), which is a parabola and is defined for all real numbers, the derivative was found to be $f'(x) = 6x + 2$, which is also defined for all real numbers. This makes sense because parabolas are continuous and smooth (meaning differentiable) curves over their entire domain and it makes sense that their derivatives would be defined and continuous over that same domain.

When x = -2, the function f(x) is changing (as x increases) at the rate $f'(-2) = 6(-2) + 2 = -10$ (y-units per x-units). So, f(x) is decreasing at the moment that x = -2 because $f'(-2) = -10$ is negative. So, the tangent line at this point is negatively sloped.

When x = -1, the function f(x) is changing (as x increases) at the rate $f'(-1) = 6(-1) + 2 = -4$ (y-units per x-units). This means that it is decreasing at the moment that x = -1, but it is decreasing at a lesser rate than it was when x = -2. The tangent line at this point is also negatively sloped, but not as steeply.

When x = 5, the function f(x) is changing (as x increases) at the rate $f'(5) = 6(5) + 2 = 32$ (y-units per x-units). This means that f(x) is increasing at the instant that x = 5. So, the tangent line at this point is positively sloped.

As an illustration of these ideas in Figure (5.54) on the next page with a different function f(x), we show the value of the derivative as the slope of the tangent line to f(x) for the three different x-values: $-\frac{3}{4}$, $-\frac{1}{3}$, and $\frac{1}{6}$.

Example (2):
We should now point out the fact that differentiation (as taking the derivative is called) is a so-called linear operation. This means that if we have two functions f(x) and g(x), along with the constant a, that:

The derivative of $(af(x)) = \lim\limits_{h \to 0} \left(\frac{af(x+h) - af(x)}{h} \right)$

$$= (a) \lim\limits_{h \to 0} \left(\frac{f(x+h) - f(x)}{h} \right)$$

$$= (a)f'(x),$$

193

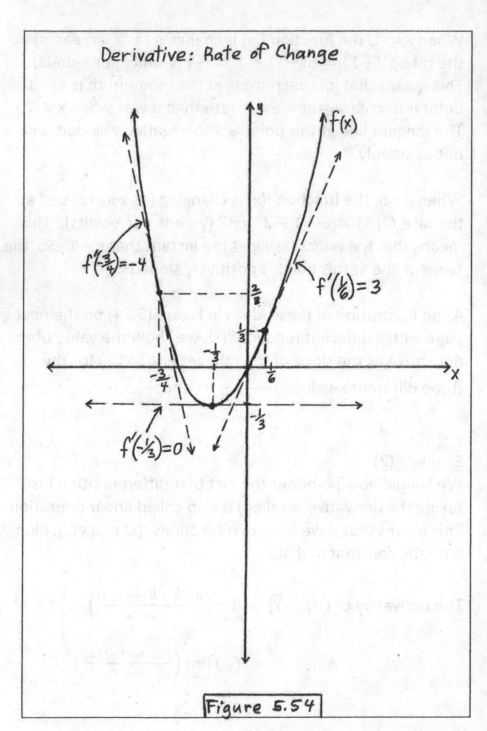

Figure 5.54

and the derivative of $(f(x) + g(x))$ is equal to:

the $\lim\limits_{h \to 0} \left(\dfrac{f(x + h) + g(x + h) - f(x) - g(x)}{h} \right)$

$= \lim\limits_{h \to 0} \left(\dfrac{f(x + h) - f(x) + g(x + h) - g(x)}{h} \right)$

$= \lim\limits_{h \to 0} \left(\dfrac{f(x + h) - f(x)}{h} \right) + \lim\limits_{h \to 0} \left(\dfrac{g(x + h) - g(x)}{h} \right)$

$= f'(x) + g'(x)$.

What this property says is that the derivative of "a constant times a function" is the constant times the derivative of the function, and that the derivative of "a sum of two functions" is the sum of the derivatives.

Further Notation and Higher Order Derivatives

Note that for a function y = f(x), we are taking the derivative of the dependent variable y with respect to the independent variable x. We use the notation $\dfrac{d}{dx}$ to mean taking the derivative with respect to x. For example, we write:

$f'(x) = \dfrac{d}{dx}(f(x)) = \dfrac{d}{dx}(y) = \dfrac{dy}{dx} = y' = y'(x)$. All of these mean the same thing.

The second derivative $f''(x)$ is simply the derivative of the first derivative $f'(x)$, and it gives us information on how the first derivative is changing. The third derivative $f'''(x)$ is the derivative of the second derivative $f''(x)$, and it gives us

195

information on how the second derivative is changing. The nth derivative $f^{(n)}(x)$ is the derivative of the $(n-1)^{st}$ derivative $f^{(n-1)}(x)$, and it gives us information on how the $(n-1)^{st}$ derivative is changing, and so on. Regarding notation:

The second derivative $f''(x) = \frac{d}{dx}(f'(x)) = \frac{d}{dx}\left(\frac{dy}{dx}\right) = \frac{d^2y}{dx^2}$.

The third derivative $f'''(x) = \frac{d}{dx}(f''(x)) = \frac{d}{dx}\left(\frac{d^2y}{dx^2}\right) = \frac{d^3y}{dx^3}$.

This pattern continues with higher order derivatives.

For most of the work that we do with derivatives, the first and second derivatives are of the most importance.

################# Exercises ##################

(1) Find the derivative using the definition above for:
   (a) f(x) = $7x^2 - 4x + 2$
   (b) f(x) = $2x^3 + x$

(2) (a) Find the derivative using the definition above for:
     f(x) = $-\frac{1}{3}x^2 + 2x - 2$
   (b) What is the rate of change for f(x) when x = 1, x = 3, and x = 6. What does the value of the derivative for each of these x-values mean?

(4.8) The Power Rule for Integer Powers

The first type of function that we will learn to differentiate are functions of the form $y = x^n$, where n is an integer. We will learn a formula which will allow us to differentiate this type of function very easily. We will derive the formula in three steps: firstly for n = 0, secondly for the case of integers n > 0, and thirdly for the case of integers n < 0.

To prove the second and third parts, we will use a proof technique called Proof by Mathematical Induction. In this book, we have introduced the reader to the technique of Proof by Contradiction, and now we will use an induction proof, which is another major proof technique that is of great use to mathematicians.

The idea is that if we desire to prove a sequence of propositions $\{P(1), P(2), P(3), \ldots \}$, which are indexed by the positive integers. What we do is this:

(A) Firstly, prove that P(1) is true. In other words, prove it is true for n = 1.
(B) Secondly, prove that if it is true for some n $\geq$ 1, then it is true for (n + 1).

From (A), we know it's true for n = 1. Given that it's true for n = 1, if we know that (B) is true, then we know that it's true for n = 2. Then, given that it's true for n = 2, if we know that (B) is true, then we know that it's true for n = 3. Then, given that it's true for n = 3, if we know that (B) is true, then we know that it's true for n = 4, and on and on the reasoning goes. This allows us to state that all of the propositions in the sequence {P(n)} are true (the proposition is true for all n). So, now we will use this to prove the power rule for integral powers.

Theorem: If n is an integer and $y = f(x) = x^n$, then $f'(x) = nx^{n-1}$. This is a very important formula in the calculus and is known as the Power Rule. It is true for any real number n, but in this section, we will prove it for integer powers n, and in a later section for rational powers n. That will be sufficient for our purposes here in this book.

Case (1): n = 0 and f(x) = x^0
If $y = x^0$, then y = 1 and we have the constant function y = f(x) = 1. So, the derivative is:

$$f'(x) = \lim_{h \to 0} \left(\frac{f(x+h) - f(x)}{h} \right) = \lim_{h \to 0} \left(\frac{1 - 1}{h} \right) = \lim_{h \to 0} \left(\frac{0}{h} \right) = 0.$$

Note that when we write $\lim_{h \to 0} \left(\frac{0}{h} \right) = 0$, this is true because "h" is taking the values of a sequence of non-zero numbers approaching 0, but never reaching 0. So $\left(\frac{0}{h} \right)$ is always 0; we don't have the indeterminate form $\frac{0}{0}$.

Our formula says that the derivative should be:
$f'(x) = (0)x^{0-1} = (0)x^{-1} = 0$. So, we have proven the proposition for this Case (1).

To generalize things, if we have any constant C and f(x) = C,

$$f'(x) = \lim_{h \to 0}\left(\frac{f(x+h)-f(x)}{h}\right) = \lim_{h \to 0}\left(\frac{C-C}{h}\right) = \lim_{h \to 0}\left(\frac{0}{h}\right) = 0.$$

Therefore, the derivative of any constant function is 0.

Our formula applied to this situation would say that if:
f(x) = C = Cx^0, then from the linearity property discussed earlier, $f'(x) = (C)[(0)x^{0-1}] = (C)(0)x^{-1} = 0$.

Case (2): Integer n > 0 and f(x) = x^n
Using a mathematical induction argument:

(A) Let n = 1. Then y = f(x) = x. So,

$$f'(x) = \lim_{h \to 0}\left(\frac{f(x+h)-f(x)}{h}\right) = \lim_{h \to 0}\left(\frac{(x+h)-(x)}{h}\right)$$
$$= \lim_{h \to 0}\left(\frac{h}{h}\right) = \lim_{h \to 0}(1) = 1.$$

Our formula says that the derivative should be:
$f'(x) = (1)x^{1-1} = (1)x^0 = 1$. So, we have proven the proposition is true for n = 1. Now for the second part of the induction argument for our Case (2):

(B) Suppose the proposition is true for some n ≥ 1, meaning that if f(x) = x^n, then $f'(x) = nx^{n-1}$. Consider the case for (n + 1), that is for f(x) = x^{n+1}:

199

$$f'(x) = \lim_{h \to 0}\left(\frac{(x+h)^{n+1}-(x)^{n+1}}{h}\right)$$

$$= \lim_{h \to 0}\left(\frac{(x+h)(x+h)^{n}-(x)(x)^{n}}{h}\right)$$

$$= \lim_{h \to 0}\left((x)\left(\frac{(x+h)^{n}-(x)^{n}}{h}\right)+(h)\left(\frac{(x+h)^{n}}{h}\right)\right)$$

$$= (x)\cdot\lim_{h \to 0}\left(\frac{(x+h)^{n}-(x)^{n}}{h}\right)+x^{n}$$

(x can be brought across the limit sign since it doesn't involve h; x is like a constant in this step of the proof)

$$= (x)(n)x^{n-1}+x^{n}$$

(because we assume the proposition is true for n)

$$= (n)x^{n}+x^{n} = (n+1)x^{n} = (n+1)x^{(n+1)-1}.$$

So, we have shown that the proposition is true for (n + 1). Therefore, by induction the proposition is true for all n in our Case (2).

Case (3): Integer n > 0 and f(x) = x^{-n}
Using a mathematical induction argument:

(A) Let n = 1. Then y = f(x) = x^{-1}. So,

$$f'(x) = \lim_{h \to 0}\left(\frac{f(x+h)-f(x)}{h}\right) = \lim_{h \to 0}\left(\frac{(x+h)^{-1}-(x)^{-1}}{h}\right)$$

$$= \lim_{h \to 0} \left(\frac{1}{(h)(x+h)} - \frac{1}{(h)(x)} \right)$$

$$= \lim_{h \to 0} \left(\frac{(x)-(x+h)}{(h)(x)(x+h)} \right) = (-1) \lim_{h \to 0} \left(\frac{1}{(x)(x+h)} \right)$$

$$= (-1)x^{-2} = (-1)x^{(-1)-1}$$

So, we have proven the proposition is true for n = 1. Now for the second part of the induction argument for our Case (3):

(B) Suppose the proposition is true for some $n \geq 1$, meaning that if $f(x) = x^{-n}$, then $f'(x) = -nx^{(-n)-1}$. Consider the case for (n + 1), that is for $f(x) = x^{-(n+1)}$:

$$f'(x) = \lim_{h \to 0} \left(\frac{(x+h)^{-(n+1)} - (x)^{-(n+1)}}{h} \right)$$

$$= \lim_{h \to 0} \left(\frac{(x+h)^{-1}(x+h)^{-n} - (x)^{-1}(x)^{-n}}{h} \right)$$

$$= \lim_{h \to 0} \left(\left(\frac{x}{(h)(x)(x+h)} \right) (x+h)^{-n} - \left(\frac{x+h}{(h)(x)(x+h)} \right) (x)^{-n} \right)$$

$$= \lim_{h \to 0} \left(\left(\frac{x}{(h)(x)(x+h)} \right) ((x+h)^{-n} - (x)^{-n}) \right) -$$

$$\lim_{h \to 0} \left(\frac{1}{(x)(x+h)} \right) x^{-n}$$

$$= \lim_{h \to 0} \left[\left(\frac{1}{(x+h)} \right) \left(\frac{(x+h)^{-n} - (x)^{-n}}{h} \right) \right] - x^{-n-2}$$

$$= \lim_{h \to 0}\left[\left(\frac{1}{(x+h)}\right)\right] \cdot \lim_{h \to 0}\left[\frac{(x+h)^{-n} - (x)^{-n}}{h}\right] - x^{-n-2}$$

$$= (x^{-1}) \cdot (-n)(x)^{(-n)-1} - (x)^{-n-2}$$
(because we assume the proposition is true for n)

$$= (-n)x^{-n-2} - x^{-n-2}$$

$$= -(n+1)x^{-(n+1)-1}$$

So, we have shown that the proposition is true for (n + 1). Therefore, by induction the proposition is true for all n in our Case (3). So, we have proven the Power Rule for all integral powers.

Example (1):

(A) If $y = x^4 + 200$, $\frac{dy}{dx} = 4x^3$.

(B) If $y = x^3 - x^8$, $\frac{dy}{dx} = 3x^2 - 8x^7$.

(C) If $y = x^{-5}$, $\frac{dy}{dx} = (-5)x^{-6} = \frac{-5}{x^6}$

(D) If $y = x^{-2} + x^{-6}$, $\frac{dy}{dx} = (-2)x^{-3} - (6)x^{-7}$

$$= \frac{-2}{x^3} - \frac{6}{x^7}.$$

(E) If $y = x^{10} - \frac{1}{x}$, Then $y = x^{10} - x^{-1}$.

So, $\frac{dy}{dx} = 10x^9 - (-1)x^{-2}$

202

$$= 10x^9 + x^{-2}$$
$$= 10x^9 + \frac{1}{x^2}$$

Example (2):

(A) If $y = 3x^2 + 10x^9 - 8x^{-3}$,

$\frac{dy}{dx} = (3)(2)x^1 + (10)(9)x^8 - (8)(-3)x^{-4}$

$\quad = 6x + 90x^8 + 24x^{-4}$

$\quad = 6x + 90x^8 + \frac{24}{x^4}$.

(B) If $y = 5x^2 + 13x^3 - 5x^4$,

$\frac{dy}{dx} = 10x + 39x^2 - 20x^3$.

(C) If $y = 2x^{-2} - 4x^{-5} + 15x + 7$,

$\frac{dy}{dx} = -4x^{-3} + 20x^{-6} + 15 + 0$

$\quad = \frac{-4}{x^3} + \frac{20}{x^6} + 15$.

(D) If $y = \frac{12}{x^3} - 10x^{20} - \frac{4}{x^{-3}}$,

Then $y = 12x^{-3} - 10x^{20} - 4x^3$.

So, $\frac{dy}{dx} = -36x^{-4} - 200x^{19} - 12x^2$

$\quad = \frac{-36}{x^4} - 200x^{19} - 12x^2$

(E) If $y = \frac{4x^3 - 12x^5 + 5x^9 - 8x}{2x}$,

Then $y = 2x^2 - 6x^4 + \left(\frac{5}{2}\right)x^8 - 4$

So, $\frac{dy}{dx} = 4x - 24x^3 + 20x^7$

(F) If $y = \dfrac{30x^{-7} - 10x^{-4} + 8x^2}{2x^{-3}}$,

Then $y = 15\left(x^{-7-(-3)}\right) - 5\left(x^{-4-(-3)}\right) + 4\left(x^{2-(-3)}\right)$

$= 15x^{-4} - 5x^{-1} + 4x^5$

So, $\dfrac{dy}{dx} = -60x^{-5} + 5x^{-2} + 20x^4$

$= \dfrac{-60}{x^5} + \dfrac{5}{x^2} + 20x^4$

################ Exercises ##################

Find the derivative for each of the functions y = f(x).

(1) $y = 3x^4 + x^2 + 9$ (2) $y = 4x^{-8} - 5x^3$

(3) $y = 2x^2 + 3x^3 + 4x^4$ (4) $y = \left(\dfrac{x^2 + x^3 + x^5}{x}\right)$

(5) $y = \left(\dfrac{x^{-2} + x^{-3} + x^{-5}}{x}\right)$ (6) $y = 10x^4 + 100x^3$

(4.9) The Chain Rule

Composite Functions

We will now consider the situation where we have functions of functions, that is functions nested within each other. This is very important for understanding the so-called Chain Rule and then Implicit Differentiation. Once we have established these results, we will be in a position to prove the Power Rule for rational powers, and much more.

Example (1):
Consider the function $y = (2x^4 - 6)^3$. In calculus we have to think of this as a nesting of two functions.
Let $u(x) = (2x^4 - 6)$. Then $y(x) = y(u(x)) = (u(x))^3$.

Example (2):
Consider the function $y = \sin(4x^4 + 2x)$.
Let $u(x) = (4x^4 + 2x)$. Then $y(x) = y(u(x)) = \sin(u(x))$.

Example (3):
Consider the function $y = \ln(\tan(3x))$.
Let $u(x) = 3x$, and $v(u) = \tan(u)$.
Then $y(x) = y(v(u(x))) = \ln(v(u(x)))$.

The Chain Rule
The chain rule is based on the following reasoning. When we have a function $y = f(x)$, we know that the derivative is defined to be:

$\frac{dy}{dx} = \lim_{\Delta x \to 0} \frac{\Delta y}{\Delta x}$. If we have a composite function $y = f(u(x))$, then we must say that:

$$\frac{dy}{dx} = \lim_{\Delta x \to 0} \frac{\Delta y}{\Delta x} = \lim_{\Delta x \to 0} \frac{\Delta y}{\Delta u} \cdot \frac{\Delta u}{\Delta x}.$$

Since $\Delta u \rightarrow 0$ as $\Delta x \rightarrow 0$, we can then say that if both of the two limits in parentheses below exist as finite numbers:

$$\frac{dy}{dx} = \lim_{\Delta x \to 0} \frac{\Delta y}{\Delta x} = \lim_{\Delta x \to 0} \frac{\Delta y}{\Delta u} \cdot \frac{\Delta u}{\Delta x} = \left(\lim_{\Delta u \to 0} \frac{\Delta y}{\Delta u} \right) \cdot \left(\lim_{\Delta x \to 0} \frac{\Delta u}{\Delta x} \right), \text{ or}$$

$$\frac{dy}{dx} = \frac{dy}{du} \cdot \frac{du}{dx} . \text{ (Chain Rule for two nested functions)}$$

This idea can be generalized to any number of nested functions. For example, as in Example (3) above, where y is a function of v, v is a function of u, and u is a function of x, we would have the result that:

Since Δv and $\Delta u \rightarrow 0$ as $\Delta x \rightarrow 0$, we must say that if each of the three limits in parentheses below exist:

$$\frac{dy}{dx} = \lim_{\Delta x \to 0} \frac{\Delta y}{\Delta x} = \lim_{\Delta x \to 0} \frac{\Delta y}{\Delta v} \cdot \frac{\Delta v}{\Delta u} \cdot \frac{\Delta u}{\Delta x}$$

$$= \left(\lim_{\Delta v \to 0} \frac{\Delta y}{\Delta v} \right) \cdot \left(\lim_{\Delta u \to 0} \frac{\Delta v}{\Delta u} \right) \cdot \left(\lim_{\Delta x \to 0} \frac{\Delta u}{\Delta x} \right), \text{ or}$$

$$\frac{dy}{dx} = \frac{dy}{dv} \cdot \frac{dv}{du} \cdot \frac{du}{dx} . \text{ (Chain Rule for three nested functions)}$$

Illustrating the chain rule with the function in Example (1) above: There we had $y = (2x^4 - 6)^3$, and we decided that $u(x) = (2x^4 - 6)$. So, $y(u(x)) = (u(x))^3$ (or $y(u) = u^3$). Then according to the chain rule:

$$\frac{dy}{dx} = \frac{dy}{du} \cdot \frac{du}{dx} = (3u^2) \cdot (8x^3) = 3(2x^4 - 6)^2 \cdot (8x^3)$$
$$= 24x^3 (2x^4 - 6)^2$$

206

Example (4):

Let $y = (10x - 90)^5$. Calculate $\frac{dy}{dx}$.

Let $u(x) = (10x - 90)$ and $y(u) = u^5$. The chain rule says that $\frac{dy}{dx} = \frac{dy}{du} \cdot \frac{du}{dx}$. So, $\frac{dy}{dx} = (5u^4) \cdot (10) = 50(10x - 90)^4$.

Example (5):

Let $y = (4x^2 + 3x)^3$. Calculate $\frac{dy}{dx}$.

Let $u(x) = (4x^2 + 3x)$ and $y(u) = u^3$. The chain rule says that $\frac{dy}{dx} = \frac{dy}{du} \cdot \frac{du}{dx} = (3u^2) \cdot (8x + 3) = 3(8x + 3)(4x^2 + 3x)^2$

With enough practice, the use of the Chain Rule becomes second nature. We just instinctively multiply the "derivative of the outer function" by the "derivative of the inner function," as in the following examples:

Example (6):

(1) The $\frac{d}{dx}(2x + 9)^6 = (6)(2x + 9)^5(2) = 12(2x + 9)^5$

(2) The $\frac{d}{dx}(4x^3 + 9x)^3 = (3)(4x^3 + 9x)^2(12x^2 + 9)$

(3) The $\frac{d}{dx}(3x^{-5} + 1)^5 = (5)(3x^{-5} + 1)^4(-15x^{-6})$
$$= (-75)(x^{-6})(3x^{-5} + 1)^4$$

(4) The $\frac{d}{dx}(12x^2 + 5)^{-3} = (-3)(12x^2 + 5)^{-4}(24x)$

$$= (-72x)(12x^2 + 5)^{-4}$$

################# Exercises #################

Find the derivative for each of the functions y = f(x).

(1) $y = (3x + 4x^2)^5$ (2) $y = (15x^3 - 1)^7$

(3) $y = (2x + 1)^3 - (13x^2 + x - 1)^5$

(4) $y = (3 - 4x^6)^3 + (10 - 2x^2)^4$

(5) $y = 1 + 2x + (3x)^2 - (4x)^3$

(6) $y = (x - 1)^3 + (3x - 1)^3$

(4.10) Implicit Differentiation

There are times when we are unable to, or choose not to, isolate y as an explicit function of x and take the derivative using our usual rules of differentiation. When this is the case, we assume that y is an implicit function of x. Then we differentiate all of the terms of the relation, on both sides of the equation, using the chain rule when it comes to the variable y. We think of the variable y as a function like the expressions which were inside the parentheses when we discussed the Chain Rule. The derivative of these implicit functions is then simply referred to as $\frac{dy}{dx}$. Perhaps we can clear up the issue of how to differentiate with respect to x when y is an implicit function of x with the following rules:

The $\frac{d}{dx}(y) = \frac{dy}{dx}$

The $\frac{d}{dx}(y^n) = ny^{n-1} \cdot \frac{dy}{dx}$ (n is an integer)

You can see how the chain rule comes into play here. For the terms involving only the variable x we differentiate them as usual. Let's consider some examples.

Example (1):
Find $\frac{dy}{dx}$ in the relation $3y^3 + 4y = x^2$.

We differentiate term by term on both sides with respect to the variable x, resulting in the following:

$9y^2 \frac{dy}{dx} + 4 \frac{dy}{dx} = 2x$, which leads to:

$(9y^2 + 4)\frac{dy}{dx} = 2x$, which says that: $\frac{dy}{dx} = \frac{2x}{(9y^2+4)}$.

Example (2):
Find $\frac{dy}{dx}$ in the relation $20y^{-4} + y + 10\sqrt{y} = x^2 - 2x^3$.

We differentiate term by term on both sides with respect to the variable x, resulting in the following:

$-80y^{-5}\frac{dy}{dx} + \frac{dy}{dx} + 5\frac{1}{\sqrt{y}}\frac{dy}{dx} = 2x - 6x^2$, which leads to:

$\left(\frac{-80}{y^5} + \frac{5}{\sqrt{y}} + 1\right)\frac{dy}{dx} = 2x - 6x^2$, which says that:

$$\frac{dy}{dx} = \frac{(2x - 6x^2)}{\left(\frac{-80}{y^5} + \frac{5}{\sqrt{y}} + 1\right)}.$$

The price that we pay in implicit differentiation is that $\frac{dy}{dx}$ is often determined for us in terms of x and y, instead of as an explicit function involving only the variable x. However, we sometimes can find it advantageous to have $\frac{dy}{dx}$ in terms of x and y, such as in the following example.

Example (3):

Find $\frac{dy}{dx}$ in the relation $x^2 + y^2 = r^2$, which is the equation of a circle centered at the origin and with the radius r.

Differentiating all terms with respect to x leads to:

$$2x + 2y\frac{dy}{dx} = 0 \quad \rightarrow \quad 2y\frac{dy}{dx} = -2x \quad \rightarrow \quad \frac{dy}{dx} = -\frac{x}{y}.$$

Note that the $\frac{d}{dx}(r^2) = 0$, since r^2 is a constant here.
If we know that the point with coordinates (a,b) is on the circle, then the slope of the tangent line to the circle at that point is $\left(-\frac{a}{b}\right)$. For example, the point with coordinates

$\left(-\frac{r}{\sqrt{2}}, \frac{r}{\sqrt{2}}\right)$ is on this circle, and the slope of the tangent line

to the circle at that point is m = $\left(-\frac{\left(-r/\sqrt{2}\right)}{\left(r/\sqrt{2}\right)}\right)$ = 1.

################ Exercises ###################

Use implicit differentiation to find $\frac{dy}{dx}$.

(1) $x^3 + y^3 = 16x^5$ (2) $3y^3 + 2y^2 + y = x + 1$

(3) $(2y^2 + 2y)^5 = 25x$ (4) $25x^3 - 25x^2 = y^3$

(5) $4y^5 - 16y + 2x = y$ (6) $(x - 1)^2 - 4(y - 3)^2 = 1$

(4.11) The Power Rule for Rational Powers

Consider the relation $y^n = x^m$, where n and m are non-zero integers. Assume that this equation represents a curve over some region of the x-y plane where y = $x^{\frac{m}{n}}$ and x = $y^{\frac{n}{m}}$ are inverse functions on that region. Differentiating both sides implicitly with respect to x, we have (since we know the Power Rule is true for integer powers):

$ny^{n-1} \cdot \frac{dy}{dx} = mx^{m-1}$, so that $\frac{dy}{dx} = \left(\frac{m}{n}\right) y^{1-n} x^{m-1}$.

211

Then, $\frac{dy}{dx} = \left(\frac{m}{n}\right)(x^{\frac{m}{n}})^{1-n}x^{m-1}$.

$$\frac{dy}{dx} = \left(\frac{m}{n}\right)x^{\frac{m}{n}-m+m-1} .$$

$$\frac{dy}{dx} = \left(\frac{m}{n}\right)x^{\frac{m}{n}-1} .$$

Therefore, the Power Rule is true for rational powers.
In other words, we can now say: If n is a rational number and $y = f(x) = x^n$, then $f'(x) = nx^{n-1}$.

Example (1):

(A) If y = f(x) = $\sqrt{x} = x^{\frac{1}{2}}$,

then $\frac{dy}{dx} = \left(\frac{1}{2}\right)x^{\frac{1}{2}-1} = \frac{1}{2}x^{-\frac{1}{2}} = \frac{1}{2x^{\frac{1}{2}}} = \frac{1}{2\sqrt{x}}$.

(B) If y = f(x) = $x^{\frac{5}{3}}$,

then $\frac{dy}{dx} = \frac{5}{3}x^{\frac{5}{3}-1} = \frac{5}{3}x^{\frac{2}{3}}$.

(C) If y = f(x) = $x^{-\frac{3}{7}}$,

then $\frac{dy}{dx} = \left(-\frac{3}{7}\right)x^{-\frac{3}{7}-1} = -\frac{3}{7}x^{-\frac{10}{7}}$.

Example (2):

(A) If y = f(x) = $(-7x^2 - 4)^{\frac{1}{3}}$, then to find $\frac{dy}{dx}$ we must use

the chain rule by letting u = $(-7x^2 - 4)$ and then y = $u^{\frac{1}{3}}$.
So, the chain rule says that $\frac{dy}{dx} = \frac{dy}{du} \cdot \frac{du}{dx}$.

So, $\dfrac{dy}{dx} = \left(\dfrac{1}{3}u^{-\frac{2}{3}}\right)\cdot(-14x)$

$$= \left(\dfrac{1}{(3)(-7x^2-4)^{\frac{2}{3}}}\right)\cdot(-14x)$$

$$= \dfrac{-14x}{(3)(-7x^2-4)^{\frac{2}{3}}} = \dfrac{-14x}{(3)\sqrt[3]{(-7x^2-4)^2}}$$

(B) If y = $\sqrt[5]{3x+3}$, then to find $\dfrac{dy}{dx}$ we must use the chain rule by letting u = $(3x + 3)$ and then y = $u^{\frac{1}{5}}$. So, the chain rule says that $\dfrac{dy}{dx} = \dfrac{dy}{du}\cdot\dfrac{du}{dx}$.

So, $\dfrac{dy}{dx} = \left(\dfrac{1}{5}u^{-\frac{4}{5}}\right)\cdot(3) = \dfrac{3}{(5)\sqrt[5]{(3x+3)^4}}$.

################# Exercises #################

Find the derivative for the functions y = f(x).

(1) $y = \dfrac{3}{2}x^{\frac{2}{3}} + 40x + 1$ (2) $y = 3x^{\frac{7}{8}} + 4x^{\frac{1}{4}}$

(3) $y = 20x^{\frac{2}{5}} + \sqrt{x^2 - 4x}$ (4) $y = \sqrt{4x + 2} - \sqrt[3]{2x - 3}$

(5) $y = 15x^{\frac{11}{2}} - 10x^{\frac{13}{7}}$ (6) $y = 5x^{\frac{1}{5}} - 6x^{\frac{2}{3}} + 7x^{\frac{1}{6}}$

213

(4.12) The Product and Quotient Rules

There are two more rules of differentiation that we need to learn. They are the product rule and the quotient rule. If U(x) and V(x) are functions of x, then these rules allow us to find the derivatives of $U(x) \cdot V(x)$ and $\frac{U(x)}{V(x)}$. In order to make the derivations a little bit easier to follow, we will refer to U(x) and V(x) as simply U and V respectively.

The Product Rule

Let's consider the derivative of P(x) = (U)·(V). In general, a change in P(x) is $\Delta P = (U + \Delta U)(V + \Delta V) - UV$. So, where the derivative $P'(x)$ exists:

$$P'(x) = \lim_{\Delta x \to 0} \frac{\Delta P}{\Delta x} = \lim_{\Delta x \to 0} \frac{(U + \Delta U)(V + \Delta V) - UV}{\Delta x}$$

$$= \lim_{\Delta x \to 0} \left(\frac{UV + U\Delta V + V\Delta U + (\Delta U)(\Delta V) - UV}{\Delta x} \right)$$

$$= \lim_{\Delta x \to 0} \left(U\left(\frac{\Delta V}{\Delta x}\right) + V\left(\frac{\Delta U}{\Delta x}\right) + (\Delta U)\left(\frac{\Delta V}{\Delta x}\right) \right)$$

$$= U\left(\lim_{\Delta x \to 0} \left(\frac{\Delta V}{\Delta x}\right) \right) + V\left(\lim_{\Delta x \to 0} \left(\frac{\Delta U}{\Delta x}\right) \right) +$$

$$\left(\lim_{\Delta x \to 0} \Delta U \right) \cdot \left(\lim_{\Delta x \to 0} \left(\frac{\Delta V}{\Delta x}\right) \right)$$

(The $\lim_{\Delta x \to 0} \Delta U = 0$ and the $\lim_{\Delta x \to 0} \left(\frac{\Delta V}{\Delta x}\right) = \frac{dV}{dx}$, which is a finite number for any x in the domain of V at which $V'(x)$ exists)

214

$$= UV'(x) + VU'(x) + (0)(V'(x))$$

$$= U(x)V'(x) + V(x)U'(x)$$

This is the so-called Product Rule.

The Quotient Rule

Let's consider the derivative of Q(x) = $\left(\frac{U}{V}\right)$. In general, a change in Q(x) is $\Delta Q = \frac{U+\Delta U}{V+\Delta V} - \frac{U}{V}$. So, where the derivative $Q'(x)$ exists:

$$Q'(x) = \lim_{\Delta x \to 0} \frac{\Delta Q}{\Delta x} = \lim_{\Delta x \to 0} \frac{\frac{U+\Delta U}{V+\Delta V} - \frac{U}{V}}{\Delta x}$$

$$= \lim_{\Delta x \to 0} \left(\frac{1}{\Delta x}\right) \cdot \left(\frac{(U+\Delta U)(V) - (U)(V+\Delta V)}{(V)(V+\Delta V)}\right)$$

$$= \lim_{\Delta x \to 0} \left(\frac{1}{\Delta x}\right) \cdot \left(\frac{UV + V\Delta U - UV - U(\Delta V)}{V(V+\Delta V)}\right)$$

$$= \frac{(V)\left(\lim_{\Delta x \to 0}\left(\frac{\Delta U}{\Delta x}\right)\right) - (U)\left(\lim_{\Delta x \to 0}\left(\frac{\Delta V}{\Delta x}\right)\right)}{(V)(V+\Delta V)}$$

$$= \frac{V(x)U'(x) - U(x)V'(x)}{(V(x))^2}, \text{ (Since } \Delta V \to 0 \text{ as } \Delta x \to 0)$$

This is the so-called Quotient Rule.

Some quick examples of the use of these two rules:

(A) The $\frac{d}{dx}(x + 9)(-3x + 1)^5$

$= (x + 9) \cdot (5)(-3x + 1)^4(-3) + (-3x + 1)^5 \cdot (1)$

$= (-15)(x + 9)(-3x + 1)^4 + (-3x + 1)^5$

$= (-3x + 1)^4 \cdot ((-15)(x + 9) + (-3x + 1))$

$= (-3x + 1)^4(-18x - 134))$

(B) The $\frac{d}{dx}(x - 1)^2(5x + 1)^8$

$= (x - 1)^2 \cdot (8)(5x + 1)^7(5) + (5x + 1)^8 \cdot (2)(x - 1)$

$= (40)(x - 1)^2(5x + 1)^7 + (2)(x - 1)(5x + 1)^8$

$= (2)(x - 1)(5x + 1)^7 \cdot ((20)(x - 1) + (5x + 1))$

$= (2)(x - 1)(5x + 1)^7 \cdot (25x - 19)$

(C) The $\frac{d}{dx}\frac{(x+3)^4}{(2x-4)^5} = \frac{(2x-4)^5(4)(x+3)^3 - (x+3)^4(5)(2x-4)^4(2)}{((2x-4)^5)^2}$

$= \frac{(4)(2x-4)^5(x+3)^3 - (10)(x+3)^4(2x-4)^4}{(2x-4)^{10}}$

$= \frac{(2)(x+3)^3(2x-4)^4 \cdot ((2)(2x-4) - (5)(x+3))}{(2x-4)^{10}}$

$$= \frac{(2)(x+3)^3(2x-4)^4 \cdot (-x-23)}{(2x-4)^{10}}$$

(D) The $\dfrac{d}{dx}\left(\dfrac{-3}{x^4}\right) = \dfrac{(x^4)(0)-(-3)(4)(x^3)}{x^8} = \dfrac{12}{x^5}$

################ Exercises ################

Find the derivative for the functions y = f(x).

(1) $y = (2x - 7)^8(\sqrt{x})$ (2) $y = (3x^2 - 1)^4(2x - 2)^5$

(3) $y = (x - 1)^4(x + 1)^4$ (4) $y = \dfrac{(x - 7)^3}{(2 - x^3)^4}$

(5) $y = \dfrac{(x^2 + x + 1)^3}{(3x + 2)^4}$ (6) $y = \dfrac{(x - 1)^4}{(x + 1)^4}$

(4.13) Increments and Differentials

When we have a smooth and continuous function y = f(x) on an open interval (a,b), increments of the variables x and y are the actual changes in x and y between some two distinct points on the graph of f(x), where the x-values of these two points are in (a,b). Usually these two points would be taken to be relatively close to each other. As we have seen, these increments are denoted as Δx and Δy.

217

We will now think again of what we mean by dx and dy. For the independent variable x, the differential dx is always considered to be equal to the increment Δx, that is, dx = Δx. However, this is not the case with the differential dy. When two distinct points are close to each other in the open interval (a,b) and we consider an x-value (x^*) between the x-values of these two points, then the following would be true:

$\frac{\Delta y}{\Delta x} \approx f'(x^*)$, where $\Delta x \neq 0$. It is valid to algebraically rearrange this approximation to get: $\Delta y \approx f'(x^*) \cdot \Delta x$.

Consider the limit of both sides of this approximation as $\Delta x \to 0$, and we get: (See Figure (5.55) on the next page)

dy = $f'(x)$dx . (We call this a differential equation)

The differentials dx and dy, up to this point, have been taken to be notation for the infinitesimal increments Δx and Δy (as $\Delta x \to 0$), where their ratio $\frac{dy}{dx}$ was just notation for $f'(x)$, rather than thinking of them as an actual quotient of non-zero quantities.

However, now consider the differential equation above to be the tangent line at a specific point, like the equation of the line y = mx, but with the two variables x and y now taken to be dx and dy respectively.

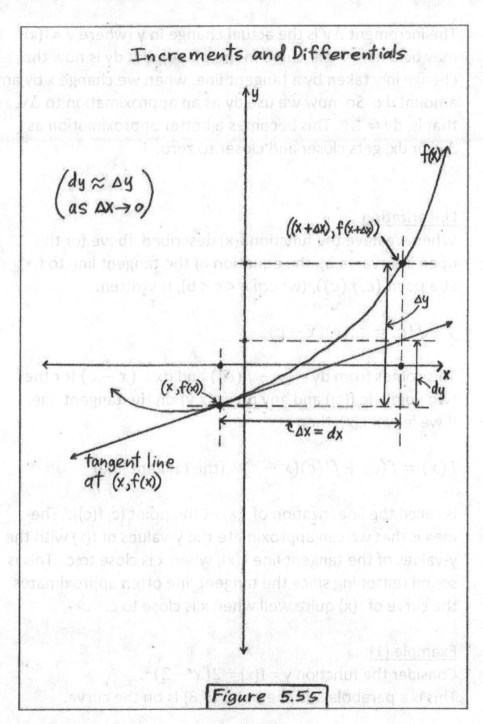

Increments and Differentials

$$\begin{pmatrix} dy \approx \Delta y \\ \text{as } \Delta x \to 0 \end{pmatrix}$$

$f(x)$

$((x+\Delta x), f(x+\Delta x))$

Δy

$(x, f(x))$

dy

$\Delta x = dx$

tangent line at $(x, f(x))$

Figure 5.55

219

The increment Δy is the actual change in y (where y = f(x) may be a non-linear curve in the plane), but dy is now the change in y taken by a tangent line, when we change x by an amount dx. So, now we use dy as an approximation to Δy, that is, dy $\approx \Delta y$. This becomes a better approximation as Δx, or dx, gets closer and closer to zero.

Linearization
When we have the function f(x) described above for the open interval (a,b), the equation of the tangent line to f(x) at a point $(c, f(c))$, (where a < c < b), is written:

$$y - f(c) = f'(c)(x - c)$$

This comes from dy = $(y - f(c))$ and dx = $(x - c)$ for the two points (c,f(c)) and any point (x,y) on the tangent line. If we let y = L(x), then

$$L(x) = f(c) + f'(c)(x - c) \quad \text{(the tangent line)}$$

is called the linearization of f(x) at the point (c, f(c)). The idea is that we can approximate the y-values of f(x) with the y-values of the tangent line L(x), when x is close to c. This is sound reasoning since the tangent line often approximates the curve of f(x) quite well when x is close to c.

Example (1):
Consider the function y = f(x) = $2(x - 3)^2$.
This is a parabola and the point (5,8) is on the curve.

$f'(x) = 4(x - 3) = 4x - 12$. The linearization at (5,8) is:

$L(x) = f(5) + f'(5)(x - 5)$, or
$L(x) = 8 + 8(x - 5)$

If we want to approximate f(5.1), we say:
$f(5.1) \approx L(5.1) = 8 + 8(5.1 - 5) = 8 + 8(0.1) = (8.8)$
as our approximation.

The actual y-value when x = 5.1 is:
$f(5.1) = 2(5.1 - 3)^2 = 8.82$. The approximation is very
good in this example, we are off the mark by only 0.02 .

Example (2):
The volume of a spherical region is $V = \frac{4}{3}\pi r^3$. Estimate the
volume V when the radius is r = 2.01 . Round all calculations
to six decimal places.

We can let c = 2. So, V(c) = V(2) = 33.510322

$V'(r) = 4\pi r^2$ (which by the way is the surface area)

If we want to estimate V(2.01) using linearization, then use:
$L(r) = V(2) + V'(2)(r - 2)$

So, $V(2.01) \approx L(2.01) = 33.510322 + (4\pi(2)^2)(2.01 - 2)$
$= 33.510322 + 0.502655 = 34.012977$

The actual volume when r = 2.01 is V = 34.015494.

221

So, for many functions, Linearization can provide us with a very good approximation in a local area. Linearization is often of good use to us in the proof of certain theorems.

Example (3):
Use linearization to estimate the $\sqrt{4.002}$.
$f(x) = \sqrt{x}$ and $f'(x) = \frac{1}{2\sqrt{x}}$. Let c = 4, so that f(c) = f(4) = 2, and $f'(c) = f'(4) = \frac{1}{2\sqrt{4}} = 0.25$. The linearization at (4,2) is: $L(x) = f(4) + f'(4)(x-4)$.

So, the $\sqrt{4.002} \approx L(4.002) = 2 + (0.25)(4.002 - 4)$
$$= 2 + 0.0005 = 2.0005 .$$

A more exact value on my scientific calculator is 2.0004999 (rounded to seven decimal places). Once again, a very good approximation.

################# Exercises #################

(1) Estimate the $\sqrt{25.01}$, by using linearization with c = 25. Compare this with the scientific calculator value.

(2) For a right circular cylinder, the volume is given by the formula $V = \pi r^2 h$. Assume h to be fixed at h = 10. So, $V = 10\pi r^2$ is a function of r alone. Use linearization to estimate the volume when r = (100.23) with c = 100. Compare this with the scientific calculator value.

(3) We have a mass m_1 = 100 kilograms and a mass

m_2 = 150 kilograms, separated by a distance of r = 20 meters. Newton's Law of Gravitation says that the force that each mass exerts on the other is given by $F = -\frac{Gm_1m_2}{r^2}$ Newtons. The negative sign is used to indicate that the force felt by each mass is toward the other. G is Newton's Gravitational Constant (we need not know its value in this example, just carry it through all of the calculations, simply denoting it by G). Estimate the gravitational force as the distance r between the masses decreases to 19.996 meters. Compare this with the scientific calculator value.

(4.14) Derivatives of Inverse Functions

Suppose there exists a closed rectangular region D of the x-y coordinate plane where y = f(x) and x = g(y) are inverse functions. Then we know that y = f(g(y)) for all y in the domain of g, and x = g(f(x)) for all x in the domain of f. The domain of f is an interval [a,b], and the domain of g is an interval [c,d], so that D = [a,b] × [c,d]. (We can restrict ourselves to a bounded region like this without any loss of generality). Then on the region D, f: [a,b] → [c,d] and the inverse function g: [c,d] → [a,b] are 1-1 and onto functions.

From y = f(g(y)), we can differentiate implicitly with respect to y to get:

$$1 = f'(g(y)) \cdot g'(y) \text{, or}$$

$$1 = f'(x) \cdot \frac{dx}{dy}, \text{ or}$$

$$1 = \frac{dy}{dx} \cdot \frac{dx}{dy} \qquad\qquad (4.14.1)$$

From x = g(f(x)), we can differentiate implicitly with respect to x to get:

$$1 = g'(f(x)) \cdot f'(x), \text{ or}$$

$$1 = g'(y) \cdot \frac{dy}{dx}, \text{ or}$$

$$1 = \frac{dx}{dy} \cdot \frac{dy}{dx} \qquad\qquad (4.14.2)$$

Equations (4.14.1) and (4.14.2) imply that:

$$\frac{dy}{dx} = \frac{1}{\left(\frac{dx}{dy}\right)} \quad \text{and} \quad \frac{dx}{dy} = \frac{1}{\left(\frac{dy}{dx}\right)}.$$

These equations define the relationship between these two derivatives for a pair of inverse functions y = f(x) and x = g(y) on D. This will be important in later sections of this chapter. See Figure (5.56) on the next page.

Example (1):
Suppose y = f(x) = $5x - 3$. Then solving for x in terms of y, we have the inverse function x = g(y) = $\frac{1}{5}y + \frac{3}{5}$.
These two functions are 1-1 and onto functions, hence inverse functions, from the reals R to the reals R.
We can see that $\frac{dy}{dx} = 5$ and $\frac{dx}{dy} = \frac{1}{5}$.

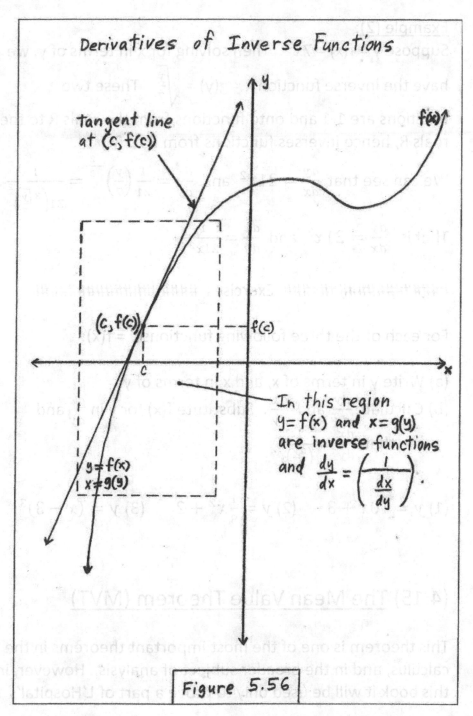

Derivatives of Inverse Functions

tangent line
at $(c, f(c))$

y

$f(x)$

$(c, f(c))$

$f(c)$

c

x

In this region
$y = f(x)$ and $x = g(y)$
are inverse functions
and $\dfrac{dy}{dx} = \left(\dfrac{1}{\dfrac{dx}{dy}}\right)$.

$y = f(x)$
$x = g(y)$

Figure 5.56

Example (2):

Suppose $y = f(x) = 7x^3$. Then solving for x in terms of y, we

have the inverse function $x = g(y) = \sqrt[3]{\dfrac{y}{7}}$. These two

functions are 1-1 and onto functions from the reals R to the reals R, hence inverses functions from R to R.

We can see that $\dfrac{dy}{dx} = 21x^2$ and $\dfrac{dx}{dy} = \dfrac{1}{21}\left(\dfrac{y}{7}\right)^{-\frac{2}{3}} = \dfrac{1}{21\left(\sqrt[3]{\dfrac{y}{7}}\right)^2}$.

That is, $\dfrac{dy}{dx} = 21x^2$ and $\dfrac{dx}{dy} = \dfrac{1}{21x^2}$.

################ Exercises ##################

For each of the three following functions y = f(x):

(a) Write y in terms of x, and x in terms of y.

(b) Calculate $\dfrac{dy}{dx}$ and $\dfrac{dx}{dy}$. Substitute f(x) for y in $\dfrac{dx}{dy}$ and

     see that $\dfrac{dx}{dy} = \dfrac{1}{\left(\dfrac{dy}{dx}\right)}$.

(1) $y = 10x + 3$ (2) $y = \dfrac{1}{8}x^2 + 2$ (3) $y = (x - 3)^3$

(4.15) The Mean Value Theorem (MVT)

This theorem is one of the most important theorems in the calculus, and in the broader subject of analysis. However, in this book it will be used only to prove a part of L'Hospital's

Rule, in the next section. L'Hospital's Rule is important for evaluating certain limits. To prove the MVT we need a preliminary result known as Rolle's theorem.

Rolle's Theorem
If we have a function f(x) that is continuous on a closed interval [a,b], and differentiable on the open interval (a,b), and in addition f(a) = f(b) = 0, then there exists at least one number c, where a < c < b, such that $f'(c) = 0$.

Proof: We can disregard the case where f(x) = 0 on [a,b]. Rolle's theorem is obvious in that case.

Case (1): The function f(x) is positive valued on at least one open interval (d,e), where (d,e) $\subset$ (a,b), and where f(d) and f(e) are 0. Since f(x) is continuous on [d,e], there must be at least one number c in (d,e) where f(x) reaches a maximum. Since f(x) is differentiable on (d,e), it must be true that $f'(c)$ is 0. There is a horizontal tangent line to f(x) at x = c.

Case (2): The function f(x) is negative valued on at least one open interval (d,e), where (d,e) $\subset$ (a,b), and where f(d) and f(e) are 0. Since f(x) is continuous on [d,e], there must be at least one number c in (d,e) where f(x) reaches a minimum. Since f(x) is differentiable on (d,e), it must be true that $f'(c)$ is 0. There is a horizontal tangent line to f(x) at x = c.
So, we have proven Rolle's theorem.

Now, to the MVT. See Figure (5.57) on the next page.

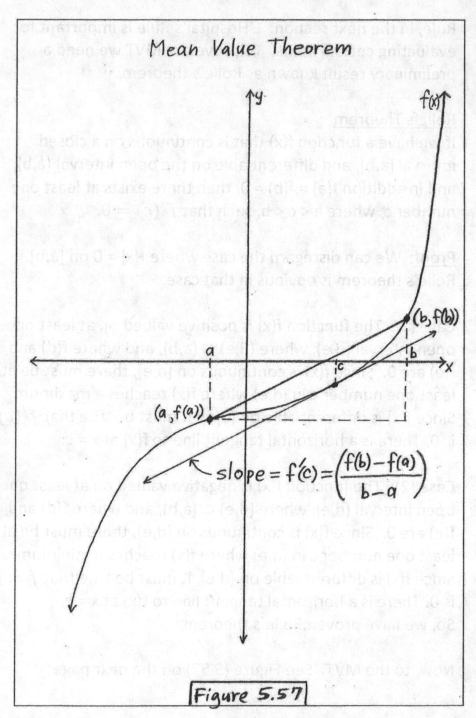

Figure 5.57

228

Mean Value Theorem (MVT)

Once again, assume that we have a function f(x) which is continuous on a closed interval [a,b] and differentiable on the open interval (a,b). Then there exists at least one number c, (a < c < b), such that $f'(c) = \left(\frac{f(b) - f(a)}{b - a}\right)$.

Proof: There exists a line through the points (a,f(a)) and (b,f(b)) with slope m = $\left(\frac{f(b) - f(a)}{b - a}\right)$. Call this linear function g(x). The equation of g(x) is:

$$g(x) = f(a) + \left(\frac{f(b) - f(a)}{b - a}\right)(x - a).$$

Let h(x) = $f(x) - g(x)$

$$= f(x) - f(a) - \left(\frac{f(b) - f(a)}{b - a}\right)(x - a).$$

Note that h(a) = h(b) = 0. Since h(x) satisfies the hypotheses of Rolle's theorem, then Rolle's theorem says that there is at least one number c (a < c < b) such that $h'(c) = 0$. This says then that $h'(c) = f'(c) - \left(\frac{f(b) - f(a)}{b - a}\right) = 0.$

Therefore, $f'(c) = \left(\frac{f(b) - f(a)}{b - a}\right)$. This proves the MVT.

################ Exercises ################

(1) Let f(x) = $4 - x^2$ on the interval [0,5]. Find some number "c" that satisfies the MVT on this interval.

(2) Let f(x) = $\sqrt[3]{x}$ on the interval [0,1]. Find some number "c" that satisfies the MVT on this interval..

(4.16) L'Hospital's Rule

This rule is very important in evaluating certain limits when we have the indeterminate form $\frac{0}{0}$ or $\frac{\infty}{\infty}$. We will prove it for the case of the indeterminate form $\frac{0}{0}$. The rule can be stated in this way: If for two functions f(x) and g(x), if the $\lim\limits_{x \to a} \frac{f(x)}{g(x)} = \frac{0}{0}$ or $\frac{\infty}{\infty}$ (where a is a number, or $-\infty$, or ∞), then

$$\lim_{x \to a} \frac{f(x)}{g(x)} = \lim_{x \to a} \frac{f'(x)}{g'(x)} .$$

(Note that this is not a situation which involves the quotient rule).

Proof: (for the case of $\frac{0}{0}$):
Suppose we have two functions f(x) and g(x) and the $\lim\limits_{x \to a} \frac{f(x)}{g(x)} = \frac{0}{0}$. Let f(a) and g(a) both be 0, $-\infty$, or ∞.

Then $\lim\limits_{x \to a} \frac{f(x)}{g(x)} = \lim\limits_{x \to a} \dfrac{\left(\frac{f(x)-f(a)}{x-a}\right)}{\left(\frac{g(x)-g(a)}{x-a}\right)}$. The MVT says that there exists two numbers c and d, such that:

230

$$\lim_{x \to a} \frac{\left(\frac{f(x) - f(a)}{x - a}\right)}{\left(\frac{g(x) - g(a)}{x - a}\right)} = \lim_{x \to a} \frac{f'(c)}{g'(d)}.$$ Since the numbers c and d are

between x and a, then it is clear that:

$$\lim_{x \to a} \frac{f(x)}{g(x)} = \lim_{x \to a} \frac{f'(x)}{g'(x)}.$$ This proves the rule for the case of $\frac{0}{0}$.

Example (1):

Evaluate the $\lim_{x \to 3} \left(\frac{x^2 - x - 6}{x^2 - 10x + 21}\right)$.

When we plug in 3 for x, we see that we have $\frac{0}{0}$. So, using
L'Hospital's Rule we have that:

$$\lim_{x \to 3} \left(\frac{x^2 - x - 6}{x^2 - 10x + 21}\right) = \lim_{x \to 3} \frac{2x - 1}{2x - 10} = -\frac{5}{4}.$$

We can verify that we have the correct result, since as we
learned in section (4.6) on limits and continuity:

$$\lim_{x \to 3} \left(\frac{x^2 - x - 6}{x^2 - 10x + 21}\right) = \lim_{x \to 3} \frac{(x - 3)(x + 2)}{(x - 3)(x - 7)} = \lim_{x \to 3} \frac{(x + 2)}{(x - 7)} = -\frac{5}{4}.$$

Example (2):

Evaluate the $\lim_{x \to 1} \left(\frac{(x - 1)^2}{x^2 + 2x - 3}\right)$.

When we plug in 1 for x, we see that we have $\frac{0}{0}$. So, using
L'Hospital's Rule we have that:

$$\lim_{x \to 1} \left(\frac{(x - 1)^2}{x^2 + 2x - 3}\right) = \lim_{x \to 1} \frac{2(x - 1)}{2x + 2} = 0.$$

We can verify that we have the correct result, since as we learned in section (4.6) on limits and continuity:

$$\lim_{x \to 1} \left(\frac{(x-1)^2}{x^2 + 2x - 3} \right) = \lim_{x \to 1} \frac{(x-1)^2}{(x-1)(x+3)} = \lim_{x \to 1} \frac{(x-1)}{(x+3)} = 0 \ .$$

Example (3):

Evaluate the $\lim_{x \to \infty} \left(\frac{3x^2 - 2x + 7}{8x - 9} \right)$.

This is an $\frac{\infty}{\infty}$ indeterminate form.

So, the $\lim_{x \to \infty} \left(\frac{3x^2 - 2x + 7}{8x - 9} \right) = \lim_{x \to \infty} \frac{6x - 2}{8} = \infty.$

Example (4):

Evaluate the $\lim_{x \to \infty} \left(\frac{7x^2 + x + 1}{2x^2 - x - 10} \right)$.

This is an $\frac{\infty}{\infty}$ indeterminate form. So, using L'Hospital's Rule,

$$\lim_{x \to \infty} \left(\frac{7x^2 + x + 1}{2x^2 - x - 10} \right) = \lim_{x \to \infty} \frac{14x + 1}{4x - 1} \ .$$

This is still an $\frac{\infty}{\infty}$ indeterminate form. So, using L'Hospital's Rule again,

$$\lim_{x \to \infty} \left(\frac{7x^2 + x + 1}{2x^2 - x - 10} \right) = \lim_{x \to \infty} \frac{14x + 1}{4x - 1} = \lim_{x \to \infty} \frac{14}{4} = \frac{7}{2}.$$

################# Exercises ##################

Find the following limits using L'Hospital's Rule.

(1) $\lim\limits_{x\to\infty} \dfrac{5x^2 + x + 1}{3x^2 - 2x + 1}$ (2) $\lim\limits_{x\to\infty} \dfrac{x + 3}{x^2 - 4}$ (3) $\lim\limits_{x\to\infty} \dfrac{x^2 + x}{x - 7}$

(4) $\lim\limits_{x\to 3} \dfrac{(x - 3)^2}{(x - 3)^{\frac{4}{3}}}$ (5) $\lim\limits_{x\to 8^+} \dfrac{\sqrt{x - 8}}{(x - 8)^3}$ (6) $\lim\limits_{x\to 2} \dfrac{3x - 6}{8x - 16}$

(7) $\lim\limits_{x\to 1} \dfrac{x^2 - 1}{x^3 - 1}$ (8) $\lim\limits_{x\to 9} \dfrac{(x - 9)^2}{(x - 9)^4}$

(4.17) Derivatives of Transcendental Functions

We now refer to the exponential, logarithmic, trig, and inverse trig functions as the transcendental functions. We want to find the derivatives of all of these transcendental functions, along with certain limits involving them.

Exponential and Logarithmic Functions
Let R be a positive real number, no matter how large, and let the interval D = (-R,R). Consider the sequence of functions:

$$\{f_n(x)\} = \left\{ \left(1 + \frac{x}{n}\right)^n \right\}, \text{ for (n = 1, 2, 3, ...), on the set D.}$$

For each n, $f_n(x)$ is a continuous and smooth function on the set D. It can be shown that this sequence of functions converges to the function $f(x) = e^x$ on the interval D. We will not derive this result, but it can be done quite easily using some advanced techniques of calculus.

233

That is, $e^x = \lim\limits_{n\to\infty}\left(1+\frac{x}{n}\right)^n$, for all x on the interval D.

We can say that the sequence of functions $\{f_n(x)\}$ does converge pointwise to f(x) on D, because the above limit definition of $f(x) = e^x$ is true for each real number x in D. It can also be shown that for every positive number ε, no matter how small, that there exists an integer N such that the $|f_n(x) - e^x| < \varepsilon$, for all x in D when n > N. So, we say that the sequence of functions $\{f_n(x)\}$ converges uniformly to the function $f(x) = e^x$ on D. Recall the discussion about uniform convergence in section 4.5.

The $\dfrac{d}{dx}(e^x) = \dfrac{d}{dx}\left(\lim\limits_{n\to\infty}\left(1+\frac{x}{n}\right)^n\right)$

$$= \lim_{n\to\infty}\left(\frac{d}{dx}\left(1+\frac{x}{n}\right)^n\right)$$

$$= \lim_{n\to\infty}\left((n)\left(1+\frac{x}{n}\right)^{n-1}\left(\frac{1}{n}\right)\right)$$

$$= \lim_{n\to\infty}\left(\left(1+\frac{x}{n}\right)^{n-1}\right)$$

$$= \frac{\lim\limits_{n\to\infty}\left(1+\frac{x}{n}\right)^n}{\lim\limits_{n\to\infty}\left(1+\frac{x}{n}\right)} = \lim_{n\to\infty}\left(1+\frac{x}{n}\right)^n = e^x$$

Therefore, the $\dfrac{d}{dx}(e^x) = e^x$.

It turns out that we need, in general, the requirement of uniform convergence of $\{f_n(x)\}$ to f(x) on D, in order to bring the derivative operation across the limit as we did in the derivation above.

We know that: $y = \ln(x)$ iff $x = e^y$. So, differentiating the equation $x = e^y$ implicitly with respect to x leads to:

$1 = e^y \cdot \frac{dy}{dx}$, which says that $\frac{dy}{dx} = \frac{1}{e^y}$. But, we know that $y = \ln(x)$ and $x = e^y$. So, we have the result that:

$$\frac{d}{dx}(\ln(x)) = \frac{1}{x} .$$

We now have the derivatives of the natural exponential and the natural logarithmic functions. We can now easily find the derivatives of more general exponential and logarithmic functions for any allowable base "a".

For $f(x) = a^x$, rewrite it as $f(x) = e^{\ln a^x} = e^{x \ln a} = e^{(\ln a)x}$. Note that $(\ln a)$ is a real number. We will use the chain rule and our derivative formula for e^x. Let $u = ((\ln a)x)$, and let $y = e^u$. Then, the chain rule says that:

$$\frac{dy}{dx} = \frac{dy}{du} \cdot \frac{du}{dx} = e^u \cdot (\ln a) = e^{(\ln a)x} \cdot (\ln a) = a^x \cdot (\ln a)$$

That is: $\frac{d}{dx}(a^x) = a^x \cdot (\ln a) .$

235

For f(x) = $\log_a x$, using our logarithm Law 5 from section (3.2) write f(x) = $\log_a x = \frac{\ln(x)}{\ln(a)}$. Then, using our derivative formula for $\ln(x)$, we have:

$$\frac{d}{dx}(\log_a x) = \frac{d}{dx}\left(\frac{\ln(x)}{\ln(a)}\right) = \frac{d}{dx}\left(\left(\frac{1}{\ln a}\right)\ln(x)\right)$$

$$= \left(\frac{1}{\ln a}\right)\frac{d}{dx}(\ln(x)) = \left(\frac{1}{\ln a}\right)\cdot\left(\frac{1}{x}\right) = \frac{1}{(\ln a)x}.$$

That is: $\frac{d}{dx}(\log_a x) = \frac{1}{(\ln a)x}$.

We need to take care of some preliminary matters before moving onward. In most books on calculus, the following two limits are verified before they are needed using simple geometric arguments. We will state them here without proving them:

(1) $\lim\limits_{h\to 0}\dfrac{\sin(h)}{h} = 1$ (2) $\lim\limits_{h\to 0}\dfrac{\cos(h)-1}{h} = 0$

Here are the nine trigonometric identities from section (4.2) that we will use in the derivation of some of the following derivative formulas. It will also be very beneficial to have these in your mind when we deal with the integral calculus.

Pythagorean Identities
$\cos^2(x) + \sin^2(x) = 1$ (4.2.1)
$1 + \tan^2(x) = \sec^2(x)$ (4.2.2)
$1 + \cot^2(x) = \csc^2(x)$ (4.2.3)

Sum and Difference Identities

$$\cos(x \pm y) = \cos(x)\cos(y) \mp \sin(x)\sin(y) \quad (4.2.4)$$
$$\sin(x \pm y) = \sin(x)\cos(y) \pm \cos(x)\sin(y) \quad (4.2.5)$$

Double Angle Formulas

$$\sin(2x) = 2\sin(x)\cos(x) \qquad\qquad (4.2.6)$$
$$\cos(2x) = \cos^2(x) - \sin^2(x) \qquad\qquad (4.2.7)$$
$$= 2\cos^2(x) - 1$$
$$= 1 - 2\sin^2(x)$$

Half Angle Formulas

$$\cos^2(x) = \frac{1 + \cos(2x)}{2} \qquad\qquad (4.2.8)$$
$$\sin^2(x) = \frac{1 - \cos(2x)}{2} \qquad\qquad (4.2.9)$$

Now to the derivatives of the remaining transcendental functions that we will see in this book.

Trigonometric Functions

(1) $f(x) = \sin(x)$

$$f'(x) = \lim_{h \to 0}\left(\frac{\sin(x+h) - \sin(x)}{h}\right)$$

$$= \lim_{h \to 0}\left(\frac{\sin(x)\cos(h) + \cos(x)\sin(h) - \sin(x)}{h}\right), \text{ using } (4.2.5)$$

$$= (\sin(x)) \cdot \lim_{h \to 0}\left(\frac{\cos(h) - 1}{h}\right) + (\cos(x)) \cdot \lim_{h \to 0}\left(\frac{\sin(h)}{h}\right)$$

(where we use both of the limits given above)

$$= [(\sin(x)) \cdot (0) + (\cos(x)) \cdot (1)]$$

$$= \cos(x).$$

(2) $f(x) = \cos(x)$

$$f'(x) = \lim_{h \to 0} \left(\frac{\cos(x+h) - \cos(x)}{h} \right)$$

$$= \lim_{h \to 0} \left(\frac{\cos(x)\cos(h) - \sin(x)\sin(h) - \cos(x)}{h} \right), \text{ using (4.2.4)}$$

$$= (\cos(x)) \cdot \lim_{h \to 0} \left(\frac{\cos(h) - 1}{h} \right) - (\sin(x)) \cdot \lim_{h \to 0} \left(\frac{\sin(h)}{h} \right)$$

$$= [\cos(x) \cdot (0) - \sin(x) \cdot (1)]$$

$$= -\sin(x).$$

(3) $f(x) = \tan(x)$ (we will use the quotient rule)

$$f'(x) = \frac{d}{dx} \left(\frac{\sin(x)}{\cos(x)} \right) = \left(\frac{\cos(x)\cos(x) - \sin(x)(-\sin(x))}{\cos^2(x)} \right)$$

$$= \left(\frac{\cos^2(x) + \sin^2(x)}{\cos^2(x)} \right) = \left(\frac{1}{\cos^2(x)} \right), \text{ using (4.2.1)}$$

$$= \sec^2(x).$$

(4) $f(x) = \cot(x)$ (we will use the quotient rule)

238

$$f'(x) = \frac{d}{dx}\left(\frac{\cos(x)}{\sin(x)}\right) = \left(\frac{\sin(x)(-\sin(x))-\cos(x)\cos(x)}{\sin^2(x)}\right)$$

$$= -\left(\frac{\sin^2(x)+\cos^2(x)}{\sin^2(x)}\right) = -\left(\frac{1}{\sin^2(x)}\right)$$

$$= -\csc^2(x).$$

(5) $f(x) = \sec(x)$

$$f'(x) = \frac{d}{dx}\left(\frac{1}{\cos(x)}\right) = \left(\frac{\cos(x)\cdot(0)-(1)\cdot(-\sin(x))}{\cos^2(x)}\right)$$

$$= \left(\frac{\sin(x)}{\cos^2(x)}\right) = \left(\frac{\sin(x)}{\cos(x)}\right)\cdot\left(\frac{1}{\cos(x)}\right)$$

$$= \sec(x)\tan(x).$$

(6) $f(x) = \csc(x)$

$$f'(x) = \frac{d}{dx}\left(\frac{1}{\sin(x)}\right) = \left(\frac{\sin(x)\cdot(0)-(1)\cdot(\cos(x))}{\sin^2(x)}\right)$$

$$= \left(\frac{-\cos(x)}{\sin^2(x)}\right) = -\left(\frac{\cos(x)}{\sin(x)}\right)\cdot\left(\frac{1}{\sin(x)}\right)$$

$$= -\csc(x)\cot(x)$$

Inverse Trigonometric Functions

(1) $f(x) = \sin^{-1}(x)$

Starting with: $y = sin^{-1}(x)$ iff $x = sin(y)$

Differentiate $x = sin(y)$ with respect to x. We get:

$1 = cos(y) \cdot \dfrac{dy}{dx}$, which says that $\dfrac{dy}{dx} = \dfrac{1}{cos\,(y)} = \dfrac{1}{\sqrt{1 - sin^2(y)}}$.

Therefore, $\dfrac{d}{dx}(sin^{-1}(x)) = \dfrac{1}{\sqrt{1 - x^2}}$.

(2) $f(x) = cos^{-1}(x)$

Starting with: $y = cos^{-1}(x)$ iff $x = cos(y)$

Differentiate $x = cos(y)$ with respect to x. We get:

$1 = -\,sin(y) \cdot \dfrac{dy}{dx}$, which says that:

$\dfrac{dy}{dx} = \dfrac{1}{-\,sin(y)} = \dfrac{-1}{\sqrt{1 - cos^2(y)}}$.

Therefore, $\dfrac{d}{dx}(cos^{-1}(x)) = \dfrac{-1}{\sqrt{1 - x^2}}$.

(3) $f(x) = tan^{-1}(x)$

Starting with: $y = tan^{-1}(x)$ iff $x = tan(y)$

Differentiate $x = tan(y)$ with respect to x. We get:

$1 = sec^2(y) \cdot \dfrac{dy}{dx}$, which says that:

$$\frac{dy}{dx} = \frac{1}{sec^2(y)} = \frac{1}{1 + tan^2(y)}.$$

Therefore, $\dfrac{d}{dx}(tan^{-1}(x)) = \dfrac{1}{1 + x^2}.$

(4) $f(x) = cot^{-1}(x)$

Starting with: $y = cot^{-1}(x)$ iff $x = cot(y)$

Differentiate $x = cot(y)$ with respect to x. We get:

$1 = -csc^2(y) \cdot \dfrac{dy}{dx}$, which says that:

$$\frac{dy}{dx} = \frac{-1}{csc^2(y)} = \frac{-1}{1 + cot^2(y)}.$$

Therefore, $\dfrac{d}{dx}(cot^{-1}(x)) = \dfrac{-1}{1 + x^2}.$

(5) $f(x) = sec^{-1}(x)$

Starting with: $y = sec^{-1}(x)$ iff $x = sec(y)$

Differentiate $x = sec(y)$ with respect to x. We get:

$1 = sec(y)tan\,(y) \cdot \dfrac{dy}{dx}$, which says that:

$$\frac{dy}{dx} = \frac{1}{sec(y)\tan(y)} = \frac{1}{sec(y)\sqrt{sec^2(y) - 1}}.$$

Therefore, $\frac{d}{dx}(sec^{-1}(x)) = \frac{1}{x\sqrt{x^2 - 1}}.$

(6) $f(x) = csc^{-1}(x)$

Starting with: $y = csc^{-1}(x)$ iff $x = csc(y)$

Differentiate $x = csc(y)$ with respect to x. We get:

$1 = -csc(y)\cot(y) \cdot \frac{dy}{dx}$, which says that:

$$\frac{dy}{dx} = \frac{-1}{csc(y)\cot(y)} = \frac{-1}{csc(y)\sqrt{csc^2(y) - 1}}.$$

Therefore, $\frac{d}{dx}(csc^{-1}(x)) = \frac{-1}{x\sqrt{x^2 - 1}}.$

Example (1):

The $\frac{d}{dx}(e^x + 3\ln(x) - 5\sin(x) + 2\tan(x))$

$\quad = e^x + \frac{3}{x} - 5\cos(x) + 2sec^2(x).$

The reader should be proficient enough in the use of the chain rule to follow along with the next 3 examples:

Example (2):

The $\frac{d}{dx}\left(e^{-x^2} + 10\log_7(2x)\right)$

$$= \left(e^{-x^2}\right) \cdot (-2x) + (10)\left(\frac{1}{(2x)(\ln 7)}\right) \cdot (2)$$

$$= \left(-2xe^{-x^2}\right) + \left(\frac{20}{(2x)(\ln 7)}\right)$$

$$= \left(-2xe^{-x^2}\right) + \left(\frac{10}{(x)(\ln 7)}\right)$$

Example (3):

The $\frac{d}{dx}\left(\ln(3x-2) + 10\cos(45x^2 - 3x)\right)$

$$= \left(\frac{1}{(3x-2)}\right)(3) + (10)(-\sin(45x^2 - 3x))(90x - 3)$$

$$= \left(\frac{3}{3x-2}\right) - 10\sin(45x^2 - 3x)(90x - 3)$$

Example (4):

The $\frac{d}{dx}\left(13\sin^3(2x+1) - \tan^{-1}(4x)\right)$

$$= \frac{d}{dx}\left((13)(\sin(2x+1))^3\right) - \frac{d}{dx}\left(\tan^{-1}(4x)\right)$$

$$= (13)(3)(\sin(2x+1))^2(\cos(2x+1))(2)$$
$$- \left(\frac{1}{1+(4x)^2}\right)(4)$$

$$= 78\sin^2(2x+1)\cos(2x+1) - \left(\frac{4}{1+16x^2}\right)$$

Example (5):
We simply stated beforehand without proof the limits:

$$\lim_{h \to 0} \frac{\sin(h)}{h} = 1 \quad \text{and} \quad \lim_{h \to 0} \frac{\cos(h) - 1}{h} = 0 \; .$$

We can now verify them using L'Hospital's Rule:

(A) $\lim_{h \to 0} \dfrac{\sin(h)}{h} = \lim_{h \to 0} \dfrac{\cos(h)}{1} = 1$, and

(B) $\lim_{h \to 0} \dfrac{\cos(h) - 1}{h} = \lim_{h \to 0} \dfrac{-\sin(h)}{1} = 0$.

We have reached the end of the discussion of the derivatives of the transcendental functions that we will consider in this book. Now, one more section for some applications of the differential calculus. Then we'll move on to the concepts and methods of the integral calculus in the remaining parts of this chapter, along with a section on infinite series.

################# Exercises #################

Find the derivative for the functions y = f(x).

(1) $y = sinx - tanx$ (2) $y = \cos(2x) + \cot(2x)$

(3) $y = \sec(x^2)$ (4) $y = tan(3x^3 - 4x)$

(5) $y = sin(cos(x^2))$ (6) $y = \frac{1}{2}e^{(-3x^2 + 2x + 1)}$

(7) $y = ln(14 - x - x^2)$ (8) $y = 2^{sinx} - csc(3x)$

(9) $y = tan^{-1}(5x)$ (10) $y = sin^{-1}(10x^2)$

(11) (a) Find the linearization of $f(x) = e^x$ when $x = 0$.
    (b) Use it to approximate $e^{(0.003)}$.
    (c) Check with the scientific calculator value.

(12) (a) Find the linearization of $f(x) = ln(x)$ when $x = 1$.
    (b) Use it to approximate $ln(0.997)$.
    (c) Check with the scientific calculator value.

(13) (a) Find the linearization of $f(x) = sinx$ when $x = 0$.
    (b) Use it to approximate $sin(0.0001)$.
    (c) Check with the scientific calculator value.

(4.18) Applications of Differentiation

In a traditional calculus course, much more effort is spent on applications of differentiation than we will do here. In this chapter, we want to simply provide the reader with a feel for how the derivative is used in certain types of math problems involving optimization (finding maximums and minimums), and with rates of change. The reader will no doubt be bringing their own applications with them from their field of study. We will present four examples.

<u>Example (1)</u>:
The first and second derivatives are very useful for analyzing and helping us graph functions in the x-y plane.

The first derivative $f'(x)$ tells us how f(x) is changing. So, it is useful for finding where f(x) is increasing or decreasing. When the derivative $f'(x) > 0$ then f(x) is increasing. When $f'(x) < 0$ then f(x) is decreasing. When $f'(x) = 0$, many times the function f(x) has reached a relative maximum or minimum (relative to other function values in the vicinity). When the first derivative is zero, that means that f(x) has a horizontal tangent line at that point. We call the x-values where $f'(x) = 0$ the critical values for f(x).

The second derivative $f''(x)$ tells us how $f'(x)$ is changing. So, it can be used for finding where $f'(x)$ is increasing or decreasing. When the second derivative $f''(x) > 0$, it means that the slopes $f'(x)$ are increasing and the function f(x) is concave up. When $f''(x) < 0$, it means that the slopes $f'(x)$ are decreasing and the function f(x) is concave down. When $f''(x) = 0$, then many times the function f(x) has reached a point where the concavity changes. We call the x-values where the second derivative $f''(x) = 0$ the inflection values for the function f(x). We will now provide an example where we will analyze a cubic polynomial. Let f(x) = $x^3 - 3x^2 - 6x + 8$. We will find $f'(x)$ and $f''(x)$ and use them to analyze f(x). We'll also consider the x and y intercepts to help us graph f(x). So,

$$f'(x) = 3x^2 - 6x - 6$$

$f''(x) = 6x - 6$

(a) Set $f'(x) = 3x^2 - 6x - 6 = 0$.
This is equivalent to when $x^2 - 2x - 2 = 0$.
Then, using the quadratic formula we have:

$$x = \frac{2 \pm \sqrt{4-(4)(1)(-2)}}{2} = \frac{2 \pm \sqrt{12}}{2} = \frac{2 \pm \sqrt{4}\sqrt{3}}{2} = \frac{2 \pm 2\sqrt{3}}{2} = 1 \pm \sqrt{3}.$$

These are our critical values and they are approximately
x = −0.73 and x = 2.73. To get the critical points (points described with a pair of coordinates), plug the critical values into the original function f(x). Then we have:

$f(-0.73) = 10.392$ and $f(2.73) = -10.392$.
So, the critical points in the x-y coordinate plane are:
(-0.73, 10.392) and (2.73, -10.392).

(b) Set $f''(x) = 6x - 6 = 0$
Then we have x = 1 as our single inflection value.
To get the inflection point, plug the inflection value back into the original function f(x). Then we have: $f(1) = 0$.
So, our single inflection point is at (1, 0).

(c) The reader can verify that the x-intercepts, the roots of f(x), are at x = −2, 1, and 4.

(d) Plugging 0 in for x in the original function f(x) provides us with the y-intercept: y = 8.

(In parts (e) and (f), use test values to verify these results):

(e) $f'(x) = 3x^2 - 6x - 6 > 0$ on the intervals:
$(-\infty, -0.73)$ and $(2.73, \infty)$.
So, f(x) is increasing on these intervals.

$f'(x) = 3x^2 - 6x - 6 < 0$ on the interval:
$(-0.73, 2.73)$.
So, f(x) is decreasing on this interval.

(f) $f''(x) = 6x - 6 < 0$ on the interval: $(-\infty, 1)$.
So, f(x) is concave down on this interval.

$f''(x) = 6x - 6 > 0$ on the interval: $(1, \infty)$.
So, f(x) is concave up on this interval.

(g) What relative extrema does f(x) have?

It has a relative maximum at the critical point (-0.73, 10.392) and a relative minimum at the critical point (2.73, -10.392). This can be determined from the results of parts (e) and (f). On its entire domain (the real numbers) it does not have an absolute minimum or an absolute maximum.

We now use these results in (a) through (g) to graph f(x).

See Figure (5.58) on the next page.

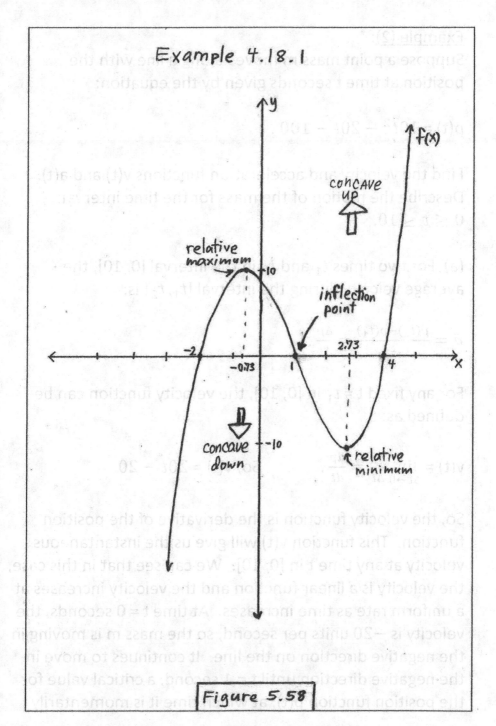

Example 4.18.1

relative maximum

concave up

inflection point

concave down

relative minimum

Figure 5.58

249

<u>Example (2):</u>
Suppose a point mass m moves along a line with the position at time t seconds given by the equation:

$$p(t) = 10t^2 - 20t - 100 .$$

Find the velocity and acceleration functions v(t) and a(t). Describe the motion of the mass for the time interval: $0 \le t \le 10$.

(a) For two times t_1 and t_2 in the interval [0, 10], the average velocity during the interval $[t_1, t_2]$ is:

$$\bar{v} = \frac{p(t_2)-p(t_1)}{t_2-t_1} = \frac{\Delta p}{\Delta t} .$$

For any fixed t = t_1 in [0, 10], the velocity function can be defined as:

$$v(t) = \lim_{\Delta t \to 0} \frac{\Delta p}{\Delta t} = \frac{dp}{dt} . \qquad \text{So, v(t) = } 20t - 20.$$

So, the velocity function is the derivative of the position function. This function v(t) will give us the instantaneous velocity at any time t in [0, 10]. We can see that in this case, the velocity is a linear function and the velocity increases at a uniform rate as time increases. At time t = 0 seconds, the velocity is −20 units per second, so the mass m is moving in the negative direction on the line. It continues to move in the negative direction until t = 1 second, a critical value for the position function p(t), at which time it is momentarily motionless. Then the mass moves in the positive direction for the next 9 seconds. The velocity continues to increase at a rate of 20 units per second per second, reaching a velocity of v(10) = 180 units per second at time t = 10.

(b) For two times t_1 and t_2 in the interval [0, 10], the average acceleration during the interval $[t_1, t_2]$ is:

$$\bar{a} = \frac{v(t_2)-v(t_1)}{t_2-t_1} = \frac{\Delta v}{\Delta t}.$$

For any fixed t = t_1 in [0, 10], the acceleration function can be defined as:

$$a(t) = \lim_{\Delta t \to 0} \frac{\Delta v}{\Delta t} = \frac{dv}{dt}. \qquad \text{So, a(t) = 20.}$$

So, the acceleration function is the derivative of the velocity function. This function a(t) will give us the instantaneous acceleration at any time t in [0, 10]. We can see that in this case, the acceleration is a constant function and causes the velocity to continually increase at a rate of 20 units per second per second, as we stated above. The position of the mass starts at p(0) = −100 units on the line. When the mass stops momentarily at time t = 1, the mass is at the position p(1) = −110 units on the line. After starting to move in the positive direction for the final 9 seconds, it reaches the point p(10) = 700 units on the line.

<u>Example (3)</u>:
An artist is building a large cardboard box for a museum exhibit, which will be filled with a large piece of styrofoam to keep it from buckling under its own weight.

It must have a width of (x) inches, a length of (x + 5) inches, and a height of (y) inches. There is the additional constraint that x + y = 500. Find the dimensions of the box that will maximize the volume.

The volume V = (x)(y)(x + 5).
The constraint x + y = 500 says that y = (500 − x). This will allow us to express the volume in terms of a single variable:

$V = (x)(500 - x)(x + 5)$, or
$V = (x^2 + 5x)(500 - x)$, or
$V = -x^3 + 495x^2 + 2500x$.

We must find the critical value by setting $\frac{dV}{dx} = 0$.
$\frac{dV}{dx} = -3x^2 + 990x + 2500 = 0$, and
$\frac{d^2V}{dx^2} = -6x + 990$.

So, using the quadratic formula:
$x = \frac{-990 \pm \sqrt{(990)^2 - (4)(-3)(2500)}}{-6}$, which says that
$x = 332.5062$ or $x = -2.5062$

Since x must be positive, we see that the critical value of interest is x = 332.5062 inches.

The $\frac{d^2V}{dx^2}$ when evaluated at x = 332.5 is -1005.0372. The important thing to consider about this is that the second derivative evaluated at x = 332.5 is negative, so we know that V(x) is concave down when x = 332.5. Therefore, V(x) is at a maximum when x = 332.5 inches.

Therefore, the dimensions of the box that maximize the volume are a width of x = 332.5 inches, a length of 337.5 inches, and a height of y = 167.5 inches.

Example (4):
This problem is called a related rates problem. A right circular cone has a volume of $V = \frac{1}{3}\pi r^2 h$, where r is the base radius and h is the height. We assume that V, r, and h are implicit functions of time t, and that at the instant that r = 20 inches and h = 100 inches, $\frac{dr}{dt} = 2 \frac{inches}{second}$, and that $\frac{dh}{dt} = 5 \frac{inches}{second}$. Find $\frac{dV}{dt}$ at that instant.

We will take the derivative of the formula for the volume assuming that all of the variables are implicit functions of time. When we do this, we have to assume that r^2 and h are two independent functions of time. Then, using the product rule and implicit differentiation,

$$\frac{dV}{dt} = \left(\frac{1}{3}\pi\right)\left((r^2)\frac{dh}{dt} + (h)\left(2r\frac{dr}{dt}\right)\right).$$

So, at the instant of time under consideration, the volume V is changing at the rate:

$$\frac{dV}{dt} = \left(\frac{1}{3}\pi\right)\left((20^2)(5) + (100)(2)(20)(2)\right)$$

$$= (10{,}471.9755)\frac{(\text{inches})^3}{\text{second}}.$$

################ Exercises #################

(1) We want to find two non-negative real numbers x and y such that the product P(x,y) = xy is maximized, subject to the constraint x + 2y = 88.

(a) Make P(x,y) a function P(x) of the one variable x, by using the constraint to eliminate y. What is P(x), $P'(x)$, and the critical values?
(b) What are the numbers x and y that maximize their product?
(c) What is the maximum value of their product?
(d) What kind of function is P(x)?
(e) Where is P(x) = 0, and what do we call these x-values?

(2) $f(x) = \frac{1}{\sqrt{2\pi}\,(a)} \cdot e^{-\frac{1}{2}\left(\frac{x-b}{a}\right)^2}$, $-\infty < x < \infty$, where a and b are real numbers.

(a) Where is f(x) = 0?
(b) What is $f'(x)$, and the critical values?
(c) What is $f''(x)$, and the inflection values?
(d) What is the $\lim\limits_{x\to-\infty} f(x)$, the $\lim\limits_{x\to\infty} f(x)$?

254

(e) Where is f(x) increasing, where is f(x) decreasing?
(f) Where is f(x) concave up, where is f(x) concave down?
(g) Does it have any extrema, relative or absolute?
(h) Sketch the graph.

(3) We have a right triangle with hypotenuse of length h and legs of length a and b. The Pythagorean theorem says that $a^2 + b^2 = h^2$. The triangle is expanding but remains a right triangle. Assume that the sides a, b, and h are implicit functions of time. If at the instant that a = 7 ft, b = 10 ft, and h = $\sqrt{149}$ ft, the rate of change of the two legs with respect to time are: $\frac{da}{dt} = 2$ $^{ft}/_{sec}$ and $\frac{db}{dt} = 3$ $^{ft}/_{sec}$. What is the rate of change of h with respect to time?

(4) A 6 ft tall man is walking directly towards a 30 ft high light pole in a level parking lot at night, with a speed of 4 ft/sec. How fast is the length of his shadow changing with respect to time?

This concludes our look at differentiation. We now have to consider the other half of the subject of calculus, called the integral calculus. We will do this in the remaining sections of this chapter except for the last. The last section deals with the important topic of infinite series.

(4.19) Indefinite Integrals

Up to this point, we have been differentiating a function f(x) to obtain the derivative $f'(x)$. We now consider the reverse process of starting with the derivative $f'(x)$ and finding the function f(x). We call this reverse process anti-differentiation or integration. We will consider indefinite integrals in this section and definite integrals in the next section along with the relation between the differential calculus and the integral calculus through the so-called Fundamental Theorem of Calculus (FTC).

<u>Antiderivatives and Indefinite Integrals</u>
For a function f(x) continuous on a closed interval [a,b], if we have a function F(x) such that $F'(x) = f(x)$, then we call F(x) an antiderivative for f(x). We have a special notation for this, which we call an indefinite integral:

F(x) = $\int f(x)\, dx$.

The elongated S is called the integral sign, f(x) is called the integrand, and dx is the differential. Think of dx as notation that tells us the variable that we are integrating with respect to (anti-differentiating with respect to). In the next section when we study definite integrals, the reader will understand more clearly the meaning of the symbolism dx.

When we have an antiderivative F(x) for f(x), more generally we write: $\int f(x)dx = F(x) + C$, where C is called an arbitrary constant.

For example, suppose we have $f(x) = x$, $F(x) = \dfrac{x^2}{2} + 10$, and $G(x) = \dfrac{x^2}{2} + 30$. Note that $F'(x) = G'(x) = f(x)$ since the derivative of a constant is 0. In other words, both F(x) and G(x) are antiderivatives of f(x). There are an infinite number of antiderivatives for a function f(x), but they all differ at most by a constant term, hence the arbitrary constant C.

Basic Formulas and Properties

(1): If $f(x) = x^n$, where n is any real number, $n \neq -1$, (we will deal with the case of n = -1 soon), then:

$$\int x^n dx = \frac{x^{n+1}}{n+1} + C.$$

We can verify that this formula is correct by differentiating:

The $\dfrac{d}{dx}\left(\dfrac{x^{n+1}}{n+1} + C\right) = \left(\left(\dfrac{n+1}{n+1}\right)x^{(n+1)-1} + 0\right) = x^n.$

This is an important result, analogous to the Power Rule of differentiation. Fortunately, it is much easier to prove.

(2): We learned that differentiation satisfies the property of linearity. Anti-differentiation does also. This means that if K is a constant and we have the functions f(x) and g(x):

(a) $\int K f(x) dx = K \int f(x) dx$, and

(b) $\int (f(x) + g(x)) dx = \int f(x) dx + \int g(x) dx$.

In other words, the integral of a "constant times a function" is the "constant times the integral of the function," and "the integral of a sum" is "the sum of the integrals."

(3): If f(x) = K, where K is a constant, then f(x) = K$\cdot x^0$.

So, $\int K dx = \int K x^0 dx = K \int x^0 dx = K \cdot \frac{x^1}{1} + C = Kx + C$.

That is: $\int K dx = Kx + C$, for any constant K.

As a special case, the $\int (0) dx = 0x + C = C$. This makes sense because we know that the derivative of any constant is 0.

(4): If we have a function Q(x), then since d(Q(x)) = $Q'(x)dx$ we have the integration formula:

$\int d(Q(x)) = \int Q'(x) dx = Q(x) + C$, that is:

$\int dQ(x) = Q(x) + C$

As a special case: $\int dx = x + C$. This makes sense since we know that the derivative of x is 1.

(5): We have an alternative notation (which will be important for us later) for the antiderivative of a function f(x). We can write: $\int f(x) dx = \int_a^x f(t) dt$.

Here, "a" and "x" are called limits. "a" is any real number and the variable x is now the upper limit. We call "t" a

258

dummy variable, since we could just as well use any symbol in place of it.

When we use this notation, then we need only remember that the $\frac{d}{dx}\left(\int_a^x f(t)dt\right) = f(x)$, just like the fact that the $\frac{d}{dx}\left(\int f(x)dx\right) = f(x)$. With this new notation, taking the derivative with respect to x of $\int_a^x f(t)dt$ simply involves the substitution of x for t, everywhere that t appears in the integrand. When taking the derivative of this form of the integral, it doesn't matter what "a" is, it should simply be a constant. The reader will understand later why the lower limit of an integral is unimportant in understanding the rate of change of the integral with respect to x. So, for example, the $\frac{d}{dx}\int_a^x (2t^3 - 6t)dt = (2x^3 - 6x)$. It is interesting to note that the chain rule can come into play in this form of the integral. As an example:

$$\frac{d}{dx}\int_a^{3x^2}(2t^3 - 6t)dt = [2(3x^2)^3 - 6(3x^2)] \cdot \frac{d}{dx}(3x^2)$$
$$= [2(3x^2)^3 - 6(3x^2)] \cdot (6x) .$$

Example (1):
We will evaluate the following integrals (find an antiderivative for the integrand):
(a) $\int(4 + x^2 - 3x^5 + 31x^{-2})dx$

$$= \int 4dx + \int x^2 dx - 3\int x^5 dx + 31\int x^{-2}dx$$

$$= 4x + \frac{x^3}{3} - \frac{3}{6}x^6 + 31\frac{x^{-1}}{-1} + C$$

$$= 4x + \frac{x^3}{3} - \frac{x^6}{2} - \frac{31}{x} + C$$

Note that we lump all the constants of integration together at the end of the process, and call it C.

(b) $\int(34x^2 - 2x^{-7} + 14x^4 + 9x^{100})dx$

$$= 34\frac{x^3}{3} - 2\frac{x^{-6}}{-6} + 14\frac{x^5}{5} + 9\frac{x^{101}}{101} + C$$

$$= \frac{34x^3}{3} + \frac{1}{3x^6} + \frac{14x^5}{5} + \frac{9x^{101}}{101} + C.$$

(c) $\int d(2x) + \int(16 - 6x)dx - \int dx$

$$= \int d(2x) + \int 16dx - 6\int xdx - \int dx$$

$$= 2x + 16x - 3x^2 - x + C.$$

(d) $\int(7x - 3x^2)dx - 12\int\frac{1}{3}xdx + \int(x^8 - 2)dx$
$$= \int 7xdx - \int 3x^2dx - 4\int xdx + \int x^8dx + \int -2dx$$

$$= \frac{7x^2}{2} - \frac{3x^3}{3} - 4\frac{x^2}{2} + \frac{x^9}{9} - 2x + C$$

$$= \frac{7x^2}{2} - x^3 - 2x^2 + \frac{x^9}{9} - 2x + C.$$

Now, we move on to the integral formulas for what we have called the transcendental functions.

Exponential and Logarithmic Formulas

(1) We know that the $\frac{d}{dx}(e^x) = e^x$, so $d(e^x) = e^x dx$.

Since the $\int d(e^x) = e^x + C$, we have the formula,

$\int e^x dx = e^x + C$.

(2) $\frac{d}{dx}(lnx) = \frac{1}{x}$ → $d(lnx) = \left(\frac{1}{x}\right) dx$

Integrating both sides gives us:

$\int \frac{1}{x} dx = \ln|x| + C$.

Note that in this formula we have $|x|$ instead of x:

(a) If x > 0, then $\int \frac{1}{x} dx = \ln(x) + C = \ln|x| + C$.

(b) If x < 0, then $\int \frac{1}{x} dx = \int \frac{1}{-x}(-1)dx$. Now let $u = -x$,
then $du = -dx$. So, $\int \frac{1}{x} dx = \int \frac{1}{-x}(-1)dx = \int \frac{1}{u} du$.
Since u > 0, $\int \frac{1}{u} dx = \ln|u| + C = \ln|-x| + C = \ln|x| + C$.
(This gives us the antiderivative of f(x) = x^n when n = -1)

261

(3) $\frac{d}{dx}(a^x) = a^x lna$ $\rightarrow$ $d(a^x) = a^x(lna)dx$

Integrating both sides gives us:

$\int a^x(lna)dx = a^x + C$.

(4) $\frac{d}{dx}(\log_a x) = \frac{1}{x(lna)}$ $\rightarrow$ $d(\log_a x) = \left(\frac{1}{x(lna)}\right)dx$

Integrating both sides gives us:

$\int \frac{1}{x(lna)}dx = \log_a x + C$.

Trigonometric Formulas

(5) $\frac{d}{dx}(sinx) = cosx$ $\rightarrow$ $d(sinx) = (cosx)dx$

Integrating both sides gives us:

$\int cosxdx = sinx + C$.

(6) $\frac{d}{dx}(cosx) = -sinx$ $\rightarrow$ $d(cosx) = -(sinx)dx$

Integrating both sides gives us:

$\int sinxdx = -cosx + C$.

262

(7) $\frac{d}{dx}(tanx) = sec^2x$ $\rightarrow$ $d(tanx) = (sec^2x)dx$

Integrating both sides gives us:

$\int sec^2xdx = tanx + C$.

(8) $\frac{d}{dx}(cotx) = -csc^2x$ $\rightarrow$ $d(cotx) = (-csc^2x)dx$

Integrating both sides gives us:

$\int csc^2xdx = -cotx + C$.

(9) $\frac{d}{dx}(secx) = secxtanx$ $\rightarrow$ $d(secx) = (secxtanx)dx$

Integrating both sides gives us:

$\int secxtanxdx = secx + C$.

(10) $\frac{d}{dx}(cscx) = -cscxcotx \rightarrow d(cscx) = (-cscxcotx)dx$

Integrating both sides gives us:

$\int cscxcotxdx = -cscx + C$.

(11) $\int secx dx = \int secx \left(\frac{secx + tanx}{secx + tanx}\right) dx$

$$= \int \left(\frac{sec^2x + secxtanx}{secx + tanx}\right) dx$$

If $u = (secx + tanx)$, $du = (sec^2x + secxtanx)dx$.
So, the formula for the integral of sec(x) is of the form
$\int \left(\frac{1}{u}\right) du = \ln|u| + C$. Therefore,

$\int secx dx = \ln|secx + tanx| + C$.

(12) $\int cscx dx = \int cscx \left(\frac{cotx - cscx}{cotx - cscx}\right) dx$

$$= \int \left(\frac{-csc^2x + cscxcotx}{cotx - cscx}\right) dx$$

If $u = (cotx - cscx)$, $du = (-csc^2x + cscxcotx)dx$.
So, the formula for the integral of csc(x) is of the form
$\int \left(\frac{1}{u}\right) du = \ln|u| + C$. Therefore,

$\int cscx dx = \ln|cotx - cscx| + C$.

(13) $\int tanx dx = \int \frac{sinx}{cosx} dx = -\int \frac{(-sinx)}{cosx} dx$.

Let $u = cosx$, $du = -sinx dx$.

So, the formula for the integral of tan(x) is of the form $-\int \left(\frac{1}{u}\right) du = -\ln|u| + C$. Therefore,

$\int tanxdx = -\ln|cosx| + C = \ln|\sec x| + C$

(14) $\int cotxdx = \int \frac{cosx}{sinx} dx = \int \frac{(cosx)}{sinx} dx$.

Let $u = sinx, \ du = cosdx$.
So, the formula for the integral of cot(x) is of the form
$\int \left(\frac{1}{u}\right) du = \ln|u| + C$. Therefore,

$\int cotxdx = \ln|sinx| + C$

Inverse Trigonometric Formulas

(15) $\frac{d}{dx}(sin^{-1}x) = \frac{1}{\sqrt{1-x^2}}$ $\rightarrow$ $d(sin^{-1}x) = \frac{1}{\sqrt{1-x^2}}dx$

Integrating both sides gives us:

$\int \frac{1}{\sqrt{1-x^2}} dx = sin^{-1}x + C$

(16) $\frac{d}{dx}(tan^{-1}x) = \frac{1}{1+x^2}$ $\rightarrow$ $d(tan^{-1}x) = \frac{1}{1+x^2}dx$

Integrating both sides gives us:

265

$$\int \frac{1}{1+x^2} dx = tan^{-1}x + C$$

(17) $\frac{d}{dx}(sec^{-1}x) = \frac{1}{x\sqrt{x^2-1}}$ → $d(sec^{-1}x) = \frac{1}{x\sqrt{x^2-1}}dx$

Integrating both sides gives us:

$$\int \frac{1}{x\sqrt{x^2-1}} dx = sec^{-1}x + C$$

The three remaining inverse trigonometric formulas are derived similarly, leading to:

(18) $\int \frac{-1}{\sqrt{1-x^2}} dx = cos^{-1}x + C$.

(19) $\int \frac{-1}{1+x^2} dx = cot^{-1}x + C$.

(20) $\int \frac{-1}{x\sqrt{x^2-1}} dx = csc^{-1}x + C$.

Example (2):
We will evaluate the following integrals:

(a) $\int (3e^x + (14)2^x - secx + cosx)dx$

$$= 3\int e^x dx + (14)\int 2^x dx - \int secxdx + \int cosxdx$$

$$= 3e^x + \frac{14}{\ln 2} 2^x - \ln|\sec x + \tan x| + \sin x + C.$$

(b) $\int \left(\frac{5}{x} + \sin x + \cos x - \tan x\right) dx$

$$= 5 \int \frac{1}{x} dx + \int \sin x dx + \int \cos x dx - \int \tan x dx$$

$$= 5 \ln|x| - \cos x + \sin x + \ln|\cos x| + C.$$

(c) $\int \csc x dx - \int \sec x \tan x dx + \int \cot x dx$

$$= \ln|\cot x - \csc x| - \sec x + \ln|\sin x| + C.$$

(d) $\int 5(\ln 3) 3^x dx + \int \frac{100}{x(\ln 10)} dx + \int \frac{3}{\cot x} dx$

$$= 5 \int (\ln 3) 3^x dx + 100 \int \frac{1}{x(\ln 10)} dx + 3 \int \tan x dx$$

$$= (5) 3^x + 100 \log_{10} x - 3 \ln|\cos x| + C.$$

(e) $\int \frac{8}{1+x^2} dx - \int \frac{25}{\sqrt{1-x^2}} dx$

$$= (8) \tan^{-1}(x) - (25) \sin^{-1}(x) + C.$$

################# Exercises #################

Evaluate the following Integrals.

(1) $\int (x^3 + x^2 + x + 1) dx$ (2) $\int dx$ (3) $\int (48) dx$

267

(4) $\int (sinx + cscx)dx$ (5) $\int (cosx + tanx)dx$

(6) $\int \left(\frac{1 - cosx}{4}\right) dx$ (7) $\int \left(\frac{1 + sinx}{7}\right) dx$

(8) $\int \left(\frac{cos^2(x) + sin^2(x)}{x}\right) dx$ (9) $\int (secxtanx + cscxcotx)dx$

(10) $\int (sec^2x + csc^2x)dx$ (11) $\int \frac{13}{\sqrt{1 - x^2}} dx$

(12) $\int \frac{25}{1 + x^2} dx$ (13) $\int \left(x^2 + \frac{1}{x\sqrt{x^2 - 1}} dx\right)$

(14) $\int (x + sinx + 3secx)dx$

(4.20) <u>Definite Integrals</u>

<u>Riemann Sums and The Definite Integral</u>
If we have a function f(x) that is continuous on the closed
interval [a,b], then often the area between the x-axis, the
graph of f(x), and between the vertical lines x = a and x = b,
has meaning. Until the calculus was developed, finding an
exact value for this area was a very difficult problem in
general. We will now explain a method for calculating this
area based on what we call a Riemann Sum (named for the
great 19th century mathematician Bernhard Riemann).
See Figure (5.61) on the next page.

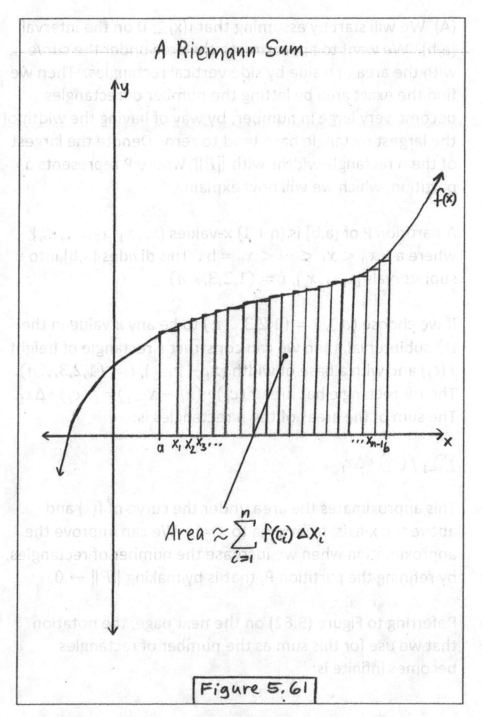

A Riemann Sum

$$\text{Area} \approx \sum_{i=1}^{n} f(c_i)\, \Delta x_i$$

Figure 5.61

(A) We will start by assuming that f(x) $\geq$ 0 on the interval [a,b]. We want to approximate the area under the curve with the area of n side by side vertical rectangles. Then we find the exact area by letting the number of rectangles become very large in number, by way of having the width of the largest rectangle base tend to zero. Denote the largest of the n rectangle widths with $\|P\|$, where P represents a partition, which we will now explain.

A partition P of [a,b] is (n + 1) x-values $\{x_0, x_1, x_2, \ldots, x_n\}$, where a = $x_0 < x_1 < \cdots < x_n = b$. This divides [a,b] into n subintervals $[x_{i-1}, x_i]$, $i = (1,2,3,\ldots n)$.

If we choose (c_i), $i = (1,2,3,\ldots n)$ to be any x-value in the ith subinterval, then we can construct a rectangle of height $f(c_i)$ and with a base of width $(x_i - x_{i-1})$, $i = (1,2,3,\ldots n)$. The ith rectangle has area $f(c_i) \cdot (x_i - x_{i-1}) = f(c_i) \cdot \Delta x_i$. The sum of the areas of the n rectangles is:

$$\sum_{i=1}^n f(c_i) \cdot \Delta x_i$$

This approximates the area under the curve of f(x) and above the x-axis, from x = a to x = b. We can improve the approximation when we increase the number of rectangles, by refining the partition P, that is by making $\|P\| \to 0$.

Referring to Figure (5.62) on the next page, the notation that we use for this sum as the number of rectangles becomes infinite is:

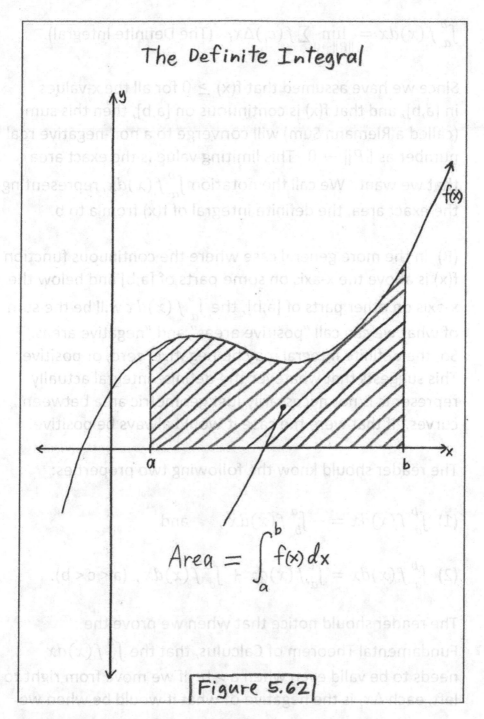

The Definite Integral

$$Area = \int_a^b f(x)\,dx$$

Figure 5.62

271

$\int_a^b f(x)dx = \lim_{\|P\| \to 0} \Sigma f(c_i)\Delta x_i$ (The Definite Integral).

Since we have assumed that f(x) ≥ 0 for all the x-values in [a,b], and that f(x) is continuous on [a,b], then this sum (called a Riemann Sum) will converge to a non-negative real number as $\|P\| \to 0$. This limiting value is the exact area that we want. We call the notation $\int_a^b f(x)dx$, representing the exact area, the definite integral of f(x) from a to b.

(B) In the more general case where the continuous function f(x) is above the x-axis on some parts of [a,b] and below the x-axis on other parts of [a,b], the $\int_a^b f(x)dx$ will be the sum of what we can call "positive areas" and "negative areas." So, the definite integral may be negative, zero, or positive. This suggests that whatever the definite integral actually represents is not necessarily just geometric area between curves. If that were the case it would always be positive.

The reader should know the following two properties:

(1) $\int_a^b f(x)dx = -\int_b^a f(x)dx$,   and

(2) $\int_a^b f(x)dx = \int_a^c f(x)dx + \int_c^b f(x)dx$, (a < c < b).

The reader should notice that when we prove the Fundamental Theorem of Calculus, that the $\int_a^b f(x)dx$ needs to be valid even when a > b. If we move from right to left, each Δx_i is the negative of what it would be when we are moving from left to right. In most applications it is usually the case for a to be less than b. It is only natural when describing things in math for us to go from left to right, to go from a lower x-value to a greater x-value. This is much like how we think of time as going from earlier times to later times. The second property should be obvious when we think about how the definite integral was defined.

****

The Fundamental Theorem of Calculus (FTC)
We will now consider a very important theorem which shows us the relation between the differential calculus and the integral calculus. It also shows us how to easily calculate definite integrals. The Theorem has two parts. In the first part we see how the indefinite integral is an antiderivative of f(x) by using the definition of the derivative. In the second part we see how a definite integral is calculated.

Part (A): For the function f(x), continuous on [a,b], and with F(x) = $\int_a^x f(t)dt$ (where $a \leq x \leq b$),

$$F'(x) = \lim_{h \to 0} \frac{F(x+h) - F(x)}{h}$$

$$= \lim_{h \to 0} \frac{\int_a^{x+h} f(t)dt - \int_a^x f(t)dt}{h}$$

$$= \lim_{h \to 0} \frac{\int_x^a f(t)dt + \int_a^{x+h} f(t)dt}{h}$$

273

$$= \lim_{h \to 0} \left(\frac{1}{h}\right) \cdot \int_x^{x+h} f(t)dt$$

The interval [x, (x + h)] is a subset of [a,b]. Therefore f(x) is continuous on [x, (x + h)]. So, it is clear that there exists a number M (which depends on x, f(x), and h) such that the

$$\int_x^{x+h} f(t)dt = (M \cdot h). \text{ (M is a mean value for the integral)}$$

So, we have:

$$F'(x) = \lim_{h \to 0} \left(\frac{M \cdot h}{h}\right) = \lim_{h \to 0} M = f(x).$$

This proves for us that the function $F(x) = \int_a^x f(t)dt$ is an antiderivative of f(x), and this represents the amount of area that has accumulated up to x. $F'(x)$ is the rate that the area is accumulating as x increases from left to right, and this is f(x). That is, the rate of accumulation of area depends on the value of f(x) at the upper limit x, and does not depend on the starting point, which we have said is any constant, in this case it is x = a. If |f(x)| is small, the rate that the area is accumulating (whatever the area represents) is low. If |f(x)| is large, the rate of accumulation of whatever the integral represents is large.

Part (B): Let c be any number in the interval [a,b], and define $F(x) = \int_c^x f(t)dt$. Then:

$$\int_a^b f(x)dx = \int_a^c f(x)dx + \int_c^b f(x)dx$$

$$= \int_c^b f(x)dx - \int_c^a f(x)dx$$

$$= F(b) - F(a).$$

This says that to evaluate the definite integral $\int_a^b f(x)dx$, we find an antiderivative F(x) for the integrand function f(x) and calculate $F(b) - F(a)$. We may use any antiderivative. To prove this, let F(x) and G(x) be any two antiderivatives of f(x). Then G(x) = F(x) + C, for some constant C. So, then:

$$\int_a^b f(x)dx = G(b) - G(a)$$

$$= (F(b) + C) - (F(a) + C)$$

$$= F(b) - F(a).$$

The constant of integration cancels out and is therefore unimportant in evaluating definite integrals. This proves the Fundamental Theorem of Calculus.

################# Exercises ##################

Evaluate the following definite integrals.

(1) $\int_1^3 (x^3 + x)dx$ (2) $\int_0^\pi \sin(x)\,dx$ (3) $\int_0^\pi \cos(x)\,dx$

(4) $\int_0^{10}(3x^2 - 9x + 1)dx$ (5) $\int_0^3 e^x dx$ (6) $\int_1^{10}\frac{1}{x}dx$

(7) $\int_{-\frac{1}{\sqrt{2}}}^{\frac{1}{\sqrt{2}}}\left(\frac{1}{\sqrt{1-x^2}}\right)dx$ (8) $\int_0^{\frac{1}{\sqrt{3}}}\left(\frac{1}{1+x^2}\right)$ (9) $\int_{-\frac{\pi}{3}}^{\frac{\pi}{4}}\sec^2(x)dx$

(10) $\int_{\frac{\pi}{2}}^{\frac{3\pi}{4}}\csc(x)\,dx$ (11) $\int_{-3}^0 e^x dx$ (12) $\int_0^1 \sqrt[3]{x}\,dx$

(4.21) Applications of Definite Integrals

Example (1):

We first want to get some practice with how to evaluate a definite integral and see the notation that is commonly used.

(a) $\int_2^6(3x - 6)dx = \left[\frac{3x^2}{2} - 6x\right]_2^6$

$$= \left(\frac{3(6)^2}{2} - 6(6)\right) - \left(\frac{3(2)^2}{2} - 6(2)\right)$$

$$= [(54 - 36) - (6 - 12)] = 24.$$

(b) $\int_0^{10}(100 - x^2)dx = \left[100x - \frac{x^3}{3}\right]_0^{10}$

$$= \left[\left(1000 - \frac{(10)^3}{3}\right) - (0)\right] = 666.\overline{666} \, .$$

(c) $\int_0^{\pi/2} \cos(x) \, dx = [\sin(x)]_0^{\pi/2}$
$$= \left[\sin(\pi/2) - \sin(0)\right] = 1 \, .$$

(d) $\int_0^2 e^x dx = [e^x]_0^2 = (e^2 - e^0)$
$$= (e^2 - 1) \approx 6.389 \, .$$

Example (2):
Our first example will be a purely geometric one, to find the area between the curves: $f(x) = x^2 - 1$ and $g(x) = -\frac{x}{2} + 2$. Refer to Figure (5.8) on the next page.

Note: when we encounter $\int_a^b f(x) dx$, with f(x) ≥ 0 on [a,b], then we can think of y = f(x) as the upper function and the x-axis or y = g(x)= 0 as the lower function, so that (f(x) − g(x)) will always be positive, and the area that we calculate will be a positive number, as it should be.

So, the first thing to do is to find the x-values for where the two functions intersect. These functions intersect where $x^2 - 1 = -\frac{1}{2}x + 2$, or the x-values for where the quadratic

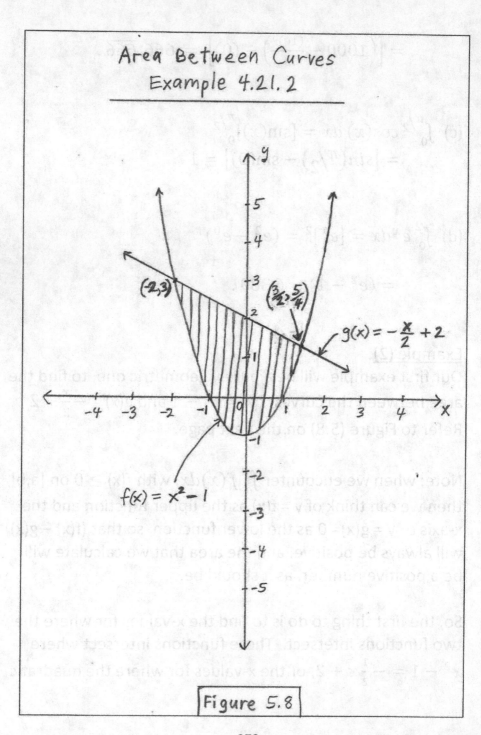

Area Between Curves
Example 4.21.2

(2,3)

$(\frac{3}{2}, \frac{5}{4})$

$g(x) = -\frac{x}{2} + 2$

$f(x) = x^2 - 1$

Figure 5.8

equation $x^2 + \frac{1}{2}x - 3 = 0$ is solved. Using the quadratic formula, we find this to be:

$$x = \frac{-\frac{1}{2} \pm \sqrt{\frac{1}{4} - (4)(1)(-3)}}{2} = \frac{-\frac{1}{2} \pm \frac{7}{2}}{2}, \text{ or } x = -2 \text{ and } \frac{3}{2}.$$

So, we can calculate the area with an integral:
For the functions that we are using in this example, function $g(x) = -\frac{1}{2}x + 2$ is actually the upper function and $f(x) = x^2 - 1$ is the lower function on $[-2, \frac{3}{2}]$.

Then, the area $= \int_{-2}^{3/2}(g(x) - f(x))dx$

$$= \int_{-2}^{3/2}\left(\left(-\frac{1}{2}x + 2\right) - (x^2 - 1)\right)dx$$

$$= \int_{-2}^{3/2}\left(-x^2 - \frac{1}{2}x + 3\right)dx$$

$$= \left[-\frac{x^3}{3} - \frac{x^2}{4} + 3x\right]_{-2}^{3/2}$$

$$= \left(-\frac{\left(\frac{3}{2}\right)^3}{3} - \frac{\left(\frac{3}{2}\right)^2}{4} + 3\left(\frac{3}{2}\right)\right) - \left(-\frac{(-2)^3}{3} - \frac{(-2)^2}{4} + 3(-2)\right)$$

$$\approx (7.1458).$$

<u>Example (3)</u>:
We have a force (measured in Newtons) that moves a mass
along the x-axis from $x_0 = 0$ meters (starting point) to $x_f = \frac{16}{3}$
meters (end point). The force is given by the function
$f(x) = \frac{3}{2}x^3 + 2x + 4$, which varies for different x-values. We
know that the amount of work done by a constant force is
defined to be the (Force) multiplied by the (distance that the
force acts on the mass) and is measured in Joules. So, in this
case with a force that varies with position x, how much work
is done?

We will divide the interval $[0, \frac{16}{3}]$ into n small subintervals
of width Δx_i , and choose a point c_i from each subinterval.
Then for the ith subinterval, we will have a rectangle with
the subinterval on the x-axis as its base and $f(c_i)$ as its
height. The area of this small rectangle $f(c_i)\Delta x_i$ will very
closely approximate the actual work done by the force on
this small sub-interval of the x-axis, as if it was a constant
force of strength f(c_i) acting through this small distance. To
approximate the total work done on $[0, \frac{16}{3}]$, we can add up
the areas of all the rectangles:

Work done $\approx \sum_{i=1}^{n} f(c_i)\Delta x_i$.

Now, refine the partition P by letting $\|P\| \to 0$. In the limit,
as the number of rectangles n goes to ∞, the limiting value
of this sum will be the exact work done by the force, and we

can calculate it because the limiting sum defines a definite integral which we can evaluate:

$$W = \lim_{\|P\| \to 0} \Sigma \left(\frac{3}{2}(c_i)^3 + 2(c_i) + 4 \right) \Delta x_i$$

$$= \int_0^{16/3} \left(\frac{3}{2}x^3 + 2x + 4 \right) dx$$

$$= \left[\frac{3x^4}{8} + x^2 + 4x \right]_0^{16/3}$$

$$= \left(\frac{3\left(\frac{16}{3}\right)^4}{8} + \left(\frac{16}{3} \right)^2 + 4\left(\frac{16}{3} \right) \right) - (0)$$

$$= \left(\frac{65,536}{216} + \frac{256}{9} + \frac{64}{3} \right) \approx (353.1851) \text{ Joules.}$$

Example (4):

Assume a Region R is perpendicular to a magnetic field. Some research physicists want to know the total magnetic flux through region R of cross section 3.75 square meters. The experiment lasts from time t = 0 to time t = 10 seconds. The magnetic field intensity through the region varies in time t with an intensity of A(t) = 5 + sin(t) Teslas for the time interval $0 \leq t \leq 10$. From physics, we know that the total flux through the region R for a magnetic field of constant intensity A(t) in a time interval Δt is defined to be W = (A(t))·(R)·(Δt) (measured in Webers). But, A(t) is not constant on (Δt). What is the total flux of this magnetic field, which varies in time according to the function A(t), through the region for the 10 seconds that the experiment is conducted?

We divide the time interval [0,10] into n small subintervals of time, each of duration Δt_i, and c_i is any time in the ith subinterval. Then the area of the ith rectangle with base length Δt_i and height $(A(c_i) \cdot R)$ is given by the expression $(A(c_i) \cdot R \cdot \Delta t_i)$, and this is an approximation to the total flux through the region R in the ith subinterval of time. The sum of the n rectangular areas is an approximation to the exact flux through the region R from t = 0 to t = 10 seconds. As the number of time subintervals becomes infinite by refining the partition P, then the exact value for the flux through region R is given by:

$$W = \lim_{\|P\| \to 0} \Sigma A(c_i) \cdot (R) \cdot \Delta t_i$$

$$= \lim_{\|P\| \to 0} \Sigma (5 + \sin(c_i)) (3.75)\Delta t_i$$

$$= (3.75) \int_0^{10} (5 + \sin(t)) dt$$

$$= (3.75) \cdot [5t - \cos(t)]_0^{10}$$

$$= (3.75) \cdot [(5(10) - \cos(10)) - ((5)(0) - \cos(0))]$$

$$\approx (3.75) \cdot [(50 + (0.83907153)) + (1)]$$

$$= (194.3965) \text{ Webers.}$$

Example (5):
(A) Suppose we have a function f(x) that is smooth and continuous on the closed interval [a,b], and we want to find the length of the curve of f(x) on this interval. Partition the interval [a,b] into (n + 1) points $\{x_0, x_1, x_2, \ldots x_n\}$ in the way that we have described previously in this section. Then consider the n pairs of adjacent points on the curve of our function f(x): $\{(x_{i-1}, y_{i-1})$ and $(x_i, y_i), i = 1,2,\ldots, n\}$, where y = f(x). We can consider the n straight line segments between each of these n adjacent pairs of points on f(x). The length of the ith segment, using the Pythagorean theorem, is:

$$L_i = \sqrt{(\Delta x)_i^2 + (\Delta y)_i^2} .$$ The exact length of the curve is L.

$$L \approx \sum_{i=1}^{n} L_i = \sum_{i=1}^{n} \sqrt{(\Delta x)_i^2 + (\Delta y)_i^2}$$

$$= \sum_{i=1}^{n} \sqrt{1 + \left(\frac{\Delta y}{\Delta x}\right)_i^2} \cdot (\Delta x)_i$$

From the Mean Value Theorem, there exists a number c_i in the ith subinterval such that $\left(\frac{\Delta y}{\Delta x}\right)_i^2 = (f'(c_i))^2$. Therefore,

$$L \approx \sum_{i=1}^{n} \sqrt{1 + \left(\frac{\Delta y}{\Delta x}\right)_i^2} \cdot (\Delta x)_i = \sum_{i=1}^{n} \sqrt{1 + (f'(c_i))^2} \cdot (\Delta x)_i$$

Now, if we refine the partition so that the number of line segments goes to infinity, then the exact length of the curve on [a,b] is given by:

283

$$L = \lim_{\|P\|\to 0} \Sigma \sqrt{1 + (f'(c_i))^2} \cdot (\Delta x)_i = \int_a^b \sqrt{1 + \left(\frac{dy}{dx}\right)^2}\, dx.$$

(B) Let's consider a specific case: Let $f(x) = x^{\frac{3}{2}}$, on [0,5]. Then $f'(x) = \frac{3}{2}\sqrt{x}$. The length of the curve is:

$L = \int_0^5 \sqrt{1 + \frac{9}{4}(x)}\, dx$. To evaluate this we will make a change of variable. Let $u = \left(1 + \frac{9}{4}(x)\right)$, $du = \frac{9}{4}dx$. So, $dx = \frac{4}{9}du$, and then (keeping track of the fact that the integral limits 0 and 5 correspond to the variable x):

$$L = \int_{x=0}^{x=5} \sqrt{u} \cdot \frac{4}{9}du = \left(\frac{4}{9}\right)\int_{x=0}^{x=5} u^{\frac{1}{2}}\, du = \left(\frac{4}{9}\right)\left(\frac{2}{3}\right)\left[u^{\frac{3}{2}}\right]_{x=0}^{x=5}$$

$$= \left(\frac{8}{27}\right)\left[\left(1 + \frac{9}{4}x\right)^{\frac{3}{2}}\right]_0^5 = \left(\frac{8}{27}\right)\left(\left(\frac{49}{4}\right)^{\frac{3}{2}} - 1\right) = \left(\frac{8}{27}\right)\left(\frac{335}{8}\right)$$

$\approx (12.4074).$

################## Exercises ##################

(1) From time $t_0 = 0$ to $t_f = 3$ (measured in hours), an automobile travels along a line at a velocity that varies in time according to $v(t) = 2t^3 + 10 \left(\frac{\text{miles}}{\text{hour}}\right)$. Calculate the total

distance D traveled during this time interval which is given
by the integral formula $D = \int_{t_0}^{t_f} v(t)dt$.

(2) From the time $t_0 = 0$ to $t_f = 7$, measured in seconds,
a fluid is forced through a steel pipe of cross-sectional area
$A = \frac{1}{30}$ meter^2, with a velocity that varies in time and is given
by the function $v(t) = \frac{1}{15}e^t \left(\frac{\text{meters}}{\text{second}}\right)$. Calculate the total
volume of fluid V that flows past a cross-section of the pipe
in this time interval. This is given by the integral formula
$V = \int_{t_0}^{t_f} A \cdot v(t)dt$.

(3) Suppose we want to derive the formula for the volume
of a right circular cone of base radius r and height h. Plot
the points (0,0), (0,r), and (h,0) in the x-y plane, and connect
line segments between these three points to form a right
triangle. Note that this is half of a vertical cross-section of
the cone through its apex (turned on its side). Now partition
the x-axis from x = 0 to x = h and let c_i be any point in the
i^{th} base segment. We know that the sum of the areas of
those rectangles approximates the exact area under the line
$y = r - \frac{r}{h}x$, which contains the hypotenuse (of the right
triangular half of the cone's vertical cross-section). Now,
let's consider something a bit different. Consider just one
of those rectangles and revolve it 360° around the x-axis
(keeping the base fixed on the x-axis). Then that revolved
rectangle generates a volume which we can think of as a
thin disk, the volume of which would be the side area
$\pi(f(c_i))^2$ times its thickness Δx_i. Calling this small disc the
i^{th} disc, of volume V_i, where $V_i = \pi(f(c_i))^2 \Delta x_i$, we can see

285

that the sum of the volumes of these n discs will closely approximate the volume of the cone. Let the number of rectangles become infinite by refining the partition, that is by making the norm of the partition $\|P\| \to 0$. Then the volume of the cone is:

$$\lim_{\|P\|\to 0} \pi\left(f(c_i)\right)^2 \Delta x_i \text{ which is the } \int_0^h \pi\left(f(x)\right)^2 dx,$$

where $f(x) = r - \dfrac{r}{h}x$. Calculate this integral and compare it with the volume of a cone which was given in section (3.13).

(4.22) Techniques of Integration

In this section, we will not try to provide a complete treatment of the various techniques of integration (the finding of antiderivatives) that one would get in a traditional calculus course. The goal is to provide examples of the most useful and most common techniques.

Change of Variable
The best way to proceed is through examples.

Example (1):
Evaluate $\int e^{-x^2}(x)dx$.
In the following, remember that constants can be brought back and forth across the integral sign. So, if we need to insert a constant "c" inside the integral, we have to

compensate by supplying the constant "$\left(\frac{1}{c}\right)$" outside the integral, to keep the integral in balance. The constant introduced into the inside of the integral becomes part of the differential for the new variable and the constant introduced outside the integral remains there outside the integral to be multiplied by the rest of the calculation.

In the integral above, we notice that aside from a constant, (xdx) is the differential of $(-x^2)$. When we change the variable by letting $u = -x^2$, the first thing that we have to do is calculate du. (For u = f(x), du = f'(x)dx). So, in this case, $du = -2xdx$. In the integrand we already have the xdx part, we just need to supply the (-2) part, and then compensate by multiplying $\left(-\frac{1}{2}\right)$ outside the integral. It is important to remember that we can only introduce a constant inside the integral, not anything which involves variables. So, we now have:

$$\int e^{-x^2}(x)dx = \left(-\frac{1}{2}\right)\int e^{-x^2}(-2)(x)dx$$

Since we have $e^u = e^{-x^2}$ and $du = -2xdx$, the integral is in the form $\left(-\frac{1}{2}\right)\int e^u du$. This is one of the basic forms that we derived back in section (4.19), which we have a formula for. Notice that when we change the variable, we must have our integral after the change, exactly in the form of one of those that were derived (other than the name of the variable of course). So,

$$\int e^{-x^2}(x)dx = \left(-\frac{1}{2}\right)\int e^{-x^2}(-2)(x)dx = \left(-\frac{1}{2}\right)\int e^u du .$$

Now, $\left(-\frac{1}{2}\right)\int e^u du = \left(-\frac{1}{2}\right)e^u + C = \left(-\frac{1}{2}\right)e^{-x^2} + C,$

after going back to the variable x. So, in summary,

$$\int e^{-x^2}(x)dx = \left(-\frac{1}{2}\right)e^{-x^2} + C.$$

Example (2):
Evaluate $\int (2 + x^3)^{10}(x^2)dx$.
Let $u = (2 + x^3)$, then $du = 3x^2 dx$. So, we have:

$$\int (2 + x^3)^{10}(x^2)\,dx = \left(\frac{1}{3}\right)\int (2 + x^3)^{10}(3x^2)dx.$$

Then, $\left(\frac{1}{3}\right)\int (2 + x^3)^{10}(3x^2)dx = \left(\frac{1}{3}\right)\int u^{10} du$

$$= \left(\frac{1}{3}\right)\frac{u^{11}}{11} + C = \frac{u^{11}}{33} + C = \frac{(2 + x^3)}{33} + C.$$

So, the $\int (2 + x^3)^{10}(x^2)dx = \frac{(2 + x^3)}{33} + C.$

Example (3):
Evaluate $\int sin^4(3x)\cos(3x)\,dx$.
It is probably easier to see what is going on if we write
$sin^4(3x)$ as $(\sin (3x))^4$, which is what the notation
means. So, we have to evaluate:

$$\int (\sin(3x))^4 \cos(3x)\,dx .$$

Let $u = (\sin(3x))$, then $du = (\cos(3x))(3)dx$. So,

$$\int \sin^4(3x)\cos(3x)\,dx = \int (\sin(3x))^4\cos(3x)\,dx$$

$$= \left(\tfrac{1}{3}\right)\int (\sin(3x))^4\cos(3x)(3)dx$$

$$= \left(\tfrac{1}{3}\right)\int u^4 du = \left(\tfrac{1}{3}\right)\frac{u^5}{5} + C = \frac{u^5}{15} + C$$

$$= \frac{\sin^5(3x)}{15} + C.$$

Example (4):

Evaluate $\int \left(\frac{1}{\tan(10x)}\right)\sec^2(10x)dx$.

Let u = tan(10x), then $du = \sec^2(10x)(10)dx$.

So, $\int \left(\frac{1}{\tan(10x)}\right)\sec^2(10x)dx$

$$= \left(\tfrac{1}{10}\right)\int \left(\frac{1}{\tan(10x)}\right)\sec^2(10x)(10)dx$$

$$= \left(\tfrac{1}{10}\right)\int \tfrac{1}{u}\,du = \left(\tfrac{1}{10}\right)\ln|u| + C$$

$$= \left(\tfrac{1}{10}\right)\ln|\tan(10x)| + C.$$

Example (5):

Evaluate $\int \frac{\sqrt{x}}{1+x}dx$.

Sometimes getting an integral into an integrable form

may involve a miscellaneous change of variable. This often involves some ingenuity and some trial and error.

Let $u = \sqrt{x}$, then $du = \frac{1}{2\sqrt{x}}dx$, or $dx = 2\sqrt{x}du = 2udu$.

Then $\int \frac{\sqrt{x}}{1+x}dx = \int \frac{u}{1+u^2}(2u)du = 2\int \frac{u^2}{1+u^2}du$

$$= 2\int \left(1 - \frac{1}{1+u^2}\right)du \quad \text{(after the long division)}$$

$$= 2\int du - 2\int \frac{1}{1+u^2}du = 2u - 2tan^{-1}(u) + C$$

$$= 2\sqrt{x} - 2tan^{-1}(\sqrt{x}) + C .$$

Integration by Parts
Recall the product rule for differentiating a product of two functions [u(x)·v(x)]:

$\frac{d}{dx}(uv) = u\frac{dv}{dx} + v\frac{du}{dx}$. Multiplying through by dx, we get:

$d(uv) = udv + vdu$. Then, integrating both sides gives us:

$\int d(uv) = \int udv + \int vdu$. Rearranging terms gives us the

Integration by Parts Formula: $\int udv = uv - \int vdu$.

Example (1):
Evaluate $\int xe^{-x}dx$.

We need to make a determination of what will be "u" and what will be "dv." ("dv" is everything other than what is chosen for "u", including the differential). To help in the choice, we have the following mnemonic device (ILATE):

I (Inverse Trigonometric)
L (Logarithmic)
A (Algebraic)
T (Trigonometric)
E (Exponential)

In the integral above, we have an algebraic part "x" and an exponential part "e^{-x}." Whichever is higher on the list will be our choice for u. So,

let: $u = x$ and $dv = e^{-x} dx$
Then: $du = dx$ and $v = \int dv = \int e^{-x} dx = -e^{-x}$

(we don't worry about a constant of integration when calculating "v" in an integration by parts problem).

So, according to the Integration by Parts Formula:
$\int u dv = uv - \int v du$:

$\int xe^{-x} dx = (-xe^{-x}) - \int(-e^{-x}) dx$
$\qquad\qquad = (-xe^{-x}) - e^{-x} + C$.

Example (2):
Evaluate $\int (x)(lnx) dx$.

291

We have an algebraic part "x" and a logarithmic part "lnx."
L is higher than A on the memory aid "ILATE." So,

Let $\quad u = lnx \quad$ and $\quad$ dv = (x)dx

Then $\quad du = \dfrac{1}{x}dx \quad$ and $\quad$ v = $\dfrac{x^2}{2}$

So, using the integration by parts formula:
$\int udv = uv - \int vdu,$ we have:

$$\int (x)(lnx)dx = \left(\dfrac{(x^2)\cdot lnx}{2}\right) - \dfrac{1}{2}\int (x)dx$$

$$= \left(\dfrac{x^2 \cdot lnx}{2}\right) - \dfrac{x^2}{4} + C .$$

Example (3):
Evaluate $\int (x)sinxdx$.
We have an algebraic part "x" and a trig part "sin(x)." A is
higher than T on the memory aid "ILATE." So,

Let $\quad u = x \quad$ and $\quad$ dv = sin(x)dx
Then $\quad du = dx \quad$ and $\quad$ v = $-cos(x)$

So, $\int udv = uv - \int vdu$ says that,

$$\int (x)(sinx)dx = (-(x)cosx) + \int cos(x)dx$$

$$= -xcosx + sinx + C .$$

Trigonometric Substitutions

Example (1):

Evaluate $\int \left(\frac{15}{\sqrt{1-x^2}}\right) dx$.

Here we will use: $cos^2(x) + sin^2(x) = 1$, and therefore that: $cos^2(x) = 1 - sin^2(x)$.

Let $x = sin(\theta)$, then $dx = cos\theta d\theta$.

So, $\sqrt{1-x^2} = \sqrt{1 - sin^2(\theta)} = \sqrt{cos^2(\theta)} = cos(\theta)$.

So, $\int \left(\frac{15}{\sqrt{1-x^2}}\right) dx = 15 \int \left(\frac{cos(\theta)}{cos(\theta)}\right) d\theta = 15 \int d\theta = 15\theta + C$.

Since $x = sin(\theta)$, then $\theta = sin^{-1}(x)$.

Therefore, $\int \left(\frac{15}{\sqrt{1-x^2}}\right) dx = 15 sin^{-1}(x) + C$.

Note that what we have here is an alternative method for deriving the integral formula that leads to the $sin^{-1}(x)$.

Example (2):

Evaluate $\int \left(\frac{3}{\sqrt{1+x^2}}\right) dx$.

Here we will use: $1 + tan^2(\theta) = sec^2(\theta)$.
Let $x = tan(\theta)$, then $dx = sec^2(\theta)d\theta$.
So, $\sqrt{1 + x^2} = \sqrt{1 + tan^2(\theta)} = \sqrt{sec^2(\theta)} = sec(\theta)$.

So, $\int \left(\frac{3}{\sqrt{1+x^2}}\right) dx = 3 \int \left(\frac{1}{sec(\theta)}\right) sec^2(\theta)d\theta$

$$= 3 \int \sec(\theta) \, d\theta$$

$$= 3 \ln|\sec\theta + \tan\theta| + C.$$

Since $x = tan(\theta)$, then $\theta = tan^{-1}(x)$.

So, $\int \left(\frac{3}{\sqrt{1+x^2}}\right) dx = 3 \ln|\sec(tan^{-1}(x)) + x| + C$

The fact that $\theta = tan^{-1}(x)$ says that θ is an angle such that the $tan(\theta) = x = \left(\frac{x}{1}\right)$. The reader may want to draw a right triangle as in Figure (5.2) and label the sides to see what is going on in this kind of calculation. Using the Pythagorean theorem: $\sec(\theta) = \left(\frac{\sqrt{1+x^2}}{1}\right) = \sqrt{1+x^2}$.

Therefore, $\int \left(\frac{3}{\sqrt{1+x^2}}\right) dx = 3 \ln|\sqrt{1+x^2} + x| + C$.

Example (3):
Evaluate $\int \sqrt{x^2 - 1} \, dx$.

Since $1 + tan^2(\theta) = sec^2(\theta)$,
$$tan^2(\theta) = sec^2(\theta) - 1$$

Now, let $x = \sec(\theta)$, then $dx = \sec(\theta)tan(\theta)d\theta$.

Then $\sqrt{x^2 - 1} = \sqrt{sec^2(\theta) - 1} = \sqrt{tan^2(\theta)} = tan(\theta)$.

So, $\int \sqrt{x^2 - 1} \, dx = \int \sec(\theta)\tan^2(\theta) \, d\theta$

$\qquad = \int \sec(\theta) \, (\sec^2(\theta) - 1) d\theta$

$\qquad = \int \sec^3(\theta) \, d\theta - \int \sec(\theta) \, d\theta$

Using Integration by Parts on the $\int \sec^3(\theta) d\theta$,
Let $u = \sec(\theta)$ $\qquad\qquad$ and $\qquad dv = \sec^2(\theta) d\theta$
Then, $du = \sec(\theta)\tan(\theta)d\theta$ $\quad$ and $\qquad v = \tan(\theta)$

So, $\int \sec^3(\theta) \, d\theta = \sec(\theta)\tan(\theta) - \int \sec(\theta)\tan^2(\theta) \, d\theta$

So, $\int \sec(\theta)\tan^2(\theta) \, d\theta$

$\qquad = \sec(\theta)\tan(\theta) - \int \sec(\theta)\tan^2(\theta)d\theta - \int \sec(\theta)d\theta$

From this we can see that:

$2 \int \sec(\theta)\tan^2(\theta) \, d\theta$

$\qquad\qquad = \sec(\theta)\tan(\theta) - \ln|\sec(\theta) + \tan(\theta)| + C$

So therefore, the $\int \sqrt{x^2 - 1} \, dx$

$\qquad = \frac{1}{2}(\sec(\theta)\tan(\theta)) - \frac{1}{2}(\ln|\sec(\theta) + \tan(\theta)|) + C$

$\qquad = \frac{1}{2}x\sqrt{x^2 - 1} - \frac{1}{2}(\ln|x + \sqrt{x^2 - 1}|) + C$.

295

Example (4):
(a) Evaluate $\int 26\cos^2(x)dx$.

In this kind of integral, use the half-angle trigonometric identity.

$$\int 26\cos^2(x)dx = 26 \int \cos^2(x)dx = 26 \int \left(\frac{1 + \cos(2x)}{2}\right)dx$$

$$= 26 \int \frac{1}{2}dx + 26 \int \frac{\cos(2x)}{2}dx = 13x + 13 \int \cos(2x)\,dx$$

Let u = 2x, so du = 2dx. Then we have:

$$\int 26\cos^2(x)dx = 13x + 13 \left(\frac{1}{2}\right) \int \cos(2x)\,2dx$$
$$= 13x + \frac{13}{2} \int \cos(u)\,du$$
$$= 13x + \frac{13}{2}\sin(u) + C$$
$$= 13x + \frac{13}{2}\sin(2x) + C$$

(b) Evaluate $\int \sin^2\left(\frac{x}{2}\right)dx$.

$$\int \sin^2\left(\frac{x}{2}\right)dx = \int \left(\frac{1 - \cos(x)}{2}\right)dx = \int \frac{1}{2}dx - \frac{1}{2}\int \cos(x)\,dx$$
$$= \frac{x}{2} - \frac{1}{2}\sin(x) + C$$

Partial Fractions
This technique is useful for the case where the integrand is a rational function, that is, a function which is the ratio of two polynomial functions $\frac{f(x)}{g(x)}$, where the degree of f(x) is strictly

296

less than the degree of g(x). This requirement must be true in order to use this technique. The technique's name comes from some algebraic results which we need not try to prove. They essentially say that once the denominator g(x) has been factored into linear factors like $(ax + b)^m$ and irreducible quadratic factors like $(ax^2 + b)^m$, then there is a so-called partial fraction decomposition for $\frac{f(x)}{g(x)}$.

For each linear factor of g(x) of the form $(ax + b)^m$, where "a" and "b" are real numbers and m is a positive integer, there corresponds a sum of terms:

$$\frac{A_1}{(ax + b)} + \frac{A_2}{(ax + b)^2} + \cdots + \frac{A_m}{(ax + b)^m}.$$

For each irreducible quadratic factor of g(x) of the form $(ax^2 + b)^m$, where "a" and "b" are real numbers and m is a positive integer, there corresponds a sum of terms:

$$\frac{B_1x + C_1}{(ax^2 + b)} + \frac{B_2x + C_2}{(ax^2 + b)^2} + \cdots + \frac{B_mx + C_m}{(ax^2 + b)^m}.$$

Let's consider the basic procedure for the following two cases (a) and (b):

(a) For $\frac{f(x)}{g(x)} = \frac{x^2 + 2x - 8}{x^3 - 11x^2 + 7x + 147}$, the denominator g(x) can be factored: $x^3 - 11x^2 + 7x + 147 = (x + 3)(x - 7)^2$.

So, the partial fraction decomposition for $\frac{f(x)}{g(x)}$ is:

297

$$\frac{f(x)}{g(x)} = \frac{x^2 + 2x - 8}{x^3 - 11x^2 + 7x + 147} = \frac{x^2 + 2x - 8}{(x+3)(x-7)^2}$$

$$= \frac{A}{(x+3)} + \frac{B}{(x-7)} + \frac{C}{(x-7)^2}.$$

(where A, B, and C need to be determined)

Then the next step in evaluating the integral of $\frac{f(x)}{g(x)}$, or

$\int \frac{x^2 + 2x - 8}{x^3 - 11x^2 + 7x + 147} dx$ after finding the partial fraction

decomposition $\frac{A}{(x+3)} + \frac{B}{(x-7)} + \frac{C}{(x-7)^2}$, would be to

evaluate the sum of the integrals of these three fractions.

(b) For $\frac{f(x)}{g(x)} = \frac{(x+3)}{x^3 + 5x^2 + x + 5}$, the denominator g(x) can be

factored: $x^3 + 5x^2 + x + 5 = (x+5)(x^2 + 1)$.

So, the partial fraction decomposition for $\frac{f(x)}{g(x)}$ is:

$$\frac{f(x)}{g(x)} = \frac{(x+3)}{x^3 + 5x^2 + x + 5} = \frac{(x+3)}{(x+5)(x^2+1)} = \frac{A}{(x+5)} + \frac{Bx+C}{(x^2+1)}.$$

(where A, B, and C need to be determined)

Then the next step in evaluating the integral of $\frac{f(x)}{g(x)}$, or

$\int \dfrac{(x+3)}{x^3 + 5x^2 + x + 5} \, dx$ after finding the partial fraction

decomposition $\dfrac{A}{(x+5)} + \dfrac{Bx+C}{(x^2+1)}$, would be to evaluate the

sum of the integrals of these two fractions.

Now, three examples of integration by partial fractions:

Example (1):

Evaluate $\int \dfrac{x-9}{(x+3)(x-1)} \, dx$.

Firstly, write $\dfrac{x-9}{(x+3)(x-1)} = \dfrac{A}{(x+3)} + \dfrac{B}{(x-1)}$.

Multiplying through by $(x+3)(x-1)$ yields:
$x - 9 = A(x-1) + B(x+3)$.

Let x = 1. Then $-8 = 4B$, which says that B = -2.
Let x = -3. Then $-12 = -4A$, which says that A = 3.

So, $\int \dfrac{x-9}{(x+3)(x-1)} \, dx = \int \dfrac{3}{x+3} \, dx - \int \dfrac{2}{x-1} \, dx$

$= 3 \ln|x+3| - 2 \ln|x-1| + C$.

Example (2):

Evaluate $\int \dfrac{x^2 + 2x + 9}{x^3 - 2x^2 - 8x} \, dx$.

Firstly, write $\dfrac{x^2 + 2x + 9}{x^3 - 2x^2 - 8x} = \dfrac{x^2 + 2x + 9}{x(x+2)(x-4)} = \dfrac{A}{x} + \dfrac{B}{x+2} + \dfrac{C}{x-4}$.

Multiplying through by $x(x + 2)(x - 4)$ yields:
$x^2 + 2x + 9 = A(x + 2)(x - 4) + B(x)(x - 4)$
$$+C(x)(x + 2)$$

Let x = −2. Then $9 = 12B$, which says that B = $\frac{3}{4}$.

Let x = 4. Then $33 = 24C$, which says that C = $\frac{11}{8}$.

Let x = 0. Then $9 = -8A$, which says that A = $-\frac{9}{8}$.

So, $\int \frac{x^2 + 2x + 9}{x(x + 2)(x - 4)} dx = -\frac{9}{8} \int \frac{1}{x} dx + \frac{3}{4} \int \frac{1}{x + 2} dx$
$$+\frac{11}{8} \int \frac{1}{x - 4} dx + C$$

$$= -\frac{9}{8} \ln|x| + \frac{3}{4} \ln|x + 2| + \frac{11}{8} \ln|x - 4| + C .$$

Example (3):
Evaluate $\int \frac{x + 2}{x^3 - x^2 + x - 1} dx$

Firstly, write $\frac{x + 2}{x^3 - x^2 + x - 1} = \frac{x + 2}{(x - 1)(x^2 + 1)} = \frac{A}{x - 1} + \frac{Bx + C}{x^2 + 1}$.

Multiplying through by $(x - 1)(x^2 + 1)$ yields:
$x + 2 = A(x^2 + 1) + (Bx + C)(x - 1)$, which expands to:
$x + 2 = Ax^2 + A + Bx^2 + Cx - Bx - C$, which becomes:
$x + 2 = (A + B)x^2 + (C - B)x + (A - C)$

Equating coefficients on like powers yields:
$$\begin{cases} A + B = 0 \\ C - B = 1 \\ A - C = 2 \end{cases}$$

Adding the first two equations yields:
$$\begin{cases} A + C = 1 \\ A - C = 2 \end{cases}$$

Adding these two equations together gives us:

$2A = 3$, or $A = \frac{3}{2}$. From $A + C = 1$, we get $\frac{3}{2} + C = 1$, or $C = -\frac{1}{2}$. Then from $A + B = 0$ in the first set of equations, we have $B = -\frac{3}{2}$.

So, $\int \frac{x + 2}{x^3 - x^2 + x - 1} dx = \frac{3}{2} \int \frac{1}{x - 1} dx + \int \frac{-\frac{3}{2}x - \frac{1}{2}}{x^2 + 1} dx$

$= \frac{3}{2} \int \frac{1}{x - 1} dx - \frac{3}{2} \left(\frac{1}{2} \right) \int \frac{2x}{x^2 + 1} dx - \frac{1}{2} \int \frac{1}{1 + x^2} dx$

$= \frac{3}{2} \ln|x - 1| - \frac{3}{4} \ln|x^2 + 1| - \frac{1}{2} \tan^{-1}(x) + C$.

Improper Integrals

With integrals, we often encounter the situations:

(A) In the integral $\int_a^b f(x)dx$, at the lower limit a, f(x) tends to ∞ or $-\infty$.

(B) In the integral $\int_a^b f(x)dx$, at the upper limit b, f(x) tends to ∞ or $-\infty$.

(C) In the integral $\int_a^b f(x)dx$, the lower limit $a = -\infty$.

(D) In the integral $\int_a^b f(x)dx$, the upper limit $b = \infty$.

We call these improper integrals. The best way to show how these situations are dealt with is through examples.

Example (1):

Evaluate $\int_0^2 \frac{1}{x} dx$.

The function $f(x) = \frac{1}{x}$ is asymptotic to the positive y-axis on the right side, that is $f(x)$ tends to ∞ as x approaches 0 from the right. Therefore, to evaluate this integral we need a slightly different technique, which takes into account the fact that for a definite integral both limits should be numbers. So, the correct way to evaluate this integral is to say:

$$\int_0^2 \frac{1}{x} dx = \lim_{t \to 0^+} \int_t^2 \frac{1}{x} dx = \lim_{t \to 0^+} [\ln|x|]_t^2 = \lim_{t \to 0^+} (\ln(2) - \ln|t|)$$

$$= (\ln(2) - (-\infty)) = \infty.$$

So, we say that the integral diverges because the area under $f(x) = \frac{1}{x}$, from 0 to 2, is infinite.

Example (2):

Evaluate $\int_{-\infty}^{-2} \frac{1}{x^2} dx$.

On the interval $[-\infty, -2]$, $f(x) = \frac{1}{x^2}$ is asymptotic to the x-axis from above, as $x \to -\infty$. We want to know if the area under $f(x)$ and above the x-axis, from $-\infty$ to -2, is finite or not. So, we figure that in this way:

$$\int_{-\infty}^{-2} \frac{1}{x^2} dx = \lim_{t \to -\infty} \int_{t}^{-2} \frac{1}{x^2} dx = \lim_{t \to -\infty} \left[-\frac{1}{x} \right]_{t}^{-2}$$

$$= \lim_{t \to -\infty} \left(\frac{1}{2} + \frac{1}{t} \right) = \frac{1}{2}.$$

So, we say that the integral converges because the area under $f(x) = \frac{1}{x^2}$ and above the x-axis, from $-\infty$ to -2, is finite and equals $\frac{1}{2}$.

Example (3):

Evaluate $\int_0^1 \frac{1}{\sqrt{x}} dx$.

The function $f(x) = \frac{1}{\sqrt{x}}$ is asymptotic to the y-axis and tends to ∞, as x approaches 0 from the right. So, we want to know whether the area under $f(x) = \frac{1}{\sqrt{x}}$ and above the x-axis, from 0 to 1 is finite or not. We figure this in the following way:

$$\int_0^1 \frac{1}{\sqrt{x}} dx = \lim_{t \to 0^+} \int_{t}^1 \frac{1}{\sqrt{x}} dx = \lim_{t \to 0^+} \int_{t}^1 x^{-\frac{1}{2}} dx = \lim_{t \to 0^+} \left[2\sqrt{x} \right]_{t}^1$$

$$= \lim_{t \to 0^+} (2 - 2\sqrt{t}) = 2.$$

So, the integral converges to 2. This is the area under $f(x) = \frac{1}{\sqrt{x}}$ and above the x-axis, from x = 0 to x = 1.

Example (4):

Evaluate $\int_0^{\infty} e^{-x} dx$.

The function $f(x) = e^{-x}$ starts at y = 0 when x = 0, and it decreases asymptotically to the x-axis as x increases. So, we

303

want to find the area under $f(x) = e^{-x}$ and above the x-axis, from x = 0 to ∞. We calculate it in this way:

$$\int_0^\infty e^{-x}dx = \lim_{t \to \infty}(-1)\int_0^t(-1)e^{-x}dx$$

$$= (-1)\lim_{t \to \infty}[e^{-x}]_0^t = (-1)\lim_{t \to \infty}(e^{-t}-1)$$

$$= (-1)(0-1) = 1.$$

So, the integral converges to 1. This is the area under $f(x) = e^{-x}$ and above the x-axis, from x = 0 to ∞.

This section, with change of variable, integration by parts, trigonometric substitutions, partial fractions, and improper integrals has presented to the reader the most common techniques of integration (finding antiderivatives).

################## Exercises ##################

Evaluate the following integrals:

(1) $\int(x-2)e^x dx$

(2) $\int(x^2 + x + 4)^4\left(x + \frac{1}{2}\right)dx$

(3) $\int(x)sin(x^2)dx$

(4) $\int \frac{x+1}{x}dx$

(5) $\int sec(3x^2)tan(3x^2)(x)dx$

(6) $\int tan(13x+3)\,dx$

(7) $\int \frac{1}{1+16x^2} dx$ (8) $\int \frac{1}{\sqrt{25-25x^2}} dx$

(9) $\int \cos(4x^2 + 8x)(x + 1)dx$ (10) $\int (x)\cos(x)dx$

(11) $\int \ln(x)\, dx$ (12) $\int \frac{30x}{\cos(2x^2)} dx$

(13) $\int (x + 3)e^{x+3}dx$ (14) $\int \frac{6}{x(x^2+1)} dx$

(15) $\int_{-\infty}^{3} e^x dx$ (16) $\int_{-1}^{0} \frac{1}{x^3} dx$ (17) $\int_{1}^{6} \frac{1}{x-6} dx.$

(4.23) Infinite Series

This will not be a complete treatment of infinite series. We just want to present some of the most important points and provide some important examples that a student of calculus should know.

An infinite series is an infinite sum and is written:
$\sum_{n=1}^{\infty} a_n = (a_1 + a_2 + a_3 + \cdots)$, or

$\sum_{n=0}^{\infty} a_n = (a_0 + a_1 + a_2 + \cdots)$, or similarly,

and where a_n is the nth term in the sum. So

Some examples are:

(1) $\sum_{n=1}^{\infty} \left(\frac{1}{n}\right) = \left(1 + \frac{1}{2} + \frac{1}{3} + \frac{1}{4} + \cdots\right)$

(2) $\sum_{n=1}^{\infty} \left(\frac{1}{n^2}\right) = \left(1 + \frac{1}{4} + \frac{1}{9} + \frac{1}{16} + \cdots\right)$

(3) $\sum_{n=0}^{\infty} \left(\frac{1}{2}\right)^n = \left(1 + \frac{1}{2} + \frac{1}{4} + \frac{1}{8} + \cdots\right)$

(4) $\sum_{n=1}^{\infty} \left(\frac{x^n}{n}\right) = \left(x + \frac{x^2}{2} + \frac{x^3}{3} + \frac{x^4}{4} + \cdots\right)$

(5) $\sum_{n=0}^{\infty} (-1)^n = (1 - 1 + 1 - 1 + \cdots)$.

Convergence or Divergence

We are generally concerned with when a series converges, or when it diverges. That is, when the infinite sum adds up to a specific number L, or when the infinite sum adds up to $(+\infty)$ or $(-\infty)$, or when the infinite sum does not approach any single number at all. It is certainly true that if a series converges, then the $\lim_{n \to \infty} a_n = 0$. However, the converse is not necessarily true, because there are many divergent series where $\lim_{n \to \infty} a_n = 0$. The harmonic series below (see the sub-section on p-series) is a good example of this.

The question of convergence or divergence is a question about how fast the successive terms approach zero after a certain point. So, the convergence of a series is a question about what happens in the long run with the terms of the series, not what happens with any first finite number of terms of the series.

For the series $\sum_{n=1}^{\infty} a_n$, we have the partial sums:
$$S_1 = a_1$$
$$S_2 = a_1 + a_2$$
$$S_3 = a_1 + a_2 + a_3$$
$$S_4 = a_1 + a_2 + a_3 + a_4$$
$$\vdots$$

The series will converge when the sequence of partial sums $\{s_n\}$ converges, and it will diverge when the sequence of partial sums $\{s_n\}$ diverges. That is: If the $\lim_{n \to \infty} (s_n) = L$, then the series $\sum_{n=1}^{\infty} a_n = L$. This is the most general definition of the convergence or divergence of an infinite series.

P-Series

A series of the form $\sum_{n=1}^{\infty} \frac{1}{n^p}$ is called a p-series. A test for convergence that is often useful is the integral test:

The p-series $\sum_{n=1}^{\infty} \frac{1}{n^p}$ will converge or diverge when the integral $\int_1^{\infty} \frac{1}{x^p} dx$ converges or diverges. This test leads to the result that a p-series diverges when $p \leq 1$ and it converges when $p > 1$.

The case $p = 1$ leads to the series $\sum_{n=1}^{\infty} \frac{1}{n}$ which is a divergent series. We call it the harmonic series.

Alternating Series

A series of the form $\sum_{n=0}^{\infty}(-1)^n a_n$ or $\sum_{n=1}^{\infty}(-1)^{n+1} a_n$ is called an alternating series. An alternating series will converge if:

(1) $\lim\limits_{n\to\infty}(a_n) = 0$, and

(2) $a_{n+1} \leq a_n$.

A series $\sum_{n=1}^{\infty} a_n$ is absolutely convergent if $\sum_{n=1}^{\infty} |a_n|$ converges.

A series $\sum_{n=1}^{\infty} a_n$ is conditionally convergent if $\sum_{n=1}^{\infty} a_n$ converges, but $\sum_{n=1}^{\infty} |a_n|$ diverges.

An example of a conditionally convergent series is $\sum_{n=1}^{\infty}(-1)^{n+1}\left(\frac{1}{n}\right)$, since it converges (by the above test for an alternating series), but the series $\sum_{n=1}^{\infty}\left|(-1)^{n+1}\left(\frac{1}{n}\right)\right|$ $= \sum_{n=1}^{\infty}\frac{1}{n}$ is the divergent harmonic series.

Geometric Series

A geometric series can be very useful in mathematics, and easy to work with. If a is a real number, then a geometric series is of the form:

$\sum_{n=0}^{\infty} a(r)^n$. This series converges only if $-1 < r < 1$.

Let S = $\sum_{m=0}^{n} ar^m = (a + ar + ar^2 + \cdots + ar^n)$ be the first (n + 1) terms of a geometric series.

Now, $(1 - r)S = (a + ar + ar^2 + \cdots + ar^n) -$
$$\qquad\qquad (ar + ar^2 + ar^3 + \cdots + ar^{n+1})$$
$$\qquad\quad = a(1 - r^{n+1}).$$

So, $S = \dfrac{a(1-r^{n+1})}{(1-r)}$ is the sum of the first (n + 1) terms.

Since the $\lim\limits_{n\to\infty} r^{n+1} = 0$ because $|r| < 1$,

So, the sum of a geometric series is: $\lim\limits_{n\to\infty} S = \dfrac{a}{(1-r)}$.

Note that a convergent geometric series is always absolutely convergent, because for $0 < r < 1$, the following is true:
the $|-r^n| = |(-1)^n| \cdot |r^n| = |r^n|$.

Example (1):

Find the sum of the geometric series $\sum_{n=0}^{\infty}(15)\left(\dfrac{1}{3}\right)^n$.

Using our formula: $\sum_{n=0}^{\infty}(15)\left(\dfrac{1}{3}\right)^n = \dfrac{15}{1-\frac{1}{3}} = 22.5$.

Example (2):

Find the sum of the geometric series $\sum_{n=0}^{\infty}(10)\left(-\dfrac{1}{4}\right)^n$.

Using our formula: $\sum_{n=0}^{\infty}(10)\left(-\dfrac{1}{4}\right)^n = \dfrac{10}{1+\frac{1}{4}} = 8$.

<u>Example (3)</u>:
Recall the discussion of the measure of the rational numbers
Q in section 2.7. There we stated that the measure of any
countable set within the real numbers is always 0 (in this
case countably infinite, because the term countable can
be used with finite sets). Therefore, we stated that the
measure of Q is zero. We can now easily show this to be
true using geometric series:

We will consider a so-called open cover of Q, a collection
of open sets that covers Q. Every rational number will be
inside a corresponding small open set in the cover. Certainly
then, the measure of the rational numbers will always be
less than the measure of the open cover that contains them.
Recall that Q, just like any countably infinite set, can in
principle be listed. So let $Q = \{q_0, q_1, q_2, q_3, \ldots\}$. Choose
an arbitrarily small positive number ε, $0 < \varepsilon < 1$. Now let
$\left(q_n - \frac{\varepsilon}{2}\left(\frac{1}{2}\right)^n, \ q_n + \frac{\varepsilon}{2}\left(\frac{1}{2}\right)^n\right)$ be the open cover of the nth
member of the listing of Q, (n = 0,1,2,3, ...). It is just an open
interval of real numbers that covers q_n since it contains q_n
(right in the middle of it). Note that each set in the cover
actually contains an infinite number of rationals, but we are
just concerned with the fact that it contains q_n. The width
of the cover of q_n is simply $2\frac{\varepsilon}{2}\left(\frac{1}{2}\right)^n = \varepsilon\left(\frac{1}{2}\right)^n$, for each n.
Now, obviously the measure of this open cover is greater
than the measure of Q. The measure of this cover of Q is no
more than the sum of the geometric series $\sum_{n=0}^{\infty} \varepsilon\left(\frac{1}{2}\right)^n$.

The sum of this series is $\left(\dfrac{\varepsilon}{1-\frac{1}{2}}\right) = 2\varepsilon.$ So, if we let $\varepsilon \to 0$, we can see that the magnitude of the open cover of $Q \to 0$.

Therefore, the measure of Q is 0.

Power Series
A series of the form $\sum_{n=0}^{\infty} a_n(x-c)^n$ is called a power series about x = c. This is also called a Taylor Series and it is especially useful for representing transcendental functions which are infinitely differentiable, and continuous in an open interval containing c. If c = 0, then we have a special type of series called a MacLaurin Series. A MacLaurin series will be the only type of power series that we will investigate in this section, and we will demonstrate how to find the MacLaurin power series representations for the functions f(x) = e^x, cos(x), and sin(x). A power series will, in general, converge for some values of x and diverge for others. In general, a MacLaurin power series will converge on an interval of the form $(-R, R)$, which we call its interval of convergence. $R > 0$ is called the radius of convergence.

Finding the MacLaurin Series for a function f(x):
Define $f^{(n)}(x)$ to be the nth derivative of f(x), where $f^{(0)}(x)$ = f(x) itself. The MacLaurin series for f(x) is a power series of the form $\sum_{n=0}^{\infty} a_n x^n$, and we have to find a formula for the a_n:

$$f^{(0)}(x) = (a_0 + a_1 x + a_2 x^2 + a_3 x^3 + a_4 x^4 + \cdots)$$
$$f^{(1)}(x) = ((1)a_1 + (2)a_2 x + (3)a_3 x^2 + (4)a_4 x^3 + \cdots)$$
$$f^{(2)}(x) = ((2)(1)a_2 + (3)(2)a_3 x + (4)(3)a_4 x^2 + \cdots)$$
$$f^{(3)}(x) = ((3)(2)(1)a_3 + (4)(3)(2)a_4 x + \cdots)$$
$$f^{(4)}(x) = ((4)(3)(2)(1)a_4 + (5)(4)(3)(2)a_5 x + \cdots)$$
$$\vdots$$

Therefore,
$$f^{(0)}(0) = a_0, \; f^{(1)}(0) = (1)a_1, \; f^{(2)}(0) = (2)(1)a_2,$$
$$f^{(3)}(0) = (3)(2)(1)a_3, \; f^{(4)}(0) = (4)(3)(2)(1)a_4, \; \cdots$$

From this we can see the pattern that is emerging:
$$a_0 = \frac{f^{(0)}(0)}{0!}, \; a_1 = \frac{f^{(1)}(0)}{1!}, \; a_2 = \frac{f^{(2)}(0)}{2!},$$
$$a_3 = \frac{f^{(3)}(0)}{3!}, \; a_4 = \frac{f^{(4)}(0)}{4!}, \; \cdots \; ,$$

where for an integer $n \geq 1$, $n! = (n)(n-1)\cdots(3)(2)(1)$.
0! is defined to be 1.

Therefore, for a function f(x) that is infinitely differentiable at 0, and continuous in an open interval containing 0, the MacLaurin series coefficient a_n is:

$$a_n = \frac{f^{(n)}(0)}{n!}, \text{ and we can represent f(x) as:}$$

f(x) = $\sum_{n=0}^{\infty} \left(\frac{f^{(n)}(0)}{n!} \right) x^n$, on some interval $(-R, R)$.

The Ratio Test

This test allows us to find the radius of convergence R for a MacLaurin series. We know that for a geometric series, the $\left|\frac{a_{n+1}}{a_n}\right| = \left|\frac{ar^{n+1}}{ar^n}\right| = r$, and that if $|r| < 1$, then the geometric series will be absolutely convergent. For a MacLaurin series, we determine the interval of convergence using a similar line of reasoning. We determine the set of x-values, which is the interval of convergence, by setting the following limit to be less than 1:

That is, set: $\lim\limits_{n \to \infty} \left|\frac{(n+1)^{th}\ term}{(n)^{th}\ term}\right| < 1$.

When the $\lim\limits_{n \to \infty} \left|\frac{(n+1)^{th}\ term}{(n)^{th}\ term}\right|$

$$= \lim\limits_{n \to \infty} \left|\frac{a_{n+1}x^{n+1}}{a_n x^n}\right|$$

$$= |x| \cdot \lim\limits_{n \to \infty} \left|\frac{a_{n+1}}{a_n}\right| < 1,$$

then the Maclaurin series will be absolutely convergent on a certain interval of the real numbers, in much the same way that a geometric series will be absolutely convergent when $\left|\frac{a_{n+1}}{a_n}\right| = |r| < 1$.

MacLaurin Series for e^x

For f(x) = e^x, $f^{(n)}(0) = e^0 = 1$, for n = (0, 1, 2, 3, ...).

Therefore, $a_n = \frac{f^{(n)}(0)}{n!} = \frac{1}{n!}$, for n = (0, 1, 2, 3, ...). So,

313

the MacLaurin series for e^x is:

$$e^x = \sum_{n=0}^{\infty} \frac{x^n}{n!}.$$

To find the interval of convergence, we will use the ratio test: Set $\lim_{n\to\infty} \left| \frac{x^{n+1}}{(n+1)!} \cdot \frac{(n)!}{x^n} \right| = |x| \cdot \left(\lim_{n\to\infty} \left(\frac{1}{n+1} \right) \right) < 1$. This says that: $|x| \cdot (0) < 1$, which says that $|x| < \frac{1}{0} = \infty$.

So, we say that the radius of convergence is $R = \infty$, and therefore, the interval of convergence is $(-\infty, \infty)$.

One more interesting point considering the number e. If we plug x = 1 into the series for e^x, then we have another way of determining e to any number of decimal places.

$$e = e^1 = \sum_{n=0}^{\infty} \frac{(1)^n}{n!} = \sum_{n=0}^{\infty} \frac{1}{n!} = \left(1 + 1 + \frac{1}{2} + \frac{1}{6} + \frac{1}{24} + \cdots \right)$$

MacLaurin Series for cos(x)
For f(x) = cos(x),
$f^{(0)}(x) = \cos(x)$, $f^{(1)}(x) = -\sin(x)$,
$f^{(2)}(x) = -\cos(x)$, $f^{(3)}(x) = \sin(x)$,
$f^{(4)}(x) = \cos(x)$, $\cdots$

So, we have:
$f^{(0)}(0) = \cos(0) = 1$, $f^{(1)}(0) = -\sin(0) = 0$,
$f^{(2)}(0) = -\cos(0) = -1$, $f^{(3)}(0) = \sin(0) = 0$,
$f^{(4)}(0) = \cos(0) = 1$, $\cdots$

Then, we can write the MacLaurin series for f(x) = cos(x) by looking at the pattern:

$$\cos(x) = \sum_{n=0}^{\infty} \left(\frac{f^{(n)}(0)}{n!}\right) x^n$$

$$= \sum_{n=0}^{\infty} (-1)^n \frac{x^{2n}}{(2n)!}$$

$$= \left(1 - \frac{x^2}{2!} + \frac{x^4}{4!} - \frac{x^6}{6!} + \cdots\right)$$

The interval of convergence is determined by the ratio test:

The $\lim\limits_{n\to\infty} \left|\dfrac{(-1)^{n+1}x^{2n+2}}{(2n+2)!} \cdot \dfrac{(2n)!}{(-1)^n x^{2n}}\right|$

$$= |x^2| \cdot \lim_{n\to\infty} \left|\frac{1}{(2n+1)(2n+2)}\right| = (|x|)^2 \cdot \left(\lim_{n\to\infty} \left|\frac{1}{(2n+1)(2n+2)}\right|\right)$$

$= (|x|)^2 \cdot (0)$, and we want this to be < 1.

So we're interested in where:
$|x|^2 < \frac{1}{0} = \infty$. Clearly, this is when $|x| < \infty$. So the radius of convergence $R = \infty$, and the interval of convergence is therefore $(-\infty, \infty)$.

MacLaurin Series for sin(x)
For f(x) = sin(x),
$f^{(0)}(x) = \sin(x)$, $f^{(1)}(x) = \cos(x)$,
$f^{(2)}(x) = -\sin(x)$, $f^{(3)}(x) = -\cos(x)$,
$f^{(4)}(x) = \sin(x)$, $\cdots$

So, we have:

315

$f^{(0)}(0) = \sin(0) = 0,\ f^{(1)}(0) = \cos(0) = 1,$
$f^{(2)}(0) = -\sin(0) = 0,\ f^{(3)}(0) = -\cos(0) = -1,$
$f^{(4)}(0) = \sin(0) = 0,\ \cdots$

Then, we can write the MacLaurin series for f(x) = sin(x) by looking at the pattern:

$$\sin(x) = \sum_{n=0}^{\infty} \left(\frac{f^{(n)}(0)}{n!} \right) x^n$$

$$= \sum_{n=0}^{\infty} (-1)^n \frac{x^{2n+1}}{(2n+1)!}$$

$$= \left(x - \frac{x^3}{3!} + \frac{x^5}{5!} - \frac{x^7}{7!} + \cdots \right)$$

The interval of convergence is determined by the ratio test:

The $\lim\limits_{n\to\infty} \left| \dfrac{(-1)^{n+1}x^{2n+3}}{(2n+3)!} \cdot \dfrac{(2n+1)!}{(-1)^n x^{2n+1}} \right|$

$= |x^2| \cdot \lim\limits_{n\to\infty} \left| \dfrac{1}{(2n+2)(2n+3)} \right| = (|x|)^2 \cdot \left(\lim\limits_{n\to\infty} \left| \dfrac{1}{(2n+2)(2n+3)} \right| \right)$

$= (|x|)^2 \cdot (0)$, and we want this to be < 1.

So we're interested in where:
$|x|^2 < \frac{1}{0} = \infty$. Clearly, this is when $|x| < \infty$. So the radius of convergence $R = \infty$, and the interval of convergence is therefore $(-\infty, \infty)$.

Euler's Identities

If $i = \sqrt{-1}$ (the imaginary unit) and since $f^{(n)}(x) = e^x$ for the function $f(x) = e^x$ ($n = 0, 1, 2, \ldots$), the MacLaurin series for $f(x) = e^{ix}$ (leading to the first form of Euler's identity) is:

$$e^{ix} = \sum_{n=0}^{\infty} \frac{f^{(n)}(0)}{n!}(ix)^n =$$

$$= \left(1 + i\frac{x}{1!} - \frac{x^2}{2!} - i\frac{x^3}{3!} + \frac{x^4}{4!} + i\frac{x^5}{5!} - \frac{x^6}{6!} - i\frac{x^7}{7!} + \cdots\right)$$

$$= \left(1 - \frac{x^2}{2!} + \frac{x^4}{4!} - \frac{x^6}{6!} + \cdots\right) +$$
$$i\left(x - \frac{x^3}{3!} + \frac{x^5}{5!} - \frac{x^7}{7!} + \cdots\right)$$

$$= \cos(x) + i\sin(x) .$$

It should be clear that the second form for Euler's identity: $e^{-ix} = \cos(x) - i\sin(x)$, can be derived similarly.

This proves Euler's identities, which we used in section (4.2).

Differentiation and Integration

Consider the MacLaurin series for an infinitely differentiable transcendental function $f(x)$:

$f(x) = \sum_{n=0}^{\infty}\left(\frac{f^{(n)}(0)}{n!}\right)x^n$. We know that this series is absolutely convergent on an interval of the form $(-R, R)$.

Now consider the sequence of partial sums for this series. This would be the sequence of functions $\{S_n(x)\}$ on the interval $(-R, R)$, where :

$$\{S_n(x)\} = \left\{ \sum_{m=0}^{n} \left(\frac{f^{(m)}(0)}{m!} \right) x^m \right\}.$$

It turns out that this sequence of functions can be shown to be uniformly convergent to the function f(x) on $(-R, R)$. Recall uniform convergence from section (4.5). Because of the uniform convergence it follows that:

$$\frac{d}{dx} f(x) = \frac{d}{dx} \left(\lim_{n \to \infty} S_n(x) \right) = \lim_{n \to \infty} \left(\frac{d}{dx} (S_n(x)) \right)$$

and $\int f(x) dx = \int \left(\lim_{n \to \infty} S_n(x) \right) dx = \lim_{n \to \infty} \left(\int S_n(x) dx \right),$

on $(-R, R)$. If we wish to evaluate the derivative of f(x), or calculate a definite integral of f(x), then we can find the derivative or integral of f(x) to any desired degree of accuracy by differentiating term by term or by integrating term by term a member from the sequence of functions $\{S_n(x)\}$, because this sequence converges uniformly to f(x) on $(-R, R)$.

This concludes our summary of both trigonometry and the calculus of a single variable, and brings to a conclusion our treatment of the fundamentals of modern mathematics as presented in this book. It is hoped that the discussion of the past four chapters will prepare the reader for possibly further courses in mathematics. For those more interested in statistics, by reading the rest of the book the hope is that the student will also have all the tools necessary to help them in further courses in statistics as well. Perhaps, for many who read this book in its entirety it would also help them in their research work in the sciences.

PART 2

PROBABILITY

The next two chapters present the essentials of probability theory from an intuitive and algebraic point of view. This underlies the subject of statistics in the third section of the book. Chapter 5 on combinatorics discusses the basic tools for counting possibilities, the number of ways that things can happen. This is very important in chapter 6 where we calculate probabilities and learn about the important topic of a probability distribution, which is indispensable for an understanding of the rest of the book.

(5) COMBINATORICS

(5.1) Fundamentals of Combinatorics

<u>The Fundamental Principle of Counting (FPC)</u>
This principle says that for two events A and B, if A can happen in N_A ways and B can happen in N_B ways, then events A and B can happen in $(N_A \cdot N_B)$ ways. This can be extended to more than two events.

For example, suppose a Delicatessen offers a choice of 3 types of bread, 10 different types of meat, and 7 different condiment choices, where we label these three choices events A, B, and C. Then the number of customized sandwiches that a customer could order is
$(N_A \cdot N_B \cdot N_C) = (3 \cdot 10 \cdot 7) = 210$.

<u>Permutations P(n,r)</u>
At this point we will define a quantity called (n factorial), which is denoted by n! and defined to be:
n! = (n)(n − 1)(n − 2) $\cdots$ (3)(2)(1).

For example:
1! = 1
2! = (2)(1) = 2
3! = (3)(2)(1) = 6
4! = (4)(3)(2)(1) = 24

and so on.

We define 0! = 1 (because it doesn't cause any difficulties – that kind of thing happens sometimes within mathematics).

Factorials get large very quickly as n increases. 10! is already greater than 3 million, and the largest factorial that my calculator can handle is only about 60!

Now to permutations:
Suppose we have n different objects, and we want to choose r of them at random, where $r \leq n$, and the order matters. Each set of r distinct objects chosen is called a permutation of r things chosen from n things, denoted P(n,r). We want to determine how many different such collections of r things we could choose, where the order matters. Well, imagine r different boxes that we will place the r things in (one object per box). We use the FPC in the following way to figure this out. There are n different choices for the first box, (n - 1) choices that remain for the second box, (n − 2) choices that remain for the third box, and so on. When we get to the rth box, there are (n − r + 1) remaining choices. So, the number of ways that the r objects could be chosen in this way from n objects, where order matters is:

P(n,r) = (n)(n − 1)(n − 2) $\cdots$ (n − r + 1) different ways.

Note that this can be written: $P(n,r) = \dfrac{n!}{(n-r)!}$.

Example (1):

$$P(10, 3) = \frac{10!}{(10-3)!} = \frac{10!}{7!} = \frac{(10(9)(8)(7)(6)(5)(4)(3)(2)(1)}{(7)(6)(5)(4)(3)(2)(1)}$$

$$= \frac{(10)(9)(8)\cdot(7!)}{(7!)} = (10)(9)(8) = 720.$$

So, if we have 10 objects and we randomly choose 3 of them, where order is important, there are 720 ways that this can be done.

Example (2):

$$P(9, 4) = \frac{9!}{(9-4)!} = \frac{9!}{5!} = \frac{(9)(8)(7)(6)(5)(4)(3)(2)(1)}{(5)(4)(3)(2)(1)}$$

$$= \frac{(9)(8)(7)(6)(5!)}{(5!)} = (9)(8)(7)(6) = 3024.$$

So, if we have 9 objects and we randomly choose 4 of them, where order is important, there are 3024 ways that this can be done.

Example (3):
P(10,10) is the number of permutations of 10 objects chosen from 10 objects, which is simply the number of permutations of 10 objects. This is:

$$P(10,10) = \frac{10!}{(10-10)!} = \frac{10!}{0!} = \frac{10!}{1} = 10! = 3,628,800.$$

Combinations C(n,r)

Now we want to determine the number of ways that we can choose r things from n things where order is not important, $r \le n$. We might also think of this as the number of subsets of r things that there are in a set of n things. Still another way of interpreting this is the number of ways that we could partition a set of n objects into two sets, one with r things in it and the other with (n − r) things in it. This is called the number of combinations of r things from a set of n things, denoted C(n,r). This is the most important result in this chapter.

The solution of this problem is easy if we realize that for a given one of the P(n,r) permutations, how many of the P(n,r) permutations of r things consist of the same r objects (but just in a different order). Obviously, this is (r!). So, all we have to do is divide P(n,r) by r! to arrive at the number of groups of permutations of r things chosen from n things that are the same except for order, i.e. C(n,r). So,

$$C(n,r) = \frac{P(n,r)}{r!} = \frac{n!}{(n-r)! \cdot (r)!} \; .$$

Example (1):

How many ways can we choose at random 3 people from a collection of 12 people, where the order that the three people was selected (or the particular collection of 3 people selected) is unimportant. In other words, how many subsets of 3 people are there among the 12 people. The answer is:

$$C(12,3) = \frac{12!}{(12-3)! \cdot (3)!} = \frac{12!}{9! \cdot 3!} = \frac{(12)(11)(10)}{6} = 220.$$

Example (2):

How many ways can we choose at random 5 cars from a collection of 8 cars, where the order that the 5 cars was selected (or the particular collection of 5 cars selected) is unimportant. In other words, how many subsets of 5 cars are there among the 8 cars. The answer is:

$$C(8,5) = \frac{8!}{(8-5)! \cdot (5)!} = \frac{8!}{3! \cdot 5!} = \frac{(8)(7)(6)}{6} = 56.$$

Example (3):

(A) How many ways can 0 people be chosen from 15?
(B) How many ways can 1 person be chosen from 15?
(C) How many ways can 14 people be chosen from 15?
(D) How many ways can 15 people be chosen from 15?

Solutions:

(A) $C(15,0) = \dfrac{15!}{(15-0)! \cdot (0)!} = 1$ way. We can choose nobody in one way.

(B) $C(15,1) = \dfrac{15!}{(15-1)! \cdot (1)!} = \dfrac{15!}{14! \cdot 1!} = 15$ ways. There are 15 ways to choose one person.

(C) $C(15,14) = \dfrac{15!}{(15-14)! \cdot (14)!} = \dfrac{15!}{(1)! \cdot (14)!} = 15$ ways. This amounts to figuring the number of ways that we can choose some people from the 15, leaving one person out each time.

(D) $C(15,15) = \dfrac{15!}{(15-15)! \cdot (15)!} = \dfrac{15!}{0! \cdot 15!} = 1$ way. There is only one way to choose everybody all at once.

From this last example, we can notice a symmetry with calculations of the number of combinations:

$$C(n, r) = \dfrac{n!}{(n-r)! \cdot r!} = \dfrac{n!}{(n-(n-r))! \cdot (n-r)!} = \dfrac{n!}{r! \cdot (n-r)!} = C(n, n-r).$$

For example, we see from above that C(15,0) = C(15,15), and C(15,1) = C(15,14). It is true that C(15,2) = C(15,13), C(15,7) = C(15,8), and so on.

Binomial Expansions
A binomial expansion is an expression of the form $(a + b)^n$.

$(a + b)^1 = a + b.$
$(a + b)^2 = a^2 + 2ab + b^2.$
$(a + b)^3 = a^3 + 3ab^2 + 3a^2b + b^3.$
and so on.

It can be shown that (and we will prove that this is true):

$(a+b)^n = \sum_{x=0}^{n} \binom{n}{x} a^x b^{n-x}$, for n = 1, 2, 3, ...

The symbolism $\binom{n}{x}$ is called a binomial coefficient and is quite common in books on probability and statistics. We pronounce it "n choose x." It turns out that $\binom{n}{x} = C(n,x)$; the binomial coefficients are just combinations. With some thought, one can see why this is the case. The powers on "a" and "b" in each term of the expansion are (x) and (n-x), which add to n. The binomial coefficients are just the number of ways that the term could contain "a" (x) times and "b" (n − x) times.

To prove the following theorem, we will once again use the technique of proof by mathematical induction, which we learned in chapter 4.

<u>Theorem</u>: $(a+b)^n = \sum_{x=0}^{n} \binom{n}{x} a^x b^{n-x}$.

<u>Part (A) of Proof</u>:

When n = 1: $(a+b)^1 = \binom{1}{0} a^0 b^{1-0} + \binom{1}{1} a^1 b^{1-1}$

$$= \sum_{x=0}^{1} \binom{1}{x} a^x b^{1-x} ,$$

which shows that the theorem (proposition) is true for n = 1.

<u>Part (B) of Proof</u>:
Suppose that the theorem is true for some n ≥ 1,

that is, $(a + b)^n = \sum_{x=0}^{n} \binom{n}{x} a^x b^{n-x}$, for some $n \geq 1$. Then we need to prove it is true for $(n + 1)$.

$$(a + b)^{n+1} = \left(\sum_{x=0}^{n} \binom{n}{x} a^x b^{n-x}\right) \cdot (a + b)$$

$$= \sum_{x=0}^{n} \binom{n}{x} a^{x+1} b^{n-x} + \sum_{x=0}^{n} \binom{n}{x} a^x b^{n-x+1}$$

$$= \{ \binom{n}{0} a^1 b^n + \binom{n}{1} a^2 b^{n-1} + \binom{n}{2} a^3 b^{n-2} + \cdots$$
$$+ \binom{n}{n-2} a^{n-1} b^2 + \binom{n}{n-1} a^n b^1 + \binom{n}{n} a^{n+1} b^0 \} +$$

$$\{ \binom{n}{0} a^0 b^{n+1} + \binom{n}{1} a^1 b^n + \binom{n}{2} a^2 b^{n-1} + \cdots$$
$$+ \binom{n}{n-2} a^{n-2} b^3 + \binom{n}{n-1} a^{n-1} b^2 + \binom{n}{n} a^n b^1 \}$$

$$= \{ \binom{n}{0} a^0 b^{n+1} +$$

$$\left[\binom{n}{0} + \binom{n}{1}\right] a^1 b^n + \left[\binom{n}{1} + \binom{n}{2}\right] a^2 b^{n-1} + \cdots +$$

$$\left[\binom{n}{n-2} + \binom{n}{n-1}\right] a^{n-1} b^2 + \left[\binom{n}{n-1} + \binom{n}{n}\right] a^n b^1 +$$

$$\binom{n}{n} a^{n+1} b^0 \}.$$

We need to prove another proposition at this point that will help us to simplify what we have above:

We need to prove that for $x = 1, 2, 3, \ldots, n$,

$$\binom{n+1}{x} = \binom{n}{x-1} + \binom{n}{x}$$

Or that:
$$\frac{(n+1)!}{(n+1-x)! \cdot (x)!} = \frac{n!}{(n+1-x)! \cdot (x-1)!} + \frac{n!}{(n-x)! \cdot (x)!}$$

Multiplying both sides of this equation by $[(n + 1 - x)! \cdot (x)!]$:

$(n + 1)! = (n! \cdot x) + (n! \cdot (n + 1 - x))$

$\quad\quad = (n! \cdot x) - (n! \cdot x) + (n + 1)n!$

$\quad\quad = (n + 1)!$

So, the result is true.

Now we can use this last combinatorial result to simplify what we were working on. Note that:

$$\binom{n}{0} = \binom{n+1}{0} \quad \text{and} \quad \binom{n}{n} = \binom{n+1}{n+1}.$$

$$(a + b)^{n+1} = \left\{ \binom{n+1}{0} a^0 b^{n+1} \right.$$

$$+ \left[\binom{n+1}{1} \right] a^1 b^n + \left[\binom{n+1}{2} \right] a^2 b^{n-1} + \cdots$$

$$+ \left[\binom{n+1}{n-1} \right] a^{n-1} b^2 + \left[\binom{n+1}{n} \right] a^n b^1 +$$

$$+ \binom{n+1}{n+1} a^{n+1} b^0 \Big\}$$

$$= \sum_{x=0}^{n+1} a^x b^{(n+1)-x}$$

Therefore, the proposition is true for $(n + 1)$.

Therefore, by induction the proposition is true for all n ≥ 1.

If a = 1 and b = 1, then we have:

$$2^n = \binom{n}{0} + \binom{n}{1} + \binom{n}{2} + \cdots + \binom{n}{n-2} + \binom{n}{n-1} + \binom{n}{n}.$$

So, if we have any set with n members, the right-hand side is the sum of the number of subsets with 0, 1, 2, . . . , (n – 2), (n – 1), and (n) members. In other words, the right-hand side is the number of subsets of a set of cardinality n, and it equals 2^n.

Partitioning N objects into n ≥ 3 Non-Empty Subsets
If we have a collection of N objects and we wish to partition them into n (n ≥ 3) non-empty subsets of sizes $\{x_1, x_2, \dots, x_n\}$, where N = $(x_1 + x_2 + \cdots + x_n)$. It can be done in the following way:

For purposes of illustration, choose n = 4. The number of ways that we could choose x_1 of the objects to be in the first subset is $\binom{N}{x_1}$. Then we could choose x_2 objects from the remaining (N - x_1) objects in $\binom{N - x_1}{x_2}$ ways. Then we could choose x_3 objects from the remaining $(N - x_1 - x_2)$ objects in $\binom{N - x_1 - x_2}{x_3}$ ways. Then finally, we could choose x_4 objects from the remaining $(N - x_1 - x_2 - x_3)$ objects in

$\left(\begin{array}{c} N - x_1 - x_2 - x_3 \\ x_4 \end{array}\right)$ ways. So, to figure the number of ways w of choosing the 4 subsets, multiply all of the combinations together to get:

$$w = \binom{N}{x_1} \cdot \binom{N - x_1}{x_2} \cdot \binom{N - x_1 - x_2}{x_3} \cdot \binom{N - x_1 - x_2 - x_3}{x_4}$$

ways as the answer. This is figured by writing out each of these combinations into their factorial definitions: The result would be:

$$w = \left[\frac{N!}{(N-x_1)!(x_1)!}\right]$$
$$\cdot \left[\frac{(N-x_1)!}{(N-x_1-x_2)!(x_2)!}\right]$$
$$\cdot \left[\frac{(N-x_1-x_2)!}{(N-x_1-x_2-x_3)!(x_3)!}\right]$$
$$\cdot \left[\frac{(N-x_1-x_2-x_3)!}{(N-x_1-x_2-x_3-x_4)!(x_4)!}\right] \text{ ways.}$$

(Note that $(N - x_1 - x_2 - x_3 - x_4) = 0$)
After doing all the algebraic cancellations, we have:

$$w = \frac{N!}{(x_1)!(x_2)!(x_3)!(x_4)!} \text{ ways that the N objects could be}$$

partitioned as desired into the four non-empty subsets. We have a notation for this which makes sense. We write:

$$w = \binom{N}{x_1, x_2, x_3, x_4}. \text{ This is called a multinomial}$$
coefficient.

<u>Example (1)</u>:
Suppose we have N = 12 objects and we want to partition them into 4 subsets of sizes 5, 2, 3, and 2. This can be done in:

$$w = \frac{12!}{5! \cdot 2! \cdot 3! \cdot 2!} = \frac{(479,001,600)}{(120)(2)(6)(2)} = 166,320 \text{ ways.}$$

Noting the similarity in the meaning of combinations and multinomial coefficients, we could write C(n,r) as $\binom{n}{r, n-r}$.

<u>Partitioning a Sequence of N Objects into K Subsets</u>
In this section, we want to figure the number of ways that N ordered objects (a sequence of N objects that are in a specific order) could be partitioned into K ≥ 3 subsets, where some of the subsets could be empty, and N > K.

As an example, suppose we have N = 10 objects that we can put into a horizontal list and we want to find the number of ways that we could partition that ordered list into K = 5 subsets (some possibly empty). So, the list is:

$$x_1 \quad x_2 \quad x_3 \quad x_4 \quad x_5 \quad x_6 \quad x_7 \quad x_8 \quad x_9 \quad x_{10}$$

Now, if we have (K − 1) = 4 dividers and they are put into the list as if they were members of the list, we would have a list containing 14 items, all of which we could view as the same, like these 14 circles:

o o o o o o o o o o o o o o

For the 14 circles, consider the number of ways that 4 of those circles could be selected to be dividers. Using vertical lines as dividers along with the original $x_i's$ in order, we could have:

$$| \quad | \quad x_1 \; x_2 \; x_3 \; x_4 \; x_5 \; x_6 \; | \; x_7 \; x_8 \; x_9 \; | \; x_{10}$$

or we could have:

$$x_1 \; x_2 \; | \; x_3 \; x_4 \; x_5 \; | \; x_6 \; x_7 \; x_8 \; x_9 \; | \; x_{10} \; |$$

or we could have:

$$x_1 \; | \quad x_2 \; x_3 \; | \; x_4 \; x_5 \; x_6 \; | \; x_7 \; x_8 \; | \; x_9 \; x_{10}$$

as three of the possibilities.

The K = 5 subsets are thought of this way: Subset 1 is the collection of objects to the left of the first divider. Subsets 2 through 4 would be the (K − 2) collections of things between the (K − 1) dividers, and subset 5 would be the collection of objects to the right of the last divider.

Then in the first possibility, the 5 subsets would be, using { } for the null set:

$$\{ \}, \; \{ \}, \; \{ x_1, x_2, x_3, x_4, x_5, x_6 \}, \; \{ x_7, x_8, x_9 \}, \; \{ x_{10} \} \; .$$

In the second possibility, the 5 subsets would be:

$\{x_1, x_2\}, \{x_3, x_4, x_5\}, \{x_6, x_7, x_8, x_9\}, \{x_{10}\}, \{\}$.

In the third possibility, the 5 subsets would be:

$\{x_1\}, \{x_2, x_3\}, \{x_4, x_5, x_6\}, \{x_7, x_8\}, \{x_9, x_{10}\}$.

Clearly, there are $\binom{N + K - 1}{K - 1} = \binom{10 + 5 - 1}{5 - 1} = \binom{14}{4} =$ 1001 ways that these K = 5 subsets could be created from the N = 10 ordered objects.

So, we have figured the number of ways that a sequence of N objects can be partitioned into K subsets, where N > K, and where some of the subsets could be empty.

To proceed a little further, notice that x_1 and x_{10} could not be in the same subset just by themselves. If we wanted to know the number of ways that we could have K subsets of the N objects (which includes empty subsets) for all sequences of the N objects, we just have to consider the N! = 10! permutations of the N = 10 objects:

Then the number of ways w is: $(N!) \cdot \binom{N + K - 1}{K - 1}$.

For the example above, the number of ways is:
    w = (10!) · (1001) = 3,632,428,800 ways.

################# Exercises #################

Calculate each of the following:

(1) P(10,6)   (2) P(12,2)   (3) P(8,4)

(4) $\binom{6}{2}$ (5) $\binom{10}{3}$ (6) $\binom{11}{7}$

(7) $\binom{10}{2,3,5}$ (8) $\binom{14}{6,4,4}$ (9) $\binom{15}{4,3,5,3}$

(10) $\left(\binom{20}{1} + \binom{20}{0} \right)$ (11) $\binom{15}{14}$ (12) $\binom{5}{5}$

(13) How many subsets are there for a set containing 12 objects?

(14) (a) How many subsets are there for the set {a, b, c}?
     (b) List them.

(15) How many permutations of 5 different objects are there if they are lined up side by side?

(6) <u>FUNDAMENTALS OF PROBABILITY</u>

(6.1) <u>Probability Modeling</u>

The beginning of the study of probability is some kind of "random experiment," such as the throwing of dice, the dealing of a hand of cards, the tossing of a coin, or as is more usually the case here in this book, some scientific experiment, where the result of the experiment cannot be known beforehand. Probability is a quantification of the chance of occurrence associated with all possible outcomes of the experiment.

The probability or chance of some event A associated with an experiment, denoted Pr(A), is a number from 0 to 1, or a percentage from 0% to 100%. These are the two different ways of expressing a probability.

If the Pr(A) = 0, then event A is impossible. If the Pr(A) = 1, then event A is certain to occur. For most events A that are of interest to us, the Pr(A) is somewhere between 0 and 1. When we have two events A and B, if $0 \leq Pr(A) < Pr(B) \leq 1$, we say that event A is less likely to occur than event B, or that event B is more likely to occur than event A.

<u>Sample Space, Event Space, Probability Measure</u>
The set of all outcomes of an experiment is called the
sample space S, and the members of S are not necessarily
numeric. The sample space is in reality always a finite set.
In some sample spaces the outcomes are equally likely,
but in others the outcomes are not equally likely. We will
denote a sample space S in the following way:

$$S = \{s_1, s_2, \ldots, s_M\},$$

where the cardinality of S is a finite number M. Each
outcome has a probability which we can write as p_i. We
require that each p_i (where i = 1,2,3,...,M) cannot be a
negative number, and that the sum of all of them must
equal one.

An event is any subset of S, including the null set $\varnothing$ and
the sample space S itself. Events include each of the single
outcomes of S, as well as any other subset of S. We will let E
denote all the events associated with a given sample space S
and call it the event space. The probability of every event A
in the event space E is determined by a probability measure
Pr(A), which assigns a probability to each event A in E.
Sometimes an event is a single outcome in S, but in other
cases an event consists of more than one outcome in S. In
the case that all of the outcomes in S are equally likely, then
we have the following as our probability measure Pr(A) for
the event A:

$$Pr(A) = \frac{Cardinality(A)}{Cardinality(S)}, \text{ which is also written } Pr(A) = \frac{N(A)}{N(S)}.$$

So, when the outcomes in S are equally likely, a probability just boils down to counting outcomes in S. But this is not the case in general. The determination of a probability does not always boil down to counting outcomes in S, and we will see this in later chapters where the determination of chance will involve a function where it is assumed that the number of members in S is countably or uncountably infinite. This is many times the case when M is very large. In these cases, we will see how infinity can be used to our advantage because it leads to a simplification of the situation.

The triple {S, E, Pr(A)} is called a Probability Space. Note that Pr($\emptyset$) = 0 and Pr(S) = 1, and for any other event A in E, $0 \leq Pr(A) \leq 1$. When S is an equally likely sample space, the combinatorial results discussed in Chapter 5 can figure prominently in the calculation of many probabilities.

Example (1):
Suppose we throw two dice, one of which is green and the other red. We denote an outcome by an ordered pair like (2,5), where the first number is the outcome of the green die and the second number is the outcome of the red die. So, (2,5) means that the green die landed 2 and the red die landed 5. The number of possible outcomes for each die is 6, so there are $(6) \cdot (6) = 36$ possible outcomes. This is the sample space S for this experiment, we can list the members:

S = {(1,1), (1,2), (1,3), (1,4), (1,5), (1,6),
     (2,1), (2,2), (2,3), (2,4), (2,5), (2,6),

(3,1), (3,2), (3,3), (3,4), (3,5), (3,6),
(4,1), (4,2), (4,3), (4,4), (4,5), (4,6),
(5,1), (5,2), (5,3), (5,4), (5,5), (5,6),
(6,1), (6,2), (6,3), (6,4), (6,5), (6,6)} .

Clearly, all 36 possible outcomes are equally-likely. The sets
$A_1 = \{(1,1), (2,2), (3,3), (4,4), (5,5), (6,6)\}$
$A_2 = \{(3,1), (1,3), (2,2)\}$
$A_3 = \{(6,3), (5,4), (4,5), (3,6)\}$
represent events.

A_1 is the event that the result of a toss is "doubles."
A_2 is the event that the sum of the two dice is 4.
A_3 is the event that the sum of the two dice is 9.

We can denote the number of members in the sets $\emptyset$, S,
A_1, A_2, and A_3 by N($\emptyset$), N(S), N(A_1), N(A_2), and N(A_3).
So, we are denoting the cardinality of an event A by N(A),
a whole number, which is always the case since N(S) is a
whole number.

Here, N($\emptyset$) = 0, N(S) = 36, N(A_1) = 6, N(A_2) = 3, and
N(A_3) = 4. Since all 36 outcomes are equally-likely, it is
clear that:

$$\Pr(\emptyset) = \frac{N(\emptyset)}{N(S)} = \frac{0}{36} = 0, \qquad \Pr(S) = \frac{N(S)}{N(S)} = \frac{36}{36} = 1,$$

$$\Pr(A_1) = \frac{N(A_1)}{N(S)} = \frac{6}{36} = \frac{1}{6}, \qquad \Pr(A_2) = \frac{N(A_2)}{N(S)} = \frac{3}{36} = \frac{1}{12},$$

$$Pr(A_3) = \frac{N(A_3)}{N(S)} = \frac{4}{36} = \frac{1}{9}.$$

Example (2):

Suppose we toss a fair coin three times and denote "Heads" with H and "Tails" with T. Then we can denote the sample space associated with this experiment by:

S = {HHH, HHT, HTH, HTT, THH, THT, TTH, TTT} .

This is our equally likely sample space. When we consider the tossing of a coin, the Pr(H) and the Pr(T) are both equal to $\frac{1}{2}$. So, it follows that the probability for each of the possible outcomes of an experiment of tossing a coin two times, or three times, or any number of times, is equal to that of any other outcome. Since here there are only 8 different possible outcomes, the probability of any single one is $\frac{1}{8}$. Define the collection of mutually exclusive and exhaustive sets $\{B_0, B_1, B_2, B_3\}$ as follows:

$B_0 = \{TTT\}$
$B_1 = \{HTT, THT, TTH\}$
$B_2 = \{HHT, HTH, THH\}$
$B_3 = \{HHH\}$

Since they are all subsets of S, they are members of the event space E associated with S. These events are those of (zero Heads), (one Head), (two Heads), and (three Heads), respectively. This exhausts all possibilities for the number of heads in any given outcome, hence these four events in E will contain all the outcomes in S and they will be disjoint. We can see that:

$N(B_0) = 1$, $N(B_3) = 1$, $N(B_1) = 3$, $N(B_2) = 3$, and $N(S) = 8$.

Therefore, $Pr(B_0) = \frac{N(B_0)}{N(S)} = \frac{1}{8}$, $Pr(B_3) = \frac{N(B_3)}{N(S)} = \frac{1}{8}$, and

$Pr(B_1) = \frac{N(B_1)}{N(S)} = \frac{3}{8}$, $Pr(B_2) = \frac{N(B_2)}{N(S)} = \frac{3}{8}$.

Note that the sum of the probabilities for this mutually exclusive and exhaustive set of events in E is:

$Pr(S) = \sum_{i=0}^{3} Pr(B_i) = \left(\frac{1}{8} + \frac{3}{8} + \frac{3}{8} + \frac{1}{8} \right) = 1$.

Some Probability Calculations

Combinatorial techniques are frequently used to calculate probabilities when the members of the sample space are equally-likely. This idea will be used a lot in this section. In the following examples, we will sometimes be considering the number of things in a set and at other times a definite probability.

Example (3):

A teacher has 25 kids in her second grade class. There are 15 boys and 10 girls.

(A) The teacher wants to select 7 kids at random to participate in a theatrical event, and she needs 3 boys and 4 girls. How many ways can this be done?

The number of ways that 3 boys can be chosen from 15 is $\binom{15}{3} = \left(\frac{15 \cdot 14 \cdot 13}{3!}\right) = 455$. For each of these ways for choosing the boys, there are $\binom{10}{4} = 210$ ways that the girls can be chosen. So, from the fundamental principle of counting (FPC), there are $(455) \cdot (210) = 95{,}550$ ways that she could choose 3 boys and 4 girls from her class.

(B) If the teacher doesn't care how many boys and how many girls are chosen in selecting the 7 kids, what is the probability that after choosing the 7 kids at random she would have 2 boys and 5 girls?

This probability is figured in the following way: We will call the choosing of 2 boys and 5 girls event A. The total number of ways that event A can happen, using combinations and the FPC is $\binom{15}{2} \cdot \binom{10}{5} = (105)(252) = 26{,}460$ ways. The total number of ways that 7 kids can be chosen from 25 is $\binom{25}{7} = 480{,}700$ ways. This is the number of things in the sample space S and they are equally-likely. So, the probability that we are looking for is:

$Pr(A) = \frac{N(A)}{N(S)} = \left(\frac{26{,}460}{480{,}700}\right) = (0.055) = 5.5\%$, which is a low probability. Therefore, the chance that she would randomly select 2 boys and 5 girls from her class is low. It's not very likely to happen. This makes sense because there are 15 boys and 10 girls in the class.

Example (4):
(A) The owner of a used car lot that has 43 cars wants to select a random sample of size n = 5 cars and measure the net weight of each of the 5 cars. He wants to take the sample with replacement, meaning that after a car is chosen and weighed, it is returned to the lot to be possibly chosen again. The dealer has his reasons for wanting to sample in this way, instead of sampling without replacement (which would mean that a selected car is not returned to the lot to possibly be chosen again, which would guarantee 5 different cars). If he samples with replacement, how many different samples leading to 5 weight measurements are possible?

Well, there are 43 possibilities for each of the 5 selections, so from the FPC there would be $(43)^5 = 147,008,443$ different samples and they would all be equally-likely.

(B) His employees think that the sampling should be done without replacement. They think that this would be a more sensible way of sampling and conducting the experiment, because each sample would be different and they would all still be equally-likely. How many samples would there be if we sample without replacement?

Well, there would of course be $\binom{43}{5} = 962,598$ different equally-likely samples. This is the number of different subsets of size 5 that could be chosen from the 43 cars.

Example (5):
A building has an elevator serving its ground level parking
lot and the 5 floors above containing all the offices. 8 men
(that don't know each other and don't interact in any non-
random way) get on the elevator at the ground level in a
completely random way and all 8 disembark from the
elevator to the 5 floors above in a completely random way,
including the order of disembarking (because there must be
at least one floor where more than one of them gets out).
You and a Billionaire get on at the ground level, strictly as
observers, and don't interact with the 8 men in any way.
The Billionaire has promised to give you $100,000,000 if you
can guess ahead of time where each individual man will get
out and the order of disembarkation (on any floor where
more than one gets out – which of course is not known to
you or the Billionaire, but you must be able to take a guess
at it ahead of time!) How many different ways can the 8
men disembark, taking into account the floor and the order
of disembarking for each individual man? What is the
probability that you win the $100,000,000?

The number of different ways for 8 men to be distributed
among the 5 floors is $\binom{8 + 5 - 1}{5 - 1} = \binom{12}{4} = 495$ ways. But,
if you have to take into account the floor and the order of
exiting from the elevator for each individual man, you must
multiply 495 by (8!) to get a total of (19,958,400) ways that
the situation could play itself out. So the probability that
you can guess ahead of time exactly the way that it would
happen and win the $100,000,000 is $\left(\dfrac{1}{19,958,400}\right)$, which is

(0.00000005) = (0.000005)% because all the possibilities are equally-likely. The chance is not great that you will win the money, only (.000005) of 1%, or 5 millionths of 1%.

Example (6):
(A) A group of 10 people (all members of one family consisting of mom and dad and their 8 children) will line up side by side for a picture. Mom and dad will be at the ends and they will allow their 8 children, 3 boys and 5 girls, to arrange themselves in between in any way that they wish. How many possible line-ups are there?

There are 2 ways that mom and dad can be at the ends and (8!) ways that the kids can be arranged between them, and all of them are equally-likely. So, there are (2)(8!) = 80,640 equally-likely possible line-ups for the picture.

(B) What is the probability that mom stands next to one of her daughters? Call this event A.

Mom could be in 2 possible places. For each one of these possibilities, there are 5 ways that the space next to her could be taken by one of her daughters. Then there are 7! ways that the remaining kids could fill the remaining 7 positions. So, there are (2)(5)(7!) = 50,400 equally-likely ways that event A could happen. Therefore, the probability of event A is:

$$Pr(A) = \frac{N(A)}{N(S)} = \left(\frac{50,400}{80,640}\right) = (0.625) = 62.5\% \, .$$

Example (7):
An engineer at a computer manufacturing company must have all three of the three devices A, B, and C which are inside the computer, operating simultaneously 94% of the time for the entire computer to operate correctly. Device A operates correctly 99.5% of the time. Device B operates correctly 97.93% of the time. He is in the process of designing Device C, and he needs to know the percentage of the time, call it x, that Device C must operate correctly in order to satisfy his requirements. All three devices operate completely independent of each other.

He must have: $(0.995)(0.9793)(x) \geq (0.94)$, from the product rule for independent events, which is discussed later in our section (6.3). Solving this inequality for x, he calculates that $x \geq (0.9647)$. In other words, he needs to design Device C so that it works correctly at least 96.47% of the time.

Example (8):
With a standard deck of playing cards, there are 52 cards, divided into 4 suits (Hearts, Diamonds, Clubs, and Spades), and each suit contains 13 denominations (Ace, 2, 3, 4, 5, 6, 7, 8, 9, 10, Jack, Queen, and King).

What is the probability of being dealt a Full House (3 of one denomination and 2 of another denomination) in a 5 card hand from a well-shuffled deck with the following stipulations: The three of a kind denomination must not be a 3, 7, or a Jack, and the two of a kind denomination cannot be a 2, 10, or King? Call this event A. All of these allowable Full Houses within event A are equally-likely.

For the three of a kind, there are only 10 possibilities for the denomination, and within whatever denomination it is there are $\binom{4}{3}$ choices for the 3 suits. For the two of a kind, there are only 7 possible choices for the denomination, and within whatever denomination it is there are $\binom{4}{2}$ choices for the 2 suits. There are $\binom{52}{5} = 2{,}598{,}960$ equally-likely hands that can be dealt from a well shuffled deck (the sample space S). The number of Full Houses that are within event A (the number that meet the stipulations) is:

$N(A) = (10) \cdot \binom{4}{3} \cdot (7) \cdot \binom{4}{2} = 1680$. So, we have:

$$Pr(A) = \frac{N(A)}{N(S)} = \frac{1680}{2{,}598{,}960} \approx (0.0006464) = (0.06464)\% \ .$$

This is slightly less than 65 of every 100,000 hands dealt.

Example (9):
If a woman has four pregnancies, with no twins, and no miscarriages, etc., what is the probability that there are 3 boys and 1 girl? Call this event A.

If we write B for boy, and G for girl, we can list all of the possibilities in the sample space S:

S = {BBBB, BBBG, BBGB, BBGG, BGBB, BGBG, BGGB, BGGG,
   GBBB, GBBG, GBGB, GBGG, GGBB, GGBG, GGGB,GGGG}

The event A = {BBBG, BBGB, BGBB, GBBB}

So, the Pr(A) = $\frac{N(A)}{N(S)}$ = $\frac{4}{16}$ = $\frac{1}{4}$ = 25% .

Example (10):
If 10 people are randomly selected from the population,
what is the probability that at least two people have the
same birthday? Call this event A.

We can think of the outcome of this experiment as being a
member of the sample space S = {0,2,3,4,5,6,7,8,9,10}. The
probability that we seek is the probability of getting one of
the outcomes from {2,3,4,5,6,7,8,9, or 10}, which is event A.
Since the probability that at least one of the 10 outcomes in
the sample space S occurs is 1, we want the Pr(A) which is
Pr(A) = (1 – Pr(0)) = (1 – Pr(all 10 birthdays are different)).

The first person chosen has a birthday, what it is doesn't
matter. The probability that the second person chosen has
a different birthday is $\left(\frac{364}{365}\right)$. The probability that the third
person chosen has a birthday different from the first two is
$\left(\frac{363}{365}\right)$. The probability that the fourth person chosen has a
birthday different from the first three is $\left(\frac{362}{365}\right)$, and on and
on this reasoning goes. We can figure the probability of 0
by multiplying these probabilities together, to get:

348

The Pr(0) = $\left(\frac{364}{365}\right) \cdot \left(\frac{363}{365}\right) \cdot \left(\frac{362}{365}\right) \cdots \left(\frac{357}{365}\right) \cdot \left(\frac{356}{365}\right)$

$= \left(\frac{(1.015364491) \times 10^{23}}{(1.149835678) \times 10^{23}}\right) = (0.883051822)$

So, the Pr{A} = (1 − 0.883051822)

$\qquad = (0.116948178) \approx 11.7\%$.

################# Exercises #################

(1) For the experiment of flipping a coin n = 5 times,
   (a) What is the cardinality of the sample space S?
   (b) Write in set notation the event that there are
       2 Heads, and call the event A.
   (c) What is the Pr(A)?
   (d) Write in set notation the event that there is 1 head,
       and call the event B.
   (e) What is the Pr(B)?

(2) From the set of digits {0,1,2,3,4,5,6,7,8,9}, we randomly
    select three digits a_1, a_2, and a_3 with replacement and
    construct the three digit number $(a_1 a_2 a_3)$.
   (a) What is the sample space S and what is the
       cardinality of S?
   (b) Write in set notation the event that the number
       begins and ends with 5, and call the event A.
   (c) What is the Pr(A)?
   (d) Let B = {126, 008}. What is the Pr(B)?
   (e) Let C be the event that the randomly selected three
       digit number begins with 3, 5, or 8 and ends with a
       0, 4, or 9 . What is the Pr(C)?

(3) If an opaque bag contains 6 blue and 4 white marbles, and three are taken from the bag at random without replacement, what is the probability that all three are white?

(4) Five people, 3 men and 2 women, line up side by side for a picture in a completely random way.
   (a) What is the probability that the leftmost person is male?
   (b) How many possible arrangements are there?
   (c) Let A be the event that the two leftmost people are female. What is the total number of ways that this could occur? What is the Pr(A)?
   (d) The five people are Bob, Billy, Beau, Alice, and Nancy. What is the probability that the arrangement for the picture is: (Alice, Billy, Nancy, Bob, Beau)?

(5) A license plate is constructed completely at random and it consists of 3 numbers and 4 letters in any arrangement with repeated numbers and repeated letters allowed.
   (a) How many different license plates are possible?
   (b) How many different plates are there that begin with BT5? What is the probability of getting a plate that begins with BT5?
   (c) What is the probability of getting a license plate that ends with 447?
   (d) What is the probability of getting "23RTY5U" or "23RTYH7"?

(6.2) Random Variables

A random variable X is a function from the sample space S to the real numbers. We write X: S → R. When we have a random variable X, it has a certain sample space S, with a certain event space E, and a probability measure Pr(A) associated with it. That is, the random variable X has a certain probability space (S, E, Pr(A)) associated with it. We can determine from the probabilities of each outcome in S, using the probability measure Pr(A) for X, the probabilities associated with all of the outcomes of the random variable X. The name of a random variable is a capital letter, like X, Y, or Z, and the outcomes that it can take are denoted with lower-case letters like x,y, or z.

The reader should notice that from this point onward in our treatment of the subject of statistics, we will be concerned mostly with the outcomes of random variables. Note that the outcomes of a random variable are associated with a set of mutually exclusive and exhaustive subsets from S.

Example (1):
From the first example of section (6.1), lets define the random variable X to be the sum of the outcomes of the two dice. Therefore, we have the function X: S → R defined for every outcome (x, y) in S, as $X(x, y) = (x + y)$. Clearly, this random variable X = the sum of the two dice is one of the whole numbers {2, 3, 4, 5, 6, 7, 8, 9, 10, 11, 12}, with a probability associated with each. This is the probability distribution of the random variable X:

$Pr(X = 2) = Pr\{(1,1)\} = \dfrac{1}{36}$.

$Pr(X = 3) = Pr\{(1,2), (2,1)\} = \dfrac{2}{36} = \dfrac{1}{18}$.

$Pr(X = 4) = Pr\{(1,3), (2,2), (3,1)\} = \dfrac{3}{36} = \dfrac{1}{12}$.

$Pr(X = 5) = Pr\{(1,4), (2,3), (3,2), (4,1)\} = \dfrac{4}{36} = \dfrac{1}{9}$.

$Pr(X = 6) = Pr\{(1,5), (2,4), (3,3), (4,2), (5,1)\} = \dfrac{5}{36}$.

$Pr(X = 7) = Pr\{(1,6), (2,5), (3,4), (4,3), (5,2), (6,1)\} = \dfrac{6}{36} = \dfrac{1}{6}$.

$Pr(X = 8) = Pr\{(2,6), (3,5), (4,4), (5,3), (6,2)\} = \dfrac{5}{36}$.

$Pr(X = 9) = Pr\{(3,6), (4,5), (5,4), (6,3)\} = \dfrac{4}{36} = \dfrac{1}{9}$.

$Pr(X = 10) = Pr\{(4,6), (5,5), (6,4)\} = \dfrac{3}{36} = \dfrac{1}{12}$.

$Pr(X = 11) = Pr\{(5,6), (6,5)\} = \dfrac{2}{36} = \dfrac{1}{18}$.

$Pr(X = 12) = Pr\{(6,6)\} = \dfrac{1}{36}$.

The 11 outcomes associated with this random variable X are based on a set of mutually exclusive and exhaustive sets from S. So, the sum of the probabilities associated with random variable X is: $\sum_{x=2}^{12} Pr(X = x) = Pr(S) = 1$.

Example (2):

Again, for the sample space S from above associated with the two dice, define the Random Variable Y to be $|x - y|$, that is Y: S → R is Y(x, y) = $|x - y|$ for all outcomes (x, y) in S. The set of possible outcomes for this random variable Y is {0,1,2,3,4,5}, and there is a probability associated with each, which we can determine with our probability measure on S. This is the probability distribution of the random variable Y:

$Pr(Y = 0) = Pr\{(1,1), (2,2), (3,3), (4,4), (5,5), (6,6)\} = \frac{6}{36} = \frac{1}{6}$.

$Pr(Y = 1) = Pr\{(1,2), (2,1), (2,3), (3,2), (3,4), (4,3),$
$(4,5), (5,4), (5,6), (6,5)\} = \frac{10}{36} = \frac{5}{18}$.

$Pr(Y = 2) = Pr\{(1,3), (3,1), (2,4), (4,2), (3,5), (5,3),$
$(4,6), (6,4)\} = \frac{8}{36} = \frac{2}{9}$.

$Pr(Y = 3) = Pr\{(1,4), (4,1), (2,5), (5,2), (3,6), (6,3)\} = \frac{6}{36} = \frac{1}{6}$.

$Pr(Y = 4) = Pr\{(1,5), (5,1), (2,6), (6,2)\} = \frac{4}{36} = \frac{1}{9}$.

$Pr(Y = 5) = Pr\{(1,6), (6,1)\} = \frac{2}{36} = \frac{1}{18}$.

The outcomes of this random variable Y are determined from a mutually exclusive and exhaustive collection of sets from S. So, the sum of their probabilities is:
$\sum_{y=0}^{5} Pr(Y = y) = Pr(S) = 1$.

353

Example (3):

Now, for the second example of section (6.1), we can define the random variable Z = the number of heads in an outcome from the sample space S, which is the number of heads in the tossing of a coin three times. The possible outcomes of Z are {0, 1, 2, 3}, with a probability associated with each. This is the probability distribution of the random variable Z:

$Pr(Z = 0) = Pr\{TTT\} = \frac{1}{8}$.

$Pr(Z = 1) = Pr\{HTT, THT, TTH\} = \frac{3}{8}$.

$Pr(Z = 2) = Pr\{HHT, HTH, THH\} = \frac{3}{8}$.

$Pr(Z = 3) = Pr\{HHH\} = \frac{1}{8}$.

The events from the sample space S associated with the four outcomes of the random variable Z are a mutually exclusive and exhaustive set of events from S. So, the sum of the probabilities associated with the outcomes of Z are:
$\sum_{z=0}^{3} Pr(Z = z) = Pr(S) = 1$.

Example (4):

Suppose we have an opaque bag and inside are two blue marbles and one red marble. Our experiment is to choose at random two marbles, one after the other, from the bag without replacement.

354

(A) What is the sample space S for this experiment?
Let's label one of the blue marbles as marble #1, the other
blue marble as marble #2, and the red marble as marble #3.
So, we can denote an outcome of this experiment with an
ordered pair, such as (2,3), which would mean that the first
marble pulled from the bag was marble #2, and then marble
#3 was pulled. There are six equally-likely possible
outcomes in S and we can list them:

$$S = \{(1,2), (1,3), (2,1), (2,3), (3,1), (3,2)\}$$

(B) The number of blue marbles drawn is either one or two.
One blue marble is drawn if the outcome is in event A_1:

$$A_1 = \{(1,3), (2,3), (3,1), (3,2)\}$$

Two blue marbles are drawn if the outcome is in event A_2:

$$A_2 = \{(1,2), (2,1)\}$$

Note that events A_1 and A_2 form a mutually exclusive
and exhaustive set of events from S. Using our counting
probability measure, the:

$$Pr(A_1) = \frac{N(A_1)}{N(S)} = \frac{4}{6} = \frac{2}{3}, \text{ and the}$$

$$Pr(A_2) = \frac{N(A_2)}{N(S)} = \frac{2}{6} = \frac{1}{3}.$$

(C) We define the random variable X = the number of blue
marbles drawn. It takes outcomes in the set $\{1, 2\}$.

The $Pr(X = 1) = Pr(A_1) = \frac{2}{3}$.

The $Pr(X = 2) = Pr(A_2) = \frac{1}{3}$.

So, as we would expect, the $\sum_{i=1}^{2} Pr\,(X = i) = Pr(S) = 1$.

################## Exercises ##################

(1) This exercise is derived from example (4) of this section. Let there be 4 marbles in an opaque bag, two blue and two red, labeled marbles #1, #2, #3, and #4 respectively. Once again, the experiment is to draw two marbles at random, one after the other, from the bag without replacement.

(a) What is the sample space S?

(b) the number of blue marbles pulled from the bag is now 0,1, or 2. Write out the events A_1, A_2, and A_3 which correspond to these three outcomes.

(c) Are events $\{A_1, A_2, A_3\}$ a mutually exclusive and exhaustive set of events from S. Using the uniform probability measure, determine the probability of each of these three events.

(d) If we define the random variable X = the number of blue marbles pulled from the bag, what is the set of possible outcomes? What is the probability associated with each outcome? Do the probabilities of the possible outcomes add to 1?

(2) Suppose we randomly select two digits a_1 and a_2 from the set $\{0,1,2,3,4,5,6,7,8,9\}$ with replacement and form the two-digit number $(a_1 a_2)$.

(a) What is the sample space S?
(b) What is the cardinality of S?
(c) Let A_1 be the event that the number is evenly divisible by three. Write out the set A_1 in set notation. What is the cardinality of set A_1?
(d) Let A_2 be the complement of A_1 relative to S. Write the set A_2 in set notation. What is the cardinality of set A_2?
(e) Do sets $\{A_1, A_2\}$ form a mutually exclusive and exhaustive set of events from S?
(f) Using the uniform probability measure, what are the probabilities of events A_1 and A_2?
(g) Define the random variable X to be 0 if the whole number is in event A_1, and 1 if the whole number is in A_2. What is the set of possible outcomes? What is the distribution of X?

(3) Suppose we randomly select two digits a_1 and a_2 from the set $\{0,1,2,3,4,5,6,7,8,9\}$ with replacement and form the two-digit number $(a_1 a_2)$.

(a) What is the sample space S?
(b) What is the cardinality of S?
(c) Let A_1 be the event where $a_1 = a_2$. Write out the set A_1 in set notation. What is the cardinality of set A_1?
(d) Let A_2 be the event where $a_1 \neq a_2$. Write the set A_2 in set notation. What is the cardinality of set A_2?

(e) Do sets $\{A_1, A_2\}$ form a mutually exclusive and exhaustive set of events from S?

(f) Using the uniform probability measure, what are the probabilities of events A_1 and A_2?

(g) Define the random variable X to be 0 if the whole number is in event A_1, and 1 if the whole number is in A_2. What is the set of possible outcomes? What is the distribution of X?

(6.3) Probability of Compound Events

By compound events we mean something like $(A \cup B)$, $(A \cap B)$, and so on, where A and B are events from S. We now know that the outcomes of a random variable X are associated with certain events from the sample space S. We now discuss the probability of compound events.

The General Sum Rule

We will start with a random variable X having the following equally likely sample space $\{0,1,2,3,4,5,6,7,8,9,10\}$. So, the probability of each outcome in S is $\left(\frac{1}{11}\right)$.

Let A = $\{1,2,3,4,5,6,7\}$, and B = $\{4,5,6,7,8\}$ be two events from the sample space S.

Then Pr(A) = $\frac{N(A)}{N(S)} = \frac{7}{11}$, and the Pr(B) = $\frac{N(B)}{N(S)} = \frac{5}{11}$.

The set $(A \cup B) = \{1,2,3,4,5,6,7,8\}$ and the

$$Pr(A \cup B) = \frac{N(A \cup B)}{N(S)} = \frac{8}{11}.$$

The set $(A \cap B) = \{4,5,6,7\}$ and the

$$Pr(A \cap B) = \frac{N(A \cap B)}{N(S)} = \frac{4}{11}.$$

The sets A and B are not mutually exclusive, so we cannot just say that the $Pr(A \cup B) = Pr(A) + Pr(B)$, which would equal $\frac{7}{11} + \frac{5}{11} = \frac{12}{11}$. This is not right since the probability of any event cannot be greater than 1. To figure the $Pr(A \cup B)$ correctly, we have to realize that the probabilities of certain outcomes are being double counted, once in Pr(A) and once in Pr(B). Clearly, the outcomes whose probabilities are being double counted are exactly the members in the set $(A \cap B)$, so we have to subtract the probability of $(A \cap B)$ once. This leads us to the general sum rule:

$$Pr(A \cup B) = Pr(A) + Pr(B) - Pr(A \cap B).$$

With this formula, the $Pr(A \cup B) = \left(\frac{7}{11} + \frac{5}{11} - \frac{4}{11}\right) = \frac{8}{11}$, which is the correct result.

Special Case of The Sum Rule
Now let's define the collection of outcomes E and F, with the same random variable X to be: $E = \{0,1,2,3\}$, $F = \{6,7,8\}$.

The Pr(E) = $\frac{4}{11}$, and the Pr(F) = $\frac{3}{11}$, and since $(E \cup F) = \{0,1,2,3,6,7,8\}$, the Pr($E \cup F$) = $\frac{7}{11}$.

Since $(E \cap F) = \emptyset$, the Pr($E \cap F$) = 0. In this case, we have a special case of the sum rule: If events E and F are mutually exclusive events, then:

$$Pr(E \cup F) = Pr(E) + Pr(F) .$$

Using this special case, the Pr($E \cup F$) = $\left(\frac{4}{11} + \frac{3}{11}\right) = \frac{7}{11}$, which is the correct result.

If we have a set A and its complement A^c with respect to the sample space S for the random variable X, then these are mutually exclusive and exhaustive sets and

$A \cup A^c = S$. So, the Pr(A) + Pr(A^c) = Pr(S) = 1 .

The reader should recall from chapter 2 the set statement: For any sets A and E, $A = (A \cap E) \cup (A \cap E^c)$.

Then we can make the following probability statement: For any two events A and E,

Pr(A) = Pr($A \cap E$) + Pr ($A \cap E^c$).

Since E and E^c are mutually exclusive events, the events $(A \cap E)$ and $(A \cap E^c)$ are also.

The General Product Rule and Conditional Probability

Let S = {0,1,2,3,4,5,6,7,8,9,10} be the same equally-likely sample space associated with the random variable X above. Then define the events:

$$A = \{2,3,4,5,6\}, \quad B = \{5,6,9,10\}.$$

Then $(A \cap B) = \{5,6\}$.

The Pr(A) = $\frac{5}{11}$, the Pr(B) = $\frac{4}{11}$, The Pr$(A \cap B)$ = $\frac{2}{11}$

The Product Rules are: (there are two forms for $Pr(A \cap B)$)
$$Pr(A \cap B) = Pr(A) \cdot Pr(B|A) = Pr(B) \cdot Pr(A|B).$$

These Product Rules can be rearranged to get the Conditional Probability Rules:

$$Pr(A|B) = \frac{Pr(A \cap B)}{Pr(B)} \quad \text{and} \quad Pr(B|A) = \frac{Pr(A \cap B)}{Pr(A)}.$$

This is read "Probability of event A given that event B occurs," for the first formula, and the "Probability of event B given that event A occurs" for the second formula. For example, the event B above has probability $\frac{4}{11}$ of occurrence. Given that this event B has occurred, then the probability that event A occurs is $\frac{1}{2}$ because event A occurs if the outcome is 5 or 6. But 5 and 6 are only two of the four outcomes in B. If B occurred because the outcome was a 9 or 10, then event A does not occur. So, if B occurs there is only a chance of $\frac{1}{2}$ that event A also occurs.

So, (1) $\Pr(A|B) = \frac{\Pr(A \cap B)}{\Pr(B)} = \frac{\left(\frac{2}{11}\right)}{\left(\frac{4}{11}\right)} = \frac{1}{2}$, and the

$$\Pr(A \cap B) = \Pr(B) \cdot \Pr(A|B) = \left(\frac{4}{11}\right) \cdot \left(\frac{1}{2}\right) = \frac{2}{11}.$$

(2) $\Pr(B|A) = \frac{\Pr(A \cap B)}{\Pr(A)} = \frac{\left(\frac{2}{11}\right)}{\left(\frac{5}{11}\right)} = \frac{2}{5}$, and the

$$\Pr(A \cap B) = \Pr(A) \cdot \Pr(B|A) = \left(\frac{5}{11}\right) \cdot \left(\frac{2}{5}\right) = \frac{2}{11}.$$

We see here the two forms of the product rule that give the same result, as they should.

Special Case of The Product Rule
There is an important special case of the product rule. If two events A and B are independent, then:

$$\Pr(A \cap B) = \Pr(A) \cdot \Pr(B) .$$

Considering that there were two forms of the General Product Rule, it follows that $\Pr(A|B) = \Pr(A)$, and $\Pr(B|A) = \Pr(B)$. What these two statements say is that the occurrence of B does not affect the occurrence of A, and the occurrence of A does not affect the occurrence of B.

For example, if we toss a coin twice, whether the first toss landed heads or tails has no effect on whether the second toss lands heads or tails. These are called independent events.

If we have more than two independent events $\{E_1, E_2, \ldots, E_n\}$, then the

$$Pr(E_1 \cap E_2 \cap \cdots \cap E_n) = Pr(E_1) \cdot Pr(E_2) \cdots Pr(E_n).$$

We encounter this in the subject of Statistics. We think of the outcomes of an experiment (when we take a simple random sample) as a collection of n independent events. The product of the probabilities of the n events is the probability of occurrence of the particular sample.

Bayes Theorem

There is an alternative theory and set of techniques in the subject of statistics which is based on a result known as Bayes Theorem, due to Thomas Bayes, a mathematician of the 18th century. Fortunately, the conventional statistical techniques such as those presented in this book and the Bayesian statistical techniques, usually produce very similar results. This theorem and Bayesian statistical techniques, if the reader should decide to read more about them or need to study more about them in more advanced coursework, are quite interesting and have a lot of use. However, the reader should not feel as if they are missing out on a lot of essential statistical material, the reason being that all the important and fundamental concepts of statistics are very much the same in both approaches, as they should be.

In the Bayesian point of view, the parameters of a distribution are treated as if they are random variables instead of being fixed numbers. One makes a subjective choice for the distribution of the parameter (called the prior distribution), and then uses experimental results to update the prior distribution to what is called the posterior distribution. Then the posterior distribution can be updated again with more statistical results to form yet another posterior distribution, and on and on. Bayesian statisticians make the case that they are using more of the information that they have at their disposal before the experiment, in order to make better statistical inferences.

In this discussion, we will consider the prior distribution for a parameter to be a set of n mutually exclusive and exhaustive events $\{A_1, A_2, \ldots, A_n\}$ in a sample space S. We will show how Bayes Theorem updates these prior probabilities to posterior probabilities, based on the result of some experiment, which we will call E. We have to start with n prior probabilities chosen by the experimenter:

$$\{\Pr(A_1), \Pr(A_2), \ldots, \Pr(A_n)\},$$

and we have to know the n probabilities:

$$\{\Pr(E|A_1), \Pr(E|A_2), \ldots, \Pr(E|A_n)\},$$

all of which are subjectively assumed or experimentally determined.

The ith posterior probability, $Pr(A_i|E)$, which has a probability of occurrence which depends on event E, can be determined:

$$Pr(A_i|E) = \frac{Pr(A_i \cap E)}{Pr(E)} = \frac{Pr(E|A_i)Pr(A_i)}{Pr(E)}$$ (from the product rule).

From the section on the sum rule, we saw that we can write: $E = (A_1 \cap E) \cup (A_2 \cap E) \cup \cdots \cup (A_n \cap E)$, where the n sets in this union are mutually exclusive. So, then we can write:

$$Pr(E) = \sum_{j=1}^{n} Pr(E \cap A_j).$$

From the product rule we write this: $Pr(E) = \sum_{j=1}^{n}[Pr(E|A_j) \cdot Pr(A_j)]$. So, then we have:

$$Pr(A_i|E) = \frac{Pr(E|A_i)Pr(A_i)}{\sum_{j=1}^{n}[Pr(E|A_j)Pr(A_j)]}, \text{ for } i = 1,\dots,n.$$

This is Bayes Formula for finding the posterior probabilities.

As an Example:

Suppose the members of a certain religious group are thought to be living only in the states of Vermont, New Hampshire, and Maine. Define events A_1, A_2, and A_3 to be the event that a randomly chosen member of this religious group lives in Vermont, New Hampshire, or Maine respectively. We will assume a uniform set of priors, or that:

$$Pr(A_1) = Pr(A_2) = Pr(A_3) = \frac{1}{3} \approx (0.333)$$

(We don't know what the priors actually are, but we have some kind of subjective belief that they are spread out in this way).

Now, we define the event E to be that a randomly chosen person belongs to this certain religious group, based on a randomly conducted phone survey of 540,230 people living in these three states. The results are that:

$$Pr(E|A_1) = 0.118,$$
$$Pr(E|A_2) = 0.093,$$
$$Pr(E|A_3) = 0.134$$

Now, we can use Bayes Theorem to find the posterior distribution: $\{Pr(A_1|E), \ Pr(A_2|E), \ Pr(A_3|E)\}$.

Firstly, the denominator is the $Pr(E)$
$= Pr(E|A_1)Pr(A_1) + Pr(E|A_2) \ Pr(A_2) + Pr(E|A_3)Pr(A_3)$
$= (0.118) \cdot (0.333) + (0.093) \cdot (0.333) + (0.134) \cdot (0.333)$
$= (0.039) + (0.031) + (0.045)$
$= 0.115$

$$Pr(A_1|E) = \frac{Pr(E|A_1)Pr(A_1)}{\sum_{j=1}^{n}[Pr(E|A_j)Pr(A_j)]} = \frac{(0.118) \cdot (0.333)}{(0.115)} = 0.342$$

$$Pr(A_2|E) = \frac{Pr(E|A_2)Pr(A_2)}{\sum_{j=1}^{n}[Pr(E|A_j)Pr(A_j)]} = \frac{(0.093)\cdot(0.333)}{(0.115)} = 0.269$$

$$Pr(A_3|E) = \frac{Pr(E|A_3)Pr(A_3)}{\sum_{j=1}^{n}[Pr(E|A_j)Pr(A_j)]} = \frac{(0.134)\cdot(0.333)}{(0.115)} = 0.388$$

So, the updating of the uniform prior distribution, based on the phone survey, leads us to conclude that there are about a third of this religious group living in Vermont, somewhat less than a third in New Hampshire, and somewhat more than a third in Maine.

################# Exercises #################

For questions (1) through (4):
For a uniform random variable X (chance of all outcomes equal), with S consisting of the whole numbers from 0 through 20 (21 possible outcomes), we have the events:
A = {3,7,10,11,15},   B = {6,7,8,9,10,11,18,19},
C = {4,6,12,19},   D = {8,13,20}.

(1) (a) What is the Pr(A)?
    (b) What is the Pr(B)?
    (c) What is the Pr($A \cap B$)?
    (d) What is the Pr($A \cup B$)?

(2) (a) What are the sets ($A \cap C$), ($A \cap D$), ($C \cap D$)?
    (b) What is the Pr(C)?
    (c) What is the Pr(D)?
    (d) What is the Pr($A \cup C \cup D$)?

(3) (a) What is the $\Pr(B \cap C)$?
   (b) What is the $\Pr(C|B)$?
   (c) What is the $\Pr(B|C)$?

(4) (a) What is the $\Pr(B \cap D)$?
   (b) What is the $\Pr(B|D)$?
   (c) What is the $\Pr(D|B)$?

(5) The two engines of an aircraft operate independently. One will fail with probability 0.00014 and the other will fail with probability 0.000037 on a Trans-Pacific flight. What is the Pr(both engines fail) on the flight?

(6) Events A, B, C, and D occur independently with probabilities 0.14, 0.02, 0.37, and 0.0056, respectively.
   (a) What is the $\Pr(A \cap B \cap C \cap D)$?
   (b) What is the Pr(A|D) and the Pr(B|C)?

(7) Events A and A^c occur with probabilities 0.41 and 0.59 respectively.
   (a) The $\Pr(E \cap A) = 0.37$ and the $\Pr(E \cap A^c) = 0.115$. What is the Pr(E)?
   (b) What is the $\Pr\big((E \cap A) \cap (E \cap A^c)\big)$?

(8) If the Pr(G) = 0.3 and the Pr(H|G) = 0.12, What is the $\Pr(G \cap H)$?

(9) If the Pr(B) = 0.77 and A $\subset$ B, what can we conclude about the Pr(A)?

368

(10) If an experiment is carried out one time and events A and B are possible outcomes, with the Pr(A) = 0.55 and the Pr(B) = 0.67, then:
(a) What is the minimum of the Pr(A ∩ B)?
(b) What is the maximum of the Pr(A ∩ B)?

(11) If an experiment is carried out one time and events A and B are possible outcomes, with the Pr(A) = 0.5 and Pr(B) = 0.5, then:
(a) What is the minimum of the Pr(A ∩ B)?
(b) What is the maximum of the Pr(A ∩ B)?

(12) Suppose a city employs male and female bus drivers. Event A is that the customer tends to like male drivers over female drivers, and event A^c is that the customer tends to like female drivers over male drivers. However, it is assumed that customers can still rate a particular driver as good or bad irrespective of whether the driver is male or female. It is decided that the prior probabilities for events A and A^c will be uniform in the absence of any data:

{Pr(A) = 0.50, Pr(A^c) = 0.50}.

Suppose a planned experiment, conducted over a 3 month period of time, determines how customers feel about their driver upon departure from the bus. Event E is that a customer rates the driver favorably. Event E^c is that a customer rates the driver unfavorably. The data indicates that the Pr(E|A) = .40 and the Pr(E|A^c) = .63.

(a) What is the Pr(E)?

(b) A statistician updates the prior probabilities to posterior probabilities. What are the posterior probabilities?

(c) What conclusion could one make about customer preferences as to the gender of their driver?

(6.4) Probability Mass Functions

Probability Mass Function (pmf)

When we have a random variable X that takes the outcomes $\{x_i\}$, from a set of mutually exclusive and exhaustive subsets of S, we will write $(p_i) = Pr(X = x_i)$. So, for any random variable X, the following two conditions must always hold true:

$$0 \leq (p_i) \leq 1$$
$$\sum_i Pr(X = x_i) = \sum_i (p_i) = 1.$$

We have a so-called probability mass function (pmf) for X, which we denote $p_X(x)$:

$$p_X(x) = \begin{cases} (p_1), & \text{when } x = x_1 \\ \vdots \\ (p_M), & \text{when } x = x_M \end{cases}.$$

M is the number of outcomes for the random variable X, and these M probabilities (p_i) need not be equal, however they must add to 1.

Example (1):
Suppose we have a random variable X which has the outcomes: {0, 1, 2, 3, 4, 5, 7, 8, 10, 11, 13, 15},

where each of the outcomes is equally-likely. Since there are 12 outcomes, each has the probability of occurrence equal to $\left(\frac{1}{12}\right)$. We can write the pmf for X:

$$p_X(x) = \left\{ \left(\frac{1}{12}\right), \text{ for x} = 0, 1, 2, 3, 4, 5, 7, 8, 10, 11, 13, 15 \right\}$$

Example (2):
Suppose we have a random variable X which takes the outcome "3" with probability $\frac{1}{6}$, the outcome "5" with probability $\frac{2}{5}$, and the outcome "9" with probability $\frac{13}{30}$. Then, the pmf for this random variable X is:

$$p_X(x) = \begin{cases} \left(\frac{1}{6}\right), & \text{for x} = 3 \\ \left(\frac{2}{5}\right), & \text{for x} = 5 \\ \left(\frac{13}{30}\right), & \text{for x} = 9 \end{cases}$$

Bivariate Probability Mass Function
We must often consider the joint distribution of two random variables X and Y, from which we can find so-called Marginal and Conditional distributions. We will consider this topic with a specific example, but first we must present the notation. We will write $(p_{ij}) = \Pr(X = x_i, Y = y_j)$.

So, we have a so-called joint distribution of the two random variables X and Y which we denote $p_{XY}(x, y)$:

$$p_{XY}(x, y) = \begin{cases} (p_{11}), & \text{when } (x = x_1, y = y_1) \\ \vdots \\ (p_{M_x M_y}), & \text{when } (x = x_{M_x}, y = y_{M_y}) \end{cases}.$$

M_x and M_y are the number of outcomes for the random variables X and Y respectively. These $(M_X) \cdot (M_Y)$ probabilities need not be equal, but they must add to 1.

Marginal Distributions

The Marginal distributions of X and Y are just the pmf's of X and Y respectively. We can find the Marginal distribution probabilities from the joint pmf probabilities for X and Y in the following way:

$$p_X(x_i) = \sum_{j=1}^{M_y} \text{Pr} (X = x_i, Y = y_j) = \sum_{j=1}^{M_y} (p_{ij}),$$
$$\text{for } i = 1, \ldots, M_x .$$

$$p_Y(y_j) = \sum_{i=1}^{M_x} \text{Pr} (X = x_i, Y = y_j) = \sum_{i=1}^{M_x} (p_{ij}),$$
$$\text{for } j = 1, \ldots, M_y .$$

Conditional Distributions

We have Conditional pmf's which are also found from the joint and marginal pmf's, denoted: $p_{X|Y}(x|y)$ and $p_{Y|X}(y|x)$.

The Conditional probabilities of X given that $Y = y_j$ is given by: $p_{X|Y}(x_i|Y = y_j) = \{ \frac{p_{XY}(x_i,y_j)}{p_Y(y_j)}$, for $i = 1,\ldots, M_x \}$.

The Conditional probabilities of Y given that $X = x_i$ is given by: $p_{Y|X}(y_j|X = x_i) = \{ \frac{p_{XY}(x_i,y_j)}{p_X(x_i)}$, for $j = 1,\ldots, M_y \}$.

Cumulative Distribution Function (cdf)

The cdf F(x) for a random variable X is a very important function in more advanced treatments of probability and statistics. We will not use it much, but we want to state it for completeness, and give an example.

If X is a random variable, the cdf for X is: $F(x) = \Pr(X \leq x)$, where x ranges over the entire real number line. With a little reflection, one can see that for any random variable X, because it has a finite set of outcomes, the cdf is a step function which is non-decreasing from 0 at $-\infty$ up to 1 at $+\infty$. $F(x) = 0$ for all x less than the least value in the set of possible outcomes, and $F(x) = 1$ for all x greater than or equal to the greatest value in the set of possible outcomes. It increases by an amount $p_X(x_i)$ when $x = x_i$, so it is a step function with M_x steps. As an example, suppose X is a random variable with pmf:

$$p_X(x) = \begin{cases} \frac{1}{4}, & \text{for } x = 0 \\ \frac{1}{2}, & \text{for } x = 1 \\ \frac{1}{4}, & \text{for } x = 2 \end{cases}.$$

Then the cdf for X is:

$$F_X(x) = \begin{cases} 0, & \text{for } x < 0 \\ \left(\frac{1}{4}\right), & \text{for } 0 \le x < 1 \\ \left(\frac{3}{4}\right), & \text{for } 1 \le x < 2 \\ 1, & \text{for } x \ge 2 \end{cases}.$$

Now, we will provide an example of a bivariate (joint) pmf $p_{XY}(x, y)$ and calculate the two marginal distributions, and calculate a couple conditional distributions.

Example (3):
(A) Suppose a group of Chemists have a sealed laboratory container of a pure sample of a gaseous element, and the entire vessel is subjected to a strong magnetic field. The molecules of the gas undergo random collisions with each other and the walls of the container. The Chemists are interested in two properties of the molecular motion, which are random variables X and Y, each with three components corresponding to the x, y, and z directions in space. They code the x, y, and z directions with 1, 2, and 3 respectively. It is believed that the variables X and Y are related and have a joint pmf, given in the following matrix:

$$p_{XY}(x, y) = \begin{array}{c} \\ x_1 \\ x_2 \\ x_3 \end{array} \begin{array}{ccc} y_1 & y_2 & y_3 \\ \begin{pmatrix} 0.103 & 0.112 & 0.114 \\ 0.109 & 0.107 & 0.110 \\ 0.115 & 0.111 & 0.119 \end{pmatrix} \end{array}$$

Here X takes the outcomes $\{x_1, x_2, x_3\}$ and Y takes the outcomes $\{y_1, y_2, y_3\}$.

(B) To find the marginal distribution of X, add the three entries in each row. $p_X(x_i) = \sum_{j=1}^{3}(p_{ij})$, for $i = 1, 2$, and 3. So, we have:

$$p_X(x) = \begin{pmatrix} 0.329, & \text{for } x = x_1 \\ 0.326, & \text{for } x = x_2 \\ 0.345, & \text{for } x = x_3 \end{pmatrix}$$

(C) To find the marginal distribution of Y, add the three entries in each column. $p_Y(y_j) = \sum_{i=1}^{3}(p_{ij})$, for $j = 1, 2$, and 3. So, we have:

$$p_Y(y) = \begin{pmatrix} 0.327, & \text{for } y = y_1 \\ 0.330, & \text{for } y = y_2 \\ 0.343, & \text{for } y = y_3 \end{pmatrix}$$

(D) There are six conditional distributions, but it is not necessary to figure all of them. We will figure the two corresponding to the first row and the third column of the joint pmf:

To find the conditional distribution of Y given X = 1, then $p_{Y|X=1}(y_j|X = 1) = \frac{p_{XY}(x_1, y_j)}{p_X(x_1)}$, for $j = 1, 2$, and 3. So, we have:

$$p_{Y|X=1}(y|X=1) = \begin{pmatrix} 0.313, & \text{for } y = y_1 \\ 0.340, & \text{for } y = y_2 \\ 0.347, & \text{for } y = y_3 \end{pmatrix}$$

To find the conditional distribution of X given Y = 3, then $p_{X|Y=3}(x_i|Y=3) = \frac{p_{XY}(x_i,y_3)}{p_Y(y_3)}$, for i = 1, 2, and 3. So, we have:

$$p_{X|Y=3}(x|Y=3) = \begin{pmatrix} 0.332, & \text{for } x = x_1 \\ 0.321, & \text{for } x = x_2 \\ 0.347, & \text{for } x = x_3 \end{pmatrix}.$$

################## Exercises ##################

(1) Write the pmf for the random variable Z in Example (3) of section (6.2).

(2) Write the pmf for the random variable X in Example (4) of section (6.2).

(3) Write the pmf for the random variable X in Exercise (1) of section (6.2).

(4) Write the pmf for the random variable X in Exercise (2) of section (6.2).

(5) Write the pmf for the random variable X in Exercise (3) of section (6.2).

(6.5) Probability Distribution Parameters

The expected value of a function f(X) for a random variable X combines the outcomes of f(X) and the probabilities that correspond to the outcomes of f(X). When f(X) = X and f(X) = X^2, then these expected values can be used to define some very useful things, such as the three important distribution parameters which we call the mean, variance, and the standard deviation of the distribution.

When we have the random variable X with M outcomes $\{x_i\}$ and associated probabilities $\{p_i\}$, then the expected value of the function f(X) is defined:

$$E(f(X)) = \sum_{i=1}^{M} f(x_i) \cdot (p_i)$$

$$= [f(x_1) \cdot (p_1) + f(x_2) \cdot (p_2) + \cdots + f(x_M) \cdot (p_M)] \, .$$

(1) Expected Value of a Constant
When we consider any random variable X with M outcomes and a is any constant (a real number), then

$$E(a) = \sum_{i=1}^{M}(a) \cdot (p_i)$$

$$= (a) \cdot \sum_{i=1}^{M}(p_i) = (a) \cdot (1) = (a).$$

The constant "a" can be brought across the summation sign because it does not involve the index of the summation i. It is simply multiplied by every term in the sum, and because

377

of the distributive property for real numbers, we can factor it out. The remaining summation is the sum of all the probabilities for the outcomes of any random variable X, and therefore equals 1. So, we have that $E(a) = a$.

(2) <u>Expected Value of a Multiple of a Random Variable</u>
When we have a random variable X and a real number a, then the $E(aX) = \sum_{i=1}^{M}(a)(x_i) \cdot (p_i)$

$$= (a) \cdot \sum_{i=1}^{M}(x_i)(p_i)$$

$$= (a)E(X).$$

(3) <u>Expected Value of the Sum of Two Random Variables</u>
When we have two random variables X and Y, with M_x and M_y outcomes respectively, then:

$$E(X + Y) = \sum_{i=1}^{M_x}\sum_{j=1}^{M_y}(x_i + y_j) \cdot p_{XY}(x_i, y_j)$$

$$= \sum_{i=1}^{M_x}\left(\sum_{j=1}^{M_y}(x_i)p_{XY}(x_i, y_j) + \sum_{j=1}^{M_y}(y_j)p_{XY}(x_i, y_j)\right)$$

$$= \sum_{i=1}^{M_x}(x_i)\sum_{j=1}^{M_y}p_{XY}(x_i, y_j) + \sum_{i=1}^{M_x}\sum_{j=1}^{M_y}(y_j)p_{XY}(x_i, y_j)$$

$$= \sum_{i=1}^{M_x}(x_i)p_X(x_i) + \sum_{j=1}^{M_y}\sum_{i=1}^{M_x}(y_j)p_{XY}(x_i, y_j)$$

(Note that we were able to change the order of the summations in the right-hand term because these are finite sums)

$$= E(X) + \sum_{j=1}^{M_y} (y_j) \sum_{i=1}^{M_x} p_{XY}(x_i, y_j)$$

$$= E(X) + \sum_{j=1}^{M_y} (y_j) p_Y(y_j)$$

$$= E(X) + E(Y).$$

(4) Expected Value of X
This expectation defines the important parameter of a distribution known as its mean μ. The Greek letter μ is universally used to represent the mean of any distribution. However, sometimes a subscript is used to designate the mean of a specific distribution.

The $E(X) = \sum_{i=1}^{M} (x_i)(p_i) = \mu_X$.

μ_X is a weighted average of the outcomes of the random variable X. It is a measure of the central tendency of the distribution. All of the probabilities $\{p_i\}$ associated with the outcomes $\{x_i\}$ of a random variable X are not generally equal to each other. We just require that each of the (p_i) be a number between 0 and 1, and that they all add up to 1. For purposes of simplification, we will assume that we have a random variable with M equally-likely outcomes, so the probability of each outcome is $(p_i) = \left(\frac{1}{M}\right)$. So, we will be able to use a simple formula for the mean of a distribution:

$$\mu_X = \Sigma_{i=1}^{M}(x_i)(p_i) = \Sigma_{i=1}^{M}(x_i)\left(\frac{1}{M}\right) = \left(\frac{1}{M}\right)\Sigma_{i=1}^{M}(x_i) \quad \text{or}$$

$$\mu_X = \frac{\Sigma_{i=1}^{M}(x_i)}{M} .$$

(5) The Variance of X

We use an expectation formula to define another important parameter of a distribution known as the variance of the distribution, universally designated as σ^2. The variance of the distribution X will sometimes also be written Var(X). σ_X^2 is a measure of the dispersion or variability of the outcomes of the random variable X, that is, how much the outcomes are spread out about the mean. It is a weighted average of the squared deviations of the outcomes from their mean, so $\sigma_X^2 \geq 0$. The square root of the variance is the commonly used parameter called the standard deviation σ_X of random variable X. $\sigma_X = \sqrt{\sigma_X^2}$.

The Var(X) $= E(X - E(X))^2 = E(X - \mu_X)^2 = \sigma_X^2$.
In terms of summations, the Var(X) $= \Sigma_{i=1}^{M}(x_i - \mu_X)^2 (p_i)$.

As we know, the probabilities $\{p_i\}$ are generally not all equal to each other. But for purposes of simplification, assume that we are dealing with a distribution where

$(p_i) = \left(\frac{1}{M}\right)$ for all $i = 1, 2, \ldots, M$. So, for uniform distributions, the Var(X) $= \frac{\sum_{i=1}^{M}(x_i - \mu_X)^2}{M}$.

We can derive another useful formula for computing the variance of a distribution.

The Var(X) $= \sum_{i=1}^{M}(x_i - \mu_X)^2 (p_i)$

$= \frac{\sum_{i=1}^{M}(x_i - \mu_X)^2}{M}$

$= \left(\frac{1}{M}\right) \sum_{i=1}^{M}[(x_i)^2 - 2\mu_X(x_i) + (\mu_X)^2]$

$= \left(\frac{1}{M}\right) \left(\sum_{i=1}^{M}(x_i)^2 - 2\mu_X \sum_{i=1}^{M}(x_i) + \sum_{i=1}^{M}(\mu_X)^2\right)$

$= \left(\frac{1}{M}\right) \sum_{i=1}^{M}(x_i)^2 - 2\mu_X \left(\frac{\sum_{i=1}^{M}(x_i)}{M}\right) + \left(\frac{1}{M}\right)(M)(\mu_X)^2$

$= \frac{\sum_{i=1}^{M}(x_i)^2}{M} - 2(\mu_X)^2 + (\mu_X)^2$

$= \frac{\sum_{i=1}^{M}(x_i)^2}{M} - (\mu_X)^2$

$= E(X^2) - \left(E(X)\right)^2 \;=\; E(X^2) - (\mu_X)^2 .$

(6) The Variance of a Constant

If "a" is a constant, the Var(a) $= E\left(a - E(a)\right)^2$

$$= E(a - a)^2 = E(0) = 0.$$

The expected value of a single number is just what you would expect it to be, the number itself. The variance of a number is just what you would expect it to be, 0, because a single number doesn't vary!

(7) <u>The Variance of a Multiple of a Random Variable</u>
For a real number "a", the variance of (aX) is calculated:
$$Var(aX) = E(aX - E(aX))^2 = E(aX - aE(X))^2$$

$$= E((a)(X - \mu_X))^2$$

$$= E((a^2)(X - \mu_X)^2)$$

$$= a^2 E(X - \mu_X)^2$$

$$= a^2 Var(X).$$

(8) <u>Variance of the Sum of Two Random Variables</u>
We often have to determine the variance of the sum of two or more random variables. For the case of two random variables X and Y:

$$Var(X + Y) = E\big((X + Y) - E(X + Y)\big)^2$$

$$= E(X + Y - E(X) - E(Y))^2$$

$$= E\big((X - \mu_X) + (Y - \mu_Y)\big)^2$$

$$= E(X - \mu_X)^2 + E(Y - \mu_Y)^2 + 2E(X - \mu_X)(Y - \mu_Y)$$

$$= Var(X) + Var(Y) + 2Cov(X, Y)$$

We have a new quantity on the right which we call the Covariance of random variables X and Y. It is a measure of how X and Y vary together — it may be positive or negative depending on whether an increase in X tends to lead to an increase in Y (positive covariance), or whether an increase in X tends to lead to a decrease in Y (negative covariance). The Covariance of two random variables will generally be non-zero if they have some kind of association with each other. However, if X and Y are independent:

$$Cov(X,Y) = E(X - \mu_X)(Y - \mu_Y) = E(X - \mu_X) \cdot E(Y - \mu_Y).$$

Now, for any random variable X, the expression:
$E(X - \mu_X) = (E(X) - E(\mu_X)) = (\mu_X - \mu_X) = 0$.
Therefore, if X and Y are independent, the Cov(X,Y) = 0.

So, if X and Y are independent,
Var(X + Y) = Var(X) + Var(Y).

Also, we should note that if X and Y are independent:
Var(X − Y) = Var(X + (−1)Y)

$$= Var(X) + Var((-1)Y) - 2Cov(X, Y)$$

$$= Var(X) + (-1)^2 Var(Y) - (0)$$

383

(Check the section on the Var(aX))

$$= Var(X) + Var(Y).$$

Example (1):
Suppose we have a random variable X that has 9 outcomes $\{-10, -7, -5, -1, 0, 1, 3, 5, 8\}$, and these outcomes are equally-likely (uniform distribution of probability – uniform distribution). This means the probability of occurrence is $\left(\frac{1}{9}\right)$ for each outcome x. Find the mean μ, variance σ^2, and the standard deviation σ for this random variable.

(A) The mean $\mu = E(X) = \left(\frac{1}{9}\right)\Sigma_{i=1}^{9}(x_i)$

$= \left(\frac{1}{9}\right) \cdot [(-10) + (-7) + (-5) + (-1) + (0) + (1) + (3) + (5) + (8)]$

$= \left(\frac{1}{9}\right)(-6) = -\left(\frac{2}{3}\right) \approx -(0.667).$

(B) The variance $\sigma^2 = E(X - \mu)^2 = \left(\frac{1}{9}\right)\Sigma_{i=1}^{9}(x_i - \mu)^2$

$= \left(\frac{1}{9}\right)\Sigma_{i=1}^{9}\left(x_i - \left(-\frac{2}{3}\right)\right)^2 = \left(\frac{1}{9}\right)\Sigma_{i=1}^{9}\left(x_i + \left(\frac{2}{3}\right)\right)^2.$

We could use this formula, but it may be easier to use the fact that the $E(X - \mu)^2 = E(X^2) - (\mu)^2$, since we already know μ. We just need to compute $E(X^2)$.

The $E(X^2) = \left(\frac{1}{9}\right)\Sigma_{i=1}^{9}(x_i^2)$

384

$$= \left(\frac{1}{9}\right) \cdot [100 + 49 + 25 + 1 + 0 + 1 + 9 + 25 + 64]$$
$$= \left(\frac{1}{9}\right)(274) \approx (30.444).$$

So, the variance $\sigma^2 = (30.444) - (-0.667)^2 = (29.999).$

(C) The standard deviation $\sigma = \sqrt{\sigma^2}$
$$= \sqrt{(29.999)} \approx (5.477).$$

Example (2):
Suppose we have a random variable X with sample space (set of possible outcomes) S = {2, 5, 6, 7, 9, 11, 13}. So there are M = 7 possible outcomes. Assume that these outcomes are equally likely, that is $p_i = \left(\frac{1}{7}\right)$ is the probability that any one of these M outcomes can occur. Find the mean μ, the variance σ^2, and the standard deviation σ for this random variable.

(A) The mean $\mu = E(X) = \sum_{i=1}^{7}(x_i)(p_i)$
$$= \left(\frac{1}{7}\right)(2 + 5 + 6 + 7 + 9 + 11 + 13)$$
$$= \left(\frac{1}{7}\right)(53) = 7.571$$

(B) The variance $\sigma^2 = E(X - \mu)^2 = E(X^2) - (E(X))^2$
$$= E(x^2) - \mu^2$$
$$= \left(\frac{1}{7}\right)(4 + 25 + 36 + 49 + 81 + 121 + 169) - (7.571)^2$$
$$= \left(\frac{485}{7}\right) - (57.32) = 11.9657$$

(C) The standard deviation $\sigma = \sqrt{11.9657} = 3.459$.

Example (3):
Suppose we have a random variable X whose sample space is S = {−2, 1, 2}, where $Pr(X = -2) = \frac{1}{4}$, and $Pr(X = 1) = \frac{1}{3}$, and the $Pr(X = 2) = \frac{5}{12}$. Find the values of the population parameters μ, σ^2, and σ. Note that here I'm calling the outcomes of a random variable the members of a population. In chapter 7 we will consider samples from a population. The formulas involved are similar, but not the same.

(A) $\mu = E(X) = \Sigma_{x \in S}(x)p(x)$
$$= \left[(-2)\left(\tfrac{1}{4}\right) + (1)\left(\tfrac{1}{3}\right) + (2)\left(\tfrac{5}{12}\right)\right]$$
$$= \left[-\tfrac{1}{2} + \tfrac{1}{3} + \tfrac{10}{12}\right] = \left(\tfrac{8}{12}\right) = \left(\tfrac{2}{3}\right).$$

(B) $\sigma^2 = E(X - \mu)^2 = \Sigma_{x \in S}(x - \mu)^2 p(x)$
$$= \left[\left(-2 - \tfrac{2}{3}\right)^2 \left(\tfrac{1}{4}\right) + \left(1 - \tfrac{2}{3}\right)^2 \left(\tfrac{1}{3}\right) + \left(2 - \tfrac{2}{3}\right)^2 \left(\tfrac{5}{12}\right)\right]$$
$$= \left[\left(-\tfrac{8}{3}\right)^2 \left(\tfrac{1}{4}\right) + \left(\tfrac{1}{3}\right)^2 \left(\tfrac{1}{3}\right) + \left(\tfrac{4}{3}\right)^2 \left(\tfrac{5}{12}\right)\right]$$
$$= \left[\tfrac{48}{27} + \tfrac{1}{27} + \tfrac{20}{27}\right] = \tfrac{69}{27} \approx (2.556).$$

(C) $\sigma = \sqrt{\sigma^2} = \sqrt{2.556} \approx (1.599)$.

Example (4):
We can bring several of the properties of this section together in this example: Suppose the random variables X,

Y, and Z are independent, and the random variable W is defined to be: $W = 2X - 4Y + 3Z - 9$.

then the mean of W is: $\mu_W = 2\mu_X - 4\mu_Y + 3\mu_Z - 9$, and the variance of W is: $\sigma_W^2 = 4\sigma_X^2 + 16\sigma_Y^2 + 9\sigma_Z^2$.

################# Exercises #################

(1) X is a uniform random variable that takes the outcomes
    $S = \{7, 9, 15, 20, 25, 31, 33\}$.
    (a) What is the pmf for X?
    (b) What is the mean μ, variance σ^2, and standard
        deviation σ for the distribution of X?

(2) X is a uniform random variable that takes the outcomes
    $S = \{-4, -3, -2, -1, 0, 1, 2, 3, 4, 5, 6, 7, 8\}$.
    (a) What is the pmf for X?
    (b) What is the mean μ, variance σ^2, and standard
        deviation σ for the distribution of X?

(3) If a random variable X has a distribution defined by the pmf:

$$p_X(x) = \begin{cases} 1/3, & \text{for } x = 10 \\ 1/5, & \text{for } x = 25 \\ 1/5, & \text{for } x = 30 \\ 4/15, & \text{for } x = 40 \end{cases}$$

What is the mean μ, the variance σ^2, and the standard deviation σ for the probability distribution of X?

PART 3

<u>STATISTICS</u>

In chapters 7 through 12 we present the main statistical
tools of the experimental scientist. These tools are those of
statistical inference. In statistical inference we calculate
statistics from samples. Then we use these statistics to test
between two competing hypotheses about distribution
parameters, or to get estimates of distribution parameters.
In chapter 12, we will introduce the least squares method of
finding a linear relationship between two random variables
and the related topic of the correlation between the two
random variables.

(7) SAMPLES AND STATISTICS

(7.1) Statistics and Probability Models

When we conduct an experiment, we get a collection of numbers and we want to make some kind of sense out of them. We wonder about the secrets that lay beneath the seemingly chaotic collection of numbers in our samples.

There is a wide array of descriptive techniques that we can use to get a visualization of the distribution underlying the sample data. In Figure (7.1) we have a listing of some hypothetical sample of data, and how we can display it by using a so-called histogram. This shows more clearly the pattern beneath the numbers in the sample.

We should emphasize that with a descriptive statistical technique like the histogram in Figure (7.1) and along with some statistics derived from the data like $\bar{X}$ and S^2, this is actually all we really know about the true distribution of the data. In the most commonly used methods of inference in statistics we must make one or more assumptions about the underlying distribution of the data and then make use of a probability model associated with that assumption.

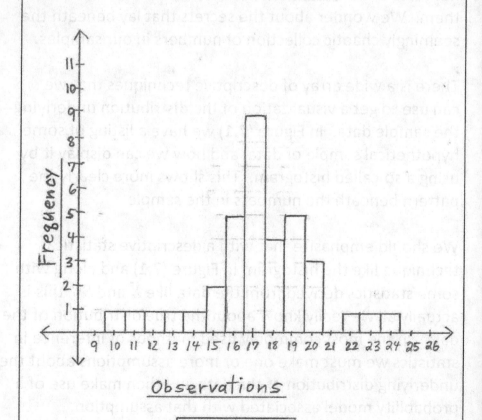

Sample of Observations:

$\{$ 16, 17, 16, 14, 17, 19, 15, 20, 23, 17, 17, 18, 19, 19, 17, 20, 12, 13, 11, 15, 16, 16, 18, 17, 19, 18, 17, 10, 24, 22, 18, 19, 16, 17, 17, 20 $\}$

Histogram for the Data:

Figure 7.1

392

A probability model is usually nothing more than a simple mathematical function that makes the analysis of the data much more simple. We can use these models to help us estimate the parameters of the assumed distribution and test hypotheses about the parameters of the assumed distribution, such as the hypothetical mean μ and the hypothetical variance σ^2. The most frequently used assumption is that the distribution that we are sampling from is a so-called normal distribution with parameters μ and σ^2. By doing so, we can make it much simpler to analyze the data. We will have much to say about these kinds of things in the remaining chapters of this book.

As we alluded to above, the histogram of Figure (7.1) is just one of many simple descriptive statistical tools that we have at our disposal. We also have so-called dot plots, box plots, stem and leaf displays, pie charts, and many more tools of this kind. Sometimes, we can recognize outliers in the data, or we can see whether the data (and hence the underlying distribution) may be unimodal or bimodal, or skewed, and so on. This is all part of the branch of statistics known as descriptive statistics, which we won't dwell upon in this book. The point of this section that we want to emphasize is the relationship between a collection of numbers (the data), and the use of a probability model to simplify our view of the data, and at the same time, make it much easier to analyze the data.

(7.2) Simple Random Sample (SRS)

Our starting point is the conducting of some experiment. For each repetition of the experiment, there is some kind of measurement of interest, which we call the random variable X. We will assume that we can conduct the experiment a large or potentially infinite number of times.

In this book we will consider large and small samples for our discussion of statistical inference. However, most of the emphasis will be on large samples in chapters 9 and 10. As the reader will see when we discuss normal distributions in the next chapter, there is a powerful result that we call the Central Limit Theorem which is very useful in a lot of areas of statistical inference when we have large samples. In most statistics books, there is the rule of thumb that a sample size of $n \geq 30$ is considered a large sample. Even though we will discuss small samples in chapter 11, we take the position that sample sizes should always be as large as possible because it makes the statistical techniques of data analysis more powerful and easier to implement.

We repeat the experiment "n" independent times to get a random sample $\{x_1, x_2, \ldots, x_n\}$ of n observations of the random variable X. The number of items in the sample is called the sample size n. In this book, we have assumed that the random variable X has M possible outcomes. When all possible samples of size n have the same chance of being selected, then we say that we have a simple random sample (SRS) of size n.

Some examples of sampling scenarios are given below. We have divided them into two different types of situations, the first is where n < M, and the second is where n > M. When we take a simple random sample of size n, usually the number of outcomes of the random variable X that appears in the sample is much fewer than the total number of possible outcomes M. However, in some situations the opposite is the case. In both cases we can still have simple random samples. We will now give some examples of both situations.

(A) The case where n < M is the most common situation. What is important in an experimental situation is that we are able to assume that the repetitions of the experiment produce independent outcomes of the random variable X, and hence representative samples. Though it is very common to have repeated outcomes in our samples, this is usually due to our inability to measure quantities with very high precision. When we get repeated outcomes, we can still assume that all the items in the sample are independent outcomes of the random variable X and we will have a representative SRS. Let's consider a few examples designed to illustrate the concepts:

Example (1):
A technician may take a sample of n = 40 measurements of the amount of soda dispensed from a soda fountain. The amount of soda dispensed from the fountain is a random variable X that varies randomly around 16 ounces, and we assume that each measurement is independent of the

others. This random variable X takes a certain number of outcomes M. The number M is very large when we can take each of the measurements with a high degree of precision. So, we assume that n < M. In this case, it is reasonable to say that that we have a SRS of size n = 40.

Example (2):
A demographics researcher measures the weight in pounds of n = 55 female students chosen at random from a large university. The random variable X is the weight of a female college student between the ages of 18 and 30 inclusive. Each of the weight measurements can be assumed to be independent measurements of the random variable X. The random variable X can take a large number M of outcomes if we could weigh the students with a high degree of precision. So, we assume that n < M. There is a very large number of female college students at this university and we can consider this to be a SRS of size n = 55.

Example (3):
A biologist may be interested in the random variable X, where X is the average speed of travel through the water for a particular species of fish living in the Atlantic Ocean (measured in meters per second). The researcher has special photographic equipment in order to make these measurements. She takes a sample of n = 100 speed measurements from fish of this species in many different locations in the Atlantic, and we can assume that each measurement is independent of the others. The random variable X can take a very large number M of outcomes if

she can measure the speeds very accurately. So, we can assume that n < M. The number of fish of this species in the ocean is very large and we can easily consider this to be a SRS of size n = 100.

(B) The less common situation is where n > M, but we can still have simple random samples that are equally-likely. We will consider some examples to illustrate the concepts.

Example (4):
Let a circular board with a pointer that can spin rapidly when set into motion, be such that it will land with equal probability on one of three $120°$ sectors, labelled sectors A, B, and C, and coded 1, 2, and 3 respectively. The outcome of the experiment is the random variable X with outcomes {1, 2, 3}. We can set the pointer spinning n = 100, n = 1000, or any number of times, and we can assume that the outcome of any given spin is independent of the others. Our sample of size n is clearly much greater than the number of outcomes M = 3. However, each such sample for a fixed n is equally-likely and we can consider it to be a SRS of size n.

Example (5):
A group of physicists conduct an experiment in a particle accelerator, where they smash together n = 100 trillion proton and anti-proton pairs. The result of each collision is a random variable X with M = 34 outcomes (one of 34 fundamental particles). The sample is a set of n = 100

trillion recorded observations, where there are only M = 34 outcomes for X. In addition, we can assume that the result of any given collision is independent of the others. Each such sample for a fixed n is equally-likely and we can consider it to be a SRS of size n.

(7.3) The Statistics $\bar{X}$ and S^2

A statistic is a function of the sample, a number which is derived from the sample, and is a random variable with a distribution. The main statistics which we will consider are the sample mean $\bar{X}$, the sample variance S^2, and the sample standard deviation S, which estimate μ, σ^2, and σ. The most important aspects of any distribution are the central tendency and the dispersion or variability of the outcomes. Before we deal further with these three statistics and the concepts of central tendency and variability, let's discuss a little more about the properties of statistics, as considered to be estimators of the parameters of a random variable X.

Mean Square Error (MSE) for a Statistic
As we have stated, generally a statistic is an estimator of a distribution parameter. It is a point estimator because when it is computed from a sample, it is a single number. Let's call $\hat{\beta}$ our statistic and β the parameter of some distribution that $\hat{\beta}$ is supposed to estimate. $\hat{\beta}$ is a random variable, so it has some distribution. For most statistics, $\hat{\beta}$ depends on the sample size n, and the $Var(\hat{\beta})$ converges to

0 as n gets larger. So, the distribution of $\hat{\beta}$ converges to a single point. To understand the following discussion, note that the $E(\hat{\beta})$ and β are numbers. Then the MSE$(\hat{\beta})$ is:

$$\text{MSE}(\hat{\beta}) = E(\hat{\beta} - \beta)^2 = E(\hat{\beta} - E(\hat{\beta}) + E(\hat{\beta}) - \beta)^2$$

$$= E(\hat{\beta} - E(\hat{\beta}))^2 + E(E(\hat{\beta}) - \beta)^2 +$$

$$E(2 \cdot (\hat{\beta} - E(\hat{\beta})) \cdot (E(\hat{\beta}) - \beta))$$

Since the $E(\hat{\beta} - E(\hat{\beta})) = 0$ and the $(E(\hat{\beta}) - \beta)$ is a number, the third term is 0. So, we have:

$$\text{MSE}(\hat{\beta}) = E(\hat{\beta} - E(\hat{\beta}))^2 + E(E(\hat{\beta}) - \beta)^2.$$

The first term is the $Var(\hat{\beta})$. The second term involves the so-called Bias$(\hat{\beta}) = (E(\hat{\beta}) - \beta)$. Therefore, the

$$\text{MSE}(\hat{\beta}) = Var(\hat{\beta}) + (Bias(\hat{\beta}))^2.$$

An estimator is unbiased if the $E(\hat{\beta}) = \beta$. This is a very desirable property of an estimator. Otherwise, the estimator is considered biased.

An estimator is called consistent if it is biased, but the bias converges to 0 as the sample size n gets larger and larger, so that in the limit as n gets larger beyond all bounds, the distribution of the estimator $\hat{\beta}$ converges to the parameter β, a single number.

Central Tendency and Dispersion

The sample average $\bar{X}$ is the most commonly used measure of central tendency, or where the center of the distribution is on the real line. If a distribution is symmetric, then the mean is at the center. If a distribution is non-symmetric, then the mean is not at the center. A non-symmetric distribution is called skewed (the outcomes are pulled more to one side or the other). A skewed distribution is either skewed left or skewed right.

The median and the mode are two other measures of central tendency that are commonly used, but especially the median. The median is usually less affected by outlying observations (or outliers) than $\bar{X}$ as a measure of the center. However, $\bar{X}$ is the most important and commonly used measure of the center of a distribution because it is an unbiased estimator of the distribution mean μ (the median is not), and it is the easiest to work with mathematically.

The sample variance S^2 and the sample standard deviation S are the two most important and commonly used measures of dispersion. S^2 is an unbiased estimator of the distribution variance σ^2, and S is a consistent estimator of the distribution standard deviation σ. There are other measures of dispersion such as the Range, which is the largest member of a sample minus the smallest value. Another measure of dispersion is the inter-quartile range

(IQR), which is the middle 50% of the sample, the 75th percentile minus the 25th percentile. Still another measure of dispersion is the mean absolute deviation (MAD), which is the average value of all the $|x_i - \bar{x}|$. The MAD is not easy to work with mathematically. So, S^2 and S are the most common measures of dispersion used in the subject of probability and statistics because they are relatively easy to work with mathematically.

The Sample Mean $\bar{X}$

The sample mean $\bar{X}$ is one of the most important statistics in the subject of statistics. For a SRS of size n, denoted by $\{x_1, x_2, \ldots, x_n\}$, which are n independent outcomes of a random variable X associated with an experiment,

$$\bar{X} = \left(\frac{1}{n}\right) \sum_{i=1}^{n}(x_i)$$

That is, the sample mean is simply the sum of the n outcomes of the random variable X, divided by the number of outcomes in the sample. It is a weighted average of the outcomes of the experiment, where the weight associated with each outcome is simply $\left(\frac{1}{n}\right)$.

(A) The statistic $\bar{X}$ is an unbiased estimator of the distribution mean μ, meaning that $E(\bar{X}) = \mu$. We can easily prove this. (Note that each item in the sample is a random variable with mean μ and variance σ^2):

$$E(\bar{X}) = E\left(\left(\tfrac{1}{n}\right)\sum_{i=1}^{n} X_i\right) = \left(\tfrac{1}{n}\right)\sum_{i=1}^{n} E(X_i)$$

(since the expected value of a sum is the sum of the expected values)

$$= \left(\tfrac{1}{n}\right)(n\mu) = \mu.$$

(B) The variance of $\bar{X}$ is an important quantity in Statistics:

$$\text{Var}(\bar{X}) = \text{Var}\left(\left(\tfrac{1}{n}\right)\sum_{i=1}^{n} X_i\right) = \left(\tfrac{1}{n^2}\right)\text{Var}(\sum_{i=1}^{n} X_i)$$

$$= \left(\tfrac{1}{n^2}\right)\left[\sum_{i=1}^{n}\text{Var}(X_i) + \sum_{i\neq j}\text{Cov}(X_i, X_j)\right]$$

(Since the n random variables $X_i, i = 1,2,\ldots,n$ are independent of each other, the $\text{Cov}(X_i, X_j) = 0$, when $i \neq j$)

$$= \left(\tfrac{1}{n^2}\right)\sum_{i=1}^{n}(\sigma^2) = \left(\tfrac{1}{n^2}\right)(n\sigma^2) = \frac{\sigma^2}{n}.$$

So, we write $\sigma_{\bar{X}}^2 = \frac{\sigma^2}{n}$. The variance of $\bar{X}$ gets smaller as n increases. This just makes sense since the more observations that one has, the smaller will be the variability of their average, or in other words, the more accurately one can estimate the true mean μ of the distribution. Both a single observation X_i and the average $\bar{X}$ of n of them have the same expected value (mean), but $\bar{X}$ has a smaller variance. It gets smaller and smaller as n increases.

(C) The standard deviation of $\bar{X}$ is called the standard error of the mean. Since the (Standard Deviation) $= \sqrt{(\text{Variance})}$ for any random variable:

$$\sigma_{\bar{X}} = \sqrt{\sigma_{\bar{X}}^2} = \sqrt{\frac{\sigma^2}{n}} = \frac{\sigma}{\sqrt{n}} .$$

The Sample Variance S^2

The sample variance S^2 is the most common estimator of the distribution variance σ^2. For a SRS of size n, denoted by $\{x_1, x_2, \dots, x_n\}$, which are n independent outcomes of a random variable X associated with an experiment, we calculate the sample variance by:

$$S^2 = \left(\frac{1}{n-1}\right) \sum_{i=1}^{n} (X_i - \bar{X})^2 .$$

Note that we have to divide by (n − 1) so that S^2 will be an unbiased estimator of σ^2. We can show this to be true:

$$E(S^2) = E\left[\left(\frac{1}{n-1}\right) \sum_{i=1}^{n} (X_i - \bar{X})^2 \right]$$

$$= \left(\frac{1}{n-1}\right) E[\sum_{i=1}^{n} (X_i - \bar{X})^2]$$

$$= \left(\frac{1}{n-1}\right) E\left(\sum_{i=1}^{n} X_i^2 - 2\bar{X} \sum_{i=1}^{n} X_i + \sum_{i=1}^{n} (\bar{X}^2)\right)$$

403

$$= \left(\frac{1}{n-1}\right) E \left(\sum_{i=1}^{n} X_i^2 - 2n(\bar{X})^2 + n(\bar{X}^2)\right)$$

(Since $\sum_{i=1}^{n} X_i = n\bar{X}$)

$$= \left(\frac{1}{n-1}\right) \left(E \left(\sum_{i=1}^{n} X_i^2\right) - nE(\bar{X}^2)\right)$$

$$= \left(\frac{1}{n-1}\right) \left(\sum_{i=1}^{n} E(X_i^2) - nE(\bar{X}^2)\right)$$

(Now, for any random variable X,
the $E(X^2) = \mu_X^2 + \sigma_X^2$)

$$= \left(\frac{1}{n-1}\right) \left(n(\mu^2 + \sigma^2) - n\left(\mu^2 + \frac{\sigma^2}{n}\right)\right)$$

$$= \left(\frac{1}{n-1}\right)(n-1)\sigma^2$$

$$= \sigma^2 .$$

The variance of S^2 can be shown to be $\left(\frac{2\sigma^4}{n-1}\right)$, when we assume that we are sampling from normal distributions (to be discussed in the next chapter). Clearly, this converges to 0 as the sample size n gets larger beyond all bounds. It is just very plausible that for distributions that we commonly sample from, that the variance of S^2 (where S^2 has an expected value of σ^2) must go to zero as n gets larger beyond all bounds.

The Sample Standard Deviation S

The sample standard deviation S is another very important measure of dispersion in a distribution, it estimates σ. We have stated that S is a consistent estimator. It seems that at least some of the time it is a biased estimator, but the bias converges to 0 as n gets larger. In other words it is the case with S that at least sometimes the $E(\sqrt{S^2}) \neq \sqrt{E(S^2)}$.

The sample standard deviation is simply the square root of the sample variance:

$$S = \sqrt{S^2} = \sqrt{\left(\left(\frac{1}{n-1}\right)\sum_{i=1}^{n}(X_i - \bar{X})^2\right)} \ .$$

The variance of S also converges to 0 for most distributions that we would sample from, as the sample size n gets larger beyond all bounds. This is just very plausible since S is the square root of S^2.

Example (1):

We have the following sample $\{-3, -4, -4, -1, 0\}$, which is a SRS of size n = 5 for some random variable X. Compute the sample mean $\bar{X}$, the sample variance S^2, and the sample standard deviation S.

(A) $\bar{X} = \frac{\sum_{i=1}^{n} X_i}{n} = \frac{(-3-4-4-1+0)}{(5)} = -\left(\frac{12}{5}\right) = -2.4$.

(B) $S^2 = \frac{\sum_{i=1}^{n}(X_i - \bar{X})^2}{n-1}$

$$= \frac{[(-3+2.4)^2+(-4+2.4)^2+(-4+2.4)^2+(-1+2.4)^2+(0+2.4)^2]}{(5-1)}$$

$$= \frac{[0.36 + 2.56 + 2.56 + 1.96 + 5.76]}{4} = 3.3 .$$

(C) $S = \sqrt{S^2} = \sqrt{3.3} \approx 1.8166$.

Example (2):

We have the following sample $\{1, 3, 5, 7, 7, 11\}$, which is a SRS of size n = 6 for some random variable X. Compute $\bar{X}$, S^2, and S.

(A) $\bar{X} = \frac{\sum_{i=1}^{n} X_i}{6} = \frac{(1+3+5+7+7+11)}{6} = \frac{34}{6} = 5.667$.

(B) $S^2 = \frac{\sum_{i=1}^{n}(X_i-\bar{X})^2}{n-1}$

$$= \frac{[(1-5.667)^2+ \cdots + (11-5.667)^2]}{6-1}$$

$$= \frac{(21.781 + \cdots + 28.441)}{5} = 12.267 .$$

(C) $S = 3.502$.

################# Exercises #################

Determine n, $\bar{X}$, S^2, and S for each of the following sets of data. Since these samples are somewhat larger than what we have seen in the previous two examples, the reader may make use of a calculator that would make short order of the process. However, the reader should be sure that they understand enough of what is actually involved in such calculations:

(1) Sample = {10,12,23,14,17,15,15,19,16,13,15}

(2) Sample = {0,1,1,1,0,1,0,0,1,1,1,1,0,0,0,1,0,1,1}

(3) Sample = {100.1, 97.6, 99.3, 101.4, 103.0, 99.8,
                  100.4, 101.9, 100.6, 102.1}

(7.4) Chebyshev's Inequality

This inequality is useful for getting conservative estimates of probabilities concerning the distribution of outcomes for a random variable X. In particular, we will use it to describe the distribution of $\bar{X}$ in the next section on the Law of Large Numbers.

Let's now derive Chebyshev's Inequality. It is applicable to all distributions, but for purposes of simplification, we will derive it for a uniform distribution. That is, for a random variable X with M outcomes, where the probability of occurrence of each possible outcome is $\left(\frac{1}{M}\right)$.

We start with: (note that we need $k \geq 1$)

$$\sigma^2 = \left(\frac{1}{M}\right) \Sigma_{i=1}^{M}(x_i - \mu)^2$$

$$= \left(\frac{1}{M}\right) \left(\Sigma_{x_i:\ |x_i-\mu|<k\sigma}(x_i - \mu)^2 + \Sigma_{x_i:\ |x_i-\mu|\geq k\sigma}(x_i - \mu)^2\right)$$

$$\geq \left(\frac{1}{M}\right)\left(\Sigma_{x_i:\ |x_i-\mu|\geq k\sigma}(x_i-\mu)^2\right)$$

$$\geq \left(\frac{1}{M}\right)\Sigma_{x_i:\ |x_i-\mu|\geq k\sigma}(k^2\sigma^2)$$

Now, dividing both sides by $(k^2\sigma^2)$, we get:

$\Sigma_{x_i:\ |x_i-\mu|\geq k\sigma}\left(\frac{1}{M}\right)\leq \frac{1}{k^2}$. The left side of this inequality is the $\Pr(|X-\mu|\geq k\sigma)$.

So, we have Chebyshev's Inequality in two equivalent forms:
$$\Pr(|X-\mu|\geq k\sigma)\leq \frac{1}{k^2}$$
$$\Pr(|X-\mu|\leq k\sigma)\geq 1-\frac{1}{k^2}$$

This inequality says that for any random variable X:

The $\Pr(|X-\mu|\leq \sigma)\geq \left(1-\frac{1}{1^2}\right)=0$.

The $\Pr(|X-\mu|\leq 2\sigma)\geq \left(1-\frac{1}{2^2}\right)=\frac{3}{4}$.

The $\Pr(|X-\mu|\leq 3\sigma)\geq \left(1-\frac{1}{3^2}\right)=\frac{8}{9}$, and so on.

These are obviously very conservative probabilities. This is the price that we pay for the fact that they apply to all possible distributions.

(7.5) The Law of Large Numbers (LLN)

Chebyshev's inequality in the form:

$\Pr(|X - \mu| \le k\sigma) \ge \left(1 - \frac{1}{k^2}\right)$ can be applied to $\bar{X}$, which

has mean μ and standard deviation $\frac{\sigma}{\sqrt{n}}$. We have:

$$\Pr\left(|\bar{X} - \mu| \le k\frac{\sigma}{\sqrt{n}}\right) \ge \left(1 - \frac{1}{k^2}\right).$$

Let ε be an arbitrarily small positive real number. Then, let $k = \frac{\varepsilon\sqrt{n}}{\sigma}$. Since k must be greater than or equal to 1, n may sometimes have to be very large. This depends on the value of ε and σ. The important point here is that in repeated sampling n will eventually be large enough so that the necessary conditions will be met. So, we have:

$$\Pr(|\bar{X} - \mu| \le \varepsilon) \ge \left(1 - \frac{1}{\left(\frac{\varepsilon^2 (n)}{\sigma^2}\right)}\right) = \left(1 - \left(\frac{\sigma^2}{(n)\varepsilon^2}\right)\right).$$

This says that the $\Pr(|\bar{X} - \mu| \le \varepsilon) \ge \left(1 - \left(\frac{\sigma^2}{(n)\varepsilon^2}\right)\right)$, which

converges to 1, as n gets larger beyond all bounds.

This is known as the Law of Large Numbers, or the LLN. In advanced statistical theory based on some high powered mathematics, there is a weak form and a strong form of the LLN. What we have derived here is the weak form of the LLN, but they both say essentially the same thing. This probability statement says that the distribution of $\bar{X}$ is a highly peaked distribution with low variance, more and more so as the sample size n gets larger beyond all bounds. When the distribution of $\bar{X}$ becomes very highly peaked with very low variability, then we can develop statistical techniques for hypothesis testing and for estimating parameters that can be very powerful and useful for the analysis of data. Though we will not refer to the LLN very much at all, what it says about the distribution of $\bar{X}$ is a big part of some of the final chapters of this book.

(8) <u>COMMON PROBABILITY DISTRIBUTIONS</u>

(8.1) <u>Bernoulli(p) Random Variables</u>

A Bernoulli (p) random variable X, or a Bernoulli trial, is one that takes only the values 0 and 1, where the $Pr(X = 1) = p$, and the $Pr(X = 0) = (1 - p)$. Sometimes we use the generic terms success and failure, coded as 1 and 0 respectively for the two outcomes. We can summarize this by writing the pmf for X:

$$p_X(x) = \begin{cases} p, & \text{for } x = 1 \\ 1 - p, & \text{for } x = 0 \end{cases}$$

The $E(X) = [(1)(p) + (0)(1 - p)] = p$
The $E(X^2) = [(1^2)(p) + (0^2)(1 - p)] = p$
The $E(X^n) = [(1^n)(p) + (0^n)(1 - p)] = p$, for all n.

Therefore, the mean $\mu = p$, and

the variance is $\sigma^2 = E(X^2) - (E(X))^2 = p - p^2$
$$= p(1 - p).$$

(8.2) Binomial(n,p) Random Variables

Binomial Random Variables and Number of Successes

Let $\{X_1, X_2, \ldots, X_n\}$ be n independent Bernoulli(p) random variables, and let

$$X = (X_1 + X_2 + \cdots + X_n).$$

Then we call X a Binomial (n,p) random variable. The random variable X takes the (n + 1) outcomes $\{0,1,2,\ldots,n\}$, which are the number of successes in the n independent trials where p = Pr(Success) is the same for each of the n trials.

$$
\begin{aligned}
\text{The } E(X) &= E(X_1 + X_2 + \cdots + X_n) \\
&= E(X_1) + E(X_2) + \cdots + E(X_n) \\
&= np.
\end{aligned}
$$

$$
\begin{aligned}
\text{The } Var(X) &= Var(X_1 + X_2 + \cdots + X_n) \\
&= Var(X_1) + Var(X_2) + \cdots + Var(X_n)
\end{aligned}
$$
(Since these n random variables are independent, the $Cov(X_i, X_j) = 0$ when $i \neq j$. So, all the covariance terms are 0)
$$= np(1-p).$$

The outcome of n Bernoulli trials is a sequence of n (0)'s and (1)'s, where (x) of them are (1)'s and (n − x) of them are (0)'s. So, there are $\binom{n}{x}$ different ways that they could be arranged when you consider that they occur

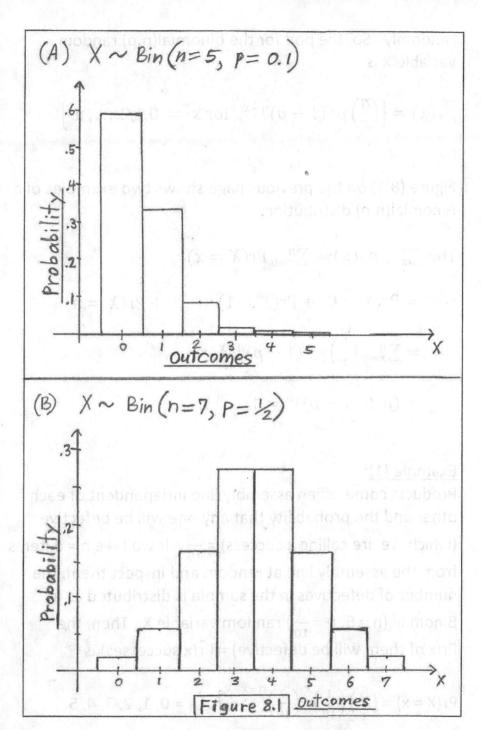

(A) $X \sim Bin(n=5, p=0.1)$

(B) $X \sim Bin(n=7, P=\frac{1}{2})$

Figure 8.1 Outcomes

413

randomly. So, the pmf for the binomial(n,p) random variable X is

$$p_X(x) = \left\{ \binom{n}{x} p^x (1-p)^{n-x}, \text{ for x } = 0,1,2,\ldots, n. \right\}.$$

Figure (8.1) on the previous page shows two examples of a Binomial(n,p) distribution.

The $\sum_{x=0}^{n} p_X(x) = \sum_{x=0}^{n} \Pr(X = x)$

$$= \Pr(X = 0) + \Pr(X = 1) + \cdots + \Pr(X = n)$$

$$= \sum_{x=0}^{n} \binom{n}{x} p^x (1-p)^{n-x}$$

$$= (p + (1-p))^n = 1$$

Example (1):
Products come off an assembly line independent of each other and the probability that any one will be defective (which we are calling a success) $= \frac{1}{10}$. If we take n = 5 items from the assembly line at random and inspect them, the number of defectives in the sample is distributed as a Binomial (n = 5, p = $\frac{1}{10}$) random variable X. Then, the Pr(x of them will be defective) = Pr(x successes) is

$$\Pr(X = x) = \binom{5}{x} \left(\frac{1}{10}\right)^x \left(\frac{9}{10}\right)^{n-x} \text{, for x = 0, 1, 2, 3, 4, 5.}$$

414

These probabilities are:

$$Pr(X = 0) = \binom{5}{0}\left(\frac{1}{10}\right)^0 \left(\frac{9}{10}\right)^{5-0} = \left(\frac{9}{10}\right)^5 = 0.59049$$

$$Pr(X = 1) = \binom{5}{1}\left(\frac{1}{10}\right)^1 \left(\frac{9}{10}\right)^{5-1} = (5)\left(\frac{1}{10}\right)^1 \left(\frac{9}{10}\right)^4 = 0.32805$$

$$Pr(X = 2) = \binom{5}{2}\left(\frac{1}{10}\right)^2 \left(\frac{9}{10}\right)^{5-2} = (10)\left(\frac{1}{10}\right)^2 \left(\frac{9}{10}\right)^3 = 0.07290$$

$$Pr(X = 3) = \binom{5}{3}\left(\frac{1}{10}\right)^3 \left(\frac{9}{10}\right)^{5-3} = (10)\left(\frac{1}{10}\right)^3 \left(\frac{9}{10}\right)^2 = 0.00810$$

$$Pr(X = 4) = \binom{5}{4}\left(\frac{1}{10}\right)^4 \left(\frac{9}{10}\right)^{5-4} = (5)\left(\frac{1}{10}\right)^4 \left(\frac{9}{10}\right)^1 = 0.00045$$

$$Pr(X = 5) = \binom{5}{5}\left(\frac{1}{10}\right)^5 \left(\frac{9}{10}\right)^{5-5} = \left(\frac{1}{10}\right)^5 = 0.00001$$

$$\sum_{x=0}^{5} Pr(X = x) = (0.59049 + 0.32805 + 0.07290$$
$$+0.00810 + 0.00045 + 0.00001) = 1$$

The expected number of defectives in a random sample of n = 5 products from the assembly line is :

$$\mu = np = (5)\left(\frac{1}{10}\right) = 0.5$$

The variance is:
$\sigma^2 = np(1 - p) = 0.45$, and the standard deviation is:
$\sigma = \sqrt{\sigma^2} = \sqrt{np(1 - p)} = \sqrt{0.45} \approx 0.671$.

Example (2):
A woman has 7 pregnancies, with only one child per pregnancy and no miscarriages. Consider a boy a success.

(A) What is the Pr(2 Boys and 5 girls)?
Considering that the outcome of the 7 pregnancies is a
Binomial random variable X with n = 7 independent trials
and with the Pr(success) = Pr(Boy) = $\frac{1}{2}$ for each trial. The

$$Pr(X = 2) = \binom{7}{2}\left(\frac{1}{2}\right)^2\left(\frac{1}{2}\right)^5 = (21)\left(\frac{1}{128}\right) = \frac{21}{128} \approx (0.164)$$

(B) What is the probability of 0 boys?
$$Pr(X = 0) = \binom{7}{0}\left(\frac{1}{2}\right)^7 = \left(\frac{1}{128}\right) \approx (0.0078)$$

(C) What is the probability of at least one boy?
The Pr(At least one boy) = Pr(1, 2, 3, 4, 5, 6, or 7 boys)
$$= 1 - Pr(0\ Boys) = (1 - 0.0078)$$
$$= (0.9922).$$

The expected number of Boys is
$$\mu = E(X) = np = (7)\left(\frac{1}{2}\right) = 3.5$$

The variance is $\sigma^2 = np(1 - p) = (7)\left(\frac{1}{2}\right)\left(\frac{1}{2}\right) = 1.75$, and
the standard deviation $\sigma = \sqrt{np(1 - p)} = \sqrt{1.75} \approx 1.323$.

Example (3):
A coin is tossed n = 10 times. Let (success) = (Heads), and
the Pr(success) = Pr(Heads) = $\frac{1}{2}$.

The number of successes X is a Binomial ($n = 10$, $p = \frac{1}{2}$) random variable if the coin is fair, meaning that the Pr(success) $= \frac{1}{2}$ on each trial, and each toss is independent of the others.

What is the probability of 3 or 4 heads?
$$Pr(X = 3 \text{ or } 4) = Pr(X = 3) + Pr(X = 4)$$
$$= \binom{10}{3}\left(\frac{1}{2}\right)^{10} + \binom{10}{4}\left(\frac{1}{2}\right)^{10}$$
$$\approx (0.1172) + (0.2051) = (0.3223)$$

The Expected number of Heads is
$$\mu = E(X) = np = (10)\left(\frac{1}{2}\right) = 5 .$$

The variance of X is $\sigma^2 = np(1 - p) = (10)\left(\frac{1}{2}\right)\left(\frac{1}{2}\right) = 2.5$,
The standard deviation of X is $\sigma = \sqrt{\sigma^2} = \sqrt{2.5} \approx 1.5811$.

Binomial Random Variables and Proportion of Successes
Another statistic that is important in connection with a Binomial(n,p) random variable X (We will see it a lot) is:

$$\bar{p} = \frac{X}{n} = \frac{\text{(number of successes)}}{\text{(number of trials)}} = \text{the proportion of successes.}$$

(Note that $0 \le \bar{p} \le 1$).
The mean of $\bar{p} = \mu_{\bar{p}} = E(\bar{p}) = E\left(\frac{X}{n}\right) = E\left(\frac{1}{n}(X)\right)$

$$= \left(\frac{1}{n}E(X)\right) = \frac{1}{n}(np) = p.$$

The Variance of $\bar{p} = \sigma_{\bar{p}}^2 = Var(\bar{p}) = Var\left(\frac{X}{n}\right) = Var\left(\frac{1}{n}X\right)$

$$= \left(\frac{1}{n^2}\right)Var(X) = \left(\frac{1}{n^2}\right)(np(1-p))$$

$$= \frac{(p)(1-p)}{n}$$

The Standard Deviation of $\bar{p} = \sigma_{\bar{p}} = \sqrt{\frac{p(1-p)}{n}}$.

As an example, a coin is tossed n = 10,000 times (with the help of a computer). Then X is a Binomial (n = 10,000, p = $\frac{1}{2}$) random variable, where X is the number of successes, which is the number of Heads. If there are X = 4980 Heads in the 10,000 tosses, then the proportion of Heads is:

$$\bar{p} = \frac{4980}{10,000} = 0.4980 \text{ , and it has variance}$$

$$\sigma_{\bar{p}}^2 = \frac{(p)(1-p)}{n} = \frac{\left(\frac{1}{2}\right)\left(\frac{1}{2}\right)}{10,000} = \frac{1}{40,000} = 0.000025$$

The standard deviation $\sigma_{\bar{p}} = \sqrt{\sigma_{\bar{p}}^2} = \sqrt{0.000025} = 0.005$

################ Exercises ################

(1) An interested television broadcaster conducts an opinion poll. She asks n = 30 people chosen at random from the population of Syracuse, New York, whether they believe that global warming is a real man-made phenomenon (which will be called a success and coded as 1), or whether they believe that global warming is just a Chinese hoax (which will be called a failure and coded as 0). The result of her poll is X = 12 successes.

(a) The experiment is modeled as the outcome of a Binomial(n = 30,p) random variable, where p is unknown. What is the sample proportion of successes $\bar{p}$? Then using this $\bar{p}$ as an estimate of the true value of p, what is the approximate standard deviation of $\bar{p}$?

(b) If indeed the true proportion p is 0.6, approximately how many standard deviations below 0.6 is the observed sample proportion $\bar{p}$?

(c) If indeed the true value of p is 0.6, how many successes from a poll of 30 people would you expect?

(2) If a fair coin is flipped 30 times,
(a) What is the expected number of Heads?
(b) What is the standard deviation?
(c) What is the interval:
     (expected number of heads) ± (2 standard deviations) ?
(d) What is the expected proportion of Heads?

419

(e) What is the standard deviation of this proportion?

(f) What is the interval:

(expected proportion) $\pm$ (2 standard deviations) ?

(8.3) The Poisson Random Variable

Poisson distributions are useful when we consider the number of occurrences of some event in a specific interval of time, where the events are occurring randomly through time, but at a constant average rate. The poisson pmf can be derived from the binomial pmf.

If we have an interval of time, and we divide it into a large number n of equal sub-intervals and we assume that the Pr(occurrence of some event in the ith sub-interval) is a small number p, where p gets smaller as n $\rightarrow \infty$, because the mean number of occurrences (np) in the larger time interval must remain constant. Furthermore, we assume that there can only be one occurrence in any one of the smaller sub-intervals. The occurrence of the event in the ith sub-interval is called a success, and the occurrence of the event in any sub-interval is independent of whatever happens in any other sub-interval. So, X = (the number of occurrences of the event in the larger time interval) is a binomial (n,p) random variable that takes the values $\{0, 1, 2, \ldots, n\}$, where n is getting infinitely large. So,

$$Pr(X = x) = \binom{n}{x} p^x (1-p)^{n-x} = \frac{[(n)(n-1)\cdots(2)(1)] \; p^x (1-p)^{n-x}}{[(n-x)(n-x-1)\cdots(2)(1)] \; (x!)}.$$

This can be written:

$$Pr(X = x) = \left(\frac{(n)\cdots(n-(x-1))\cdot(n-x)!}{(n-x)!}\right)\left(\frac{(n^x)(p^x)}{(n^x)(1-p)^x}\right)\frac{(1-p)^n}{x!}$$

$$= \left(\left(\frac{n}{n-np}\right)\cdots\left(\frac{n-(x-1)}{n-np}\right)\right)\cdot\left(1-\frac{(np)}{n}\right)^n\cdot\frac{(np)^x}{x!}$$

Now, (np) remains constant and small compared to n, as n gets larger. Also, (x) remains small compared to n, as n gets larger. Therefore, the large factor on the left converges to 1 as n gets larger. There is a result in mathematics that says $\left(1-\frac{(np)}{n}\right)^n \rightarrow e^{-(np)}$, as n gets larger. So, let the average number of occurrences (np) be denoted by μ. Then as n gets larger, the random variable X which was binomial (n,p), becomes a poisson random variable, which takes an infinite number of outcomes, with pmf:

The $Pr(X = x) = \left\{e^{-\mu}\left(\frac{\mu^x}{x!}\right), \text{ for } x = 0, 1, 2, 3, \ldots\right\}$.

Here we have a countably infinite sample space. Of course, not all of the infinite number of outcomes could occur in the real world, but the use of this pmf shows how infinity can be used to our advantage in calculating certain probabilities.

The sum of the probabilities of all the outcomes is:

$$\sum_{x=0}^{\infty}(e^{-\mu})\left(\frac{\mu^x}{x!}\right) = (e^{-\mu})\sum_{x=0}^{\infty}\frac{\mu^x}{x!}.$$

For those that have studied calculus, the infinite sum on

the right is (e^{μ}), so the sum of the probabilities for all the infinite number of possible outcomes is $(e^{-\mu})(e^{\mu})$ which equals 1.

The mean of X, where X is a Poisson (μ) random variable, is the average number of occurrences μ in the time interval under consideration. To calculate the probability of the outcomes of the Poisson (μ) random variable, we need to be given the average number of events μ in the time interval.

As an example, let the random variable X be the number of radioactive decays of some element in a given hour. If the average number of radioactive decays per hour is known to be 5, then the random variable X is distributed as a poisson $(\mu = 5)$ random variable. We can calculate the probability of 3 decays in an hour as:

$$Pr(X = 3) = (e^{-5}) \left(\frac{5^3}{3!}\right) \approx (0.006738)(20.833) = (0.1404).$$

The probability of 0 radioactive decays in an hour is:
$$Pr(X = 0) = (e^{-5}) \left(\frac{5^0}{0!}\right) \approx (0.006738).$$

So, the probability of at least 1 radioactive decay in an hour is the probability of (1 or more decays in an hour) equals $Pr(X \geq 1) = 1 - Pr(X = 0) \approx 1 - (0.006738) = 0.9933$.

################# Exercises ###################

(1) A machine for making widgets breaks down at random and at an average rate of $\mu = 3$ times per day. Modeling the random variable X = the number of breakdowns per day as a Poisson($\mu = 3$) random variable:
(a) What is the Pr(At least two breakdowns per day)?
(b) What is the Pr(0 breakdowns in a single day)?

(8.4) The Normal Distribution

Probability Density Functions

Throughout this book we have stated that the outcomes that a random variable X can take is a finite set, called the sample space S with M outcomes. In a situation where M can be very large, we will assume that the set of possible outcomes is uncountably infinite on some part of the real line. It turns out that this assumption is common and very useful. The sample space S becomes an uncountably infinite set of real numbers despite the reality that the number of outcomes M is actually very large and finite in number. So once again we will be using infinity to our advantage. This allows us to simplify things and make certain probabilities much easier to calculate, by utilizing continuous functions.

In many situations in the real world, the outcomes of a random variable are concentrated around some center and trail off in density on the real line as the distance from the center increases. In other words, a lot of distributions are mound-shaped. We can model the so-called probability density for the set of outcomes with a continuous curve, and we call it a probability density function, a pdf f(x).

So, the pmf p(x) is replaced by a pdf f(x). This pdf is a continuous curve on some part of the real line, which may be a bounded interval [a, b], or some unbounded part of the real line such as [0,∞) or (−∞, ∞). When we model a distribution in this way, we are forced to say that the probability of any single outcome is 0 rather than some very small positive number (like the height of the curve) because the sum of an uncountably infinite set of very small positive numbers would be ∞. So, we have a situation where the events with non-zero probability are intervals of the real line rather than single points. If (c, d) is a subset of the domain, then the interval (c, d) is an event, and the $\Pr(X \in (c, d))$ is the area under the pdf and above the real number interval (c, d). The area under the curve and over the entire domain on the real line is 1. This just says that the Pr(S) = 1, as the case must always be. The pdf f(x) is a way of modeling the density of probability for a random variable over its entire domain with a continuous function. This turns out to be very useful mathematically. As we will now see, one of the most common and important models that we use in this way is the so-called normal distribution.

Normal Distributions

For normal distributions, the pdf is what we commonly call the bell-shaped curve. The pdf of a normal distribution with the two parameters μ and σ, is given by the function:

$$f(x) = \frac{1}{\sqrt{2\pi}\,\sigma} \cdot e^{-\frac{1}{2}\left(\frac{x-\mu}{\sigma}\right)^2}, \quad -\infty < x < \infty.$$

See Figure (13) and Figure (13.5) on the next two pages. Note that the set of possible outcomes is $(-\infty, \infty)$ instead of a bounded interval [a, b]. The normal curve is mound shaped in the middle and quickly trails off on either side, getting infinitesimally close to the real line as it extends to $-\infty$ on the left and to $+\infty$ on the right. It is symmetrically shaped with the center at $x = \mu$ (the mean), and its spread is determined by its other parameter σ^2 (the variance). If the variance σ^2 is small, then the hump will be more highly peaked and the distribution is concentrated tightly about μ. If σ^2 is large, then the hump is lower and the distribution is more spread out about the mean μ.

To calculate areas (probabilities) under the pdf f(x) of a normal random variable requires the integral calculus and the assistance of a computer. We need to use a computer because there is no antiderivative for the function f(x) given above, which models the probability density for all normal distributions. We do not expect the reader to know the calculus. To calculate probabilities corresponding to areas under a normal distribution curve, we must use tables, which have been tabulated with a computer.

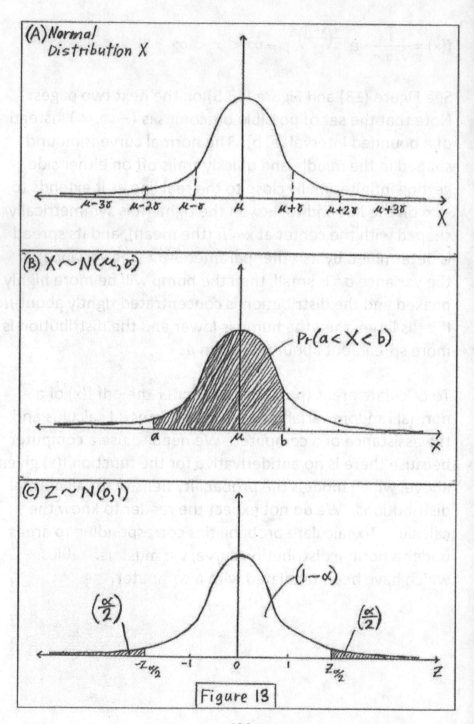

(A) Normal
    Distribution X

$\mu-3\sigma$ $\mu-2\sigma$ $\mu-\sigma$ μ $\mu+\sigma$ $\mu+2\sigma$ $\mu+3\sigma$ X

(B) $X \sim N(\mu, \sigma)$

$Pr(a < X < b)$

a μ b X

(C) $Z \sim N(0,1)$

$\left(\dfrac{\alpha}{2}\right)$ $(1-\alpha)$ $\left(\dfrac{\alpha}{2}\right)$

$-Z_{\alpha/2}$ -1 0 1 $Z_{\alpha/2}$ Z

Figure 13

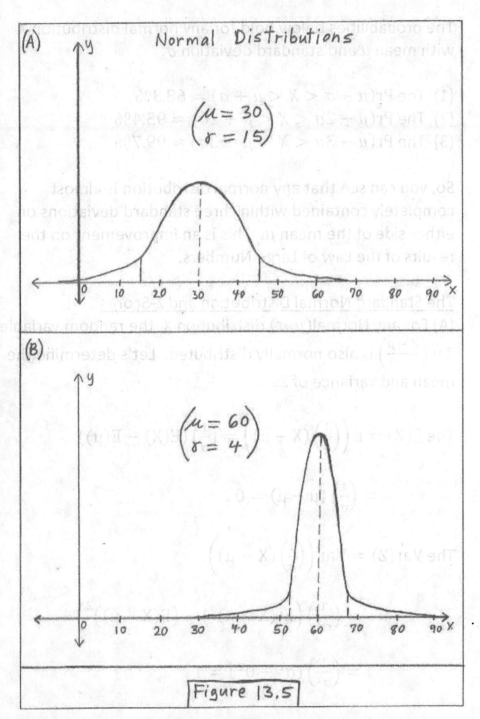

Figure 13.5

427

The probabilities below hold for any normal distribution X with mean μ and standard deviation σ:

(1) The $\Pr(\mu - \sigma < X < \mu + \sigma) \approx 68.3\%$
(2) The $\Pr(\mu - 2\sigma < X < \mu + 2\sigma) \approx 95.4\%$
(3) The $\Pr(\mu - 3\sigma < X < \mu + 3\sigma) \approx 99.7\%$

So, you can see that any normal distribution is almost completely contained within three standard deviations on either side of the mean μ. This is an improvement on the results of the Law of Large Numbers.

The Standard Normal Distribution and Z-Scores
(A) For any Normal(μ, σ) distribution X, the random variable $Z = \left(\dfrac{X - \mu}{\sigma}\right)$ is also normally distributed. Let's determine the mean and variance of Z:

The $E(Z) = E\left(\left(\dfrac{1}{\sigma}\right)(X - \mu)\right) = \left(\dfrac{1}{\sigma}\right)(E(X) - E(\mu))$

$$= \left(\dfrac{1}{\sigma}\right)(\mu - \mu) = 0.$$

The $\text{Var}(Z) = \text{Var}\left(\left(\dfrac{1}{\sigma}\right)(X - \mu)\right)$

$$= \left(\dfrac{1}{\sigma^2}\right)\left(E((X - \mu)^2) - (E(X - \mu))^2\right)$$

$$= \left(\dfrac{1}{\sigma^2}\right)(\sigma^2 - 0^2) = 1.$$

We say that Z has a standard normal distribution or a Z distribution, and we write: $Z \sim N(0,1)$. This is a very important normal distribution in the subject of probability and statistics.

(B) For a normally distributed random variable X with mean μ and standard deviation σ, we have an X-score, and a Z-score, where

$$Z = \left(\frac{X - \mu}{\sigma}\right) \leftrightarrow X = \mu + \sigma Z .$$

A Z-score for X is the distance between X and its mean μ in terms of the standard deviation σ of X.

We can think of the standard deviation of a random variable X as a typical deviation of X from its mean.

For example, if X has mean $\mu = 60$ and standard deviation $\sigma = 5$, then the Z-scores for X = 62, 55, 50, and 68 are:

$$Z = \left(\frac{62 - 60}{5}\right) = 0.40 \qquad Z = \left(\frac{55 - 60}{5}\right) = -1.00$$

$$Z = \left(\frac{50 - 60}{5}\right) = -2.00 \qquad Z = \left(\frac{68 - 60}{5}\right) = 1.60$$

X = 62 and 68 are (0.40) and (1.60) standard deviations above their mean respectively.
X = 50 and 55 are (2.00) and (1.00) standard deviations below their mean respectively.

The larger a Z-score is, the more significantly an X-value is separated from its mean. X-values that deviate greatly from their mean are considered to be a less-likely occurrence.

Calculations Concerning Normal Distributions
For normal distributions X, we find the $\Pr(X \in (c, d))$ from a table of the cdf of the standard normal distribution (which we call the Z distribution). The reader should recall the cdf for a distribution from section 6.4. In appendix A, we have tabulated the probabilities that $Z \leq z$, for every value of z from -3.49 to +3.49 in increments of (.01), that is the cdf for the distribution of Z for all z-values from -3.49 standard deviations below the mean to 3.49 standard deviations above the mean. It turns out that this range of z-values is essentially all that is needed to find a probability of interest for any normally distributed random variable X. The table values in Appendix A will be denoted by $\Phi(z)$. Once again, to belabor the point, Appendix A provides the cumulative probability for the standard normal curve, denoted $\Phi(z)$, for values of Z up to z (-3.49 $\leq z \leq$ 3.49), or in other words:

$$\Phi(z) = \Pr(Z \leq z), \text{ where } (-3.49 \leq z \leq 3.49),.$$

There are three different types of calculations that we will ever need to make with normal distributions. Let a and b be real numbers, where a is less than b, which are outcomes of a normally distributed random variable X with mean μ and standard deviation σ.

430

(1) $\Pr(X \le a) = \Pr\left(Z \le \left(\frac{a-\mu}{\sigma}\right)\right) = \Phi\left(\frac{a-\mu}{\sigma}\right).$

(2) $\Pr(a \le X \le b) = \Pr\left(Z \le \left(\frac{b-\mu}{\sigma}\right)\right) - \Pr\left(Z \le \left(\frac{a-\mu}{\sigma}\right)\right)$

$$= \Phi\left(\frac{b-\mu}{\sigma}\right) - \Phi\left(\frac{a-\mu}{\sigma}\right)$$

(3) $\Pr(X \ge a) = \Pr\left(Z \ge \left(\frac{a-\mu}{\sigma}\right)\right) = 1 - \Pr\left(Z \le \left(\frac{a-\mu}{\sigma}\right)\right)$

$$= 1 - \Phi\left(\frac{a-\mu}{\sigma}\right).$$

Example (1):

Applicants to a medical school must take an entrance exam designed by the school. The minimum score is 20 and the maximum score is 180. Historical data shows that the exam scores X are approximately normally distributed with mean $\mu = 105$ and standard deviation $\sigma = 9$. The school will admit you if you score 110 or higher.

(A) What is the probability that you will be denied admission?

The Pr(rejection) = Pr(X < 110)

$$= \Pr\left(Z < \left(\frac{110 - 105}{9}\right)\right)$$

$$= \Pr(Z < 0.56)$$

$$= \Phi(0.56) \approx (0.7123) = 71.23\%$$

$\Phi(0.56)$ was found from the table for the standard normal distribution in Appendix A.

(B) Given that you get accepted, what is the probability that you got a score of 130 or higher?

This is a conditional probability.

Let A be the event that you scored 130 or higher.
Let B be the event that you scored 110 or higher.
We seek the $\Pr(A|B) = \dfrac{\Pr(A \cap B)}{\Pr(B)}$.

In terms of the random variable X, we want to find the

$$\frac{\Pr(X \geq 130)}{\Pr(X \geq 110)} = \frac{1 - \Pr(X \leq 130)}{1 - \Pr(X \leq 110)} = \frac{1 - \Pr\left(Z \leq \left(\frac{130 - 105}{9}\right)\right)}{1 - \Pr\left(Z \leq \left(\frac{110 - 105}{9}\right)\right)}$$

$$= \frac{1 - \Pr(Z \leq 2.78)}{1 - \Pr(Z \leq 0.56)} = \frac{1 - \Phi(2.78)}{1 - \Phi(0.56)} = \frac{1 - (0.9973)}{1 - (0.7123)} = \frac{0.0027}{0.2877}$$

$$\approx (0.0094) = 0.94\%. \text{ (Just about 1\%)}$$

Example (2):
Average lifetime human body temperature X actually varies a little bit from person to person. It is now believed to be approximately normally distributed with mean $\mu = 98.6$ and $\sigma = 0.13$.

(A) What is the Pr(98 < X < 99)?

The Pr(98 < X < 99) = $Pr\left(\left(\frac{98-98.6}{0.13}\right) \leq Z \leq \left(\frac{99-98.6}{0.13}\right)\right)$

$= Pr(-4.62 \leq Z \leq 3.08)$

$= \text{Pr}(Z \leq 3.08) - \text{Pr}(Z \leq -4.62)$

$= \Phi(3.08) - \Phi(-4.62)$

$\approx (0.9990) - (0.0000)$

$= (0.9990)$, which is just about 100%.

(B) Doctors are concerned about average lifetime body temperatures for humans. They say that those with average lifetime body temperatures greater than 98.77 degrees are less likely to live beyond age 80 than those with lower average temperatures. What percentage of humans have an average body temperature greater than 98.77 degrees?

They want to know the Pr(X ≥ 98.77).
The Pr(X ≥ 98.77) = $1 - Pr(X \leq 98.77)$

$$= 1 - \Pr\left(Z \leq \left(\frac{98.77-98.6}{0.13}\right)\right)$$

$$= 1 - \Pr(Z \leq 1.31)$$

$$= 1 - \Phi(1.31)$$

$$\approx 1 - (0.9049)$$

$$= (0.0951) = 9.51\%$$

Example (3):
Human IQ scores X are believed to be approximately normally distributed with mean $\mu = 100$ and $\sigma = 16$. By one account, Einstein had an IQ of 160. What is the Pr(any given person has an IQ $\geq$ 160)?

We seek the $\Pr(X \geq 160) = 1 - \Pr(X \leq 160)$

$$= 1 - Pr\left(Z \leq \left(\frac{160-100}{16}\right)\right)$$

$$= 1 - \Pr(Z \leq 3.75)$$

$$= 1 - \Phi(3.75)$$

$$\approx (0.0000) \text{ (very much less than 1 \%)}$$

<u>Example (4):</u>

We can often use a normal distribution to more easily calculate certain binomial distribution probabilities. If X is distributed binomial (n,p) with mean μ = np and standard deviation $\sigma = \sqrt{n(p)(1-p)}$, then if the practical interval $\left(np - 3\sqrt{np(1-p)}, \ np + 3\sqrt{np(1-p)}\right)$ is a subset of the interval (0, n), then we can use the normal distribution to calculate certain probabilities for the random variable X. For example, assume X is distributed binomially with parameters n = 40 and p = .4. Then the mean of X is μ = np = 16 and its standard deviation $\sigma = \sqrt{np(1-p)}$ = 3.098.

The interval $\left(np - 3\sqrt{np(1-p)}, \ np + 3\sqrt{np(1-p)}\right)$ is (6.705, 25.295) $\subset$ (0, 40). So, we can use the normal distribution with parameters μ = 16 and σ = 3.098 to find, for example, the Pr(13 $\leq X \leq$ 19). Note we use here what statisticians call a continuity correction:

The Pr(13 $\leq X \leq$ 19) = Pr(12.5 $\leq X \leq$ 19.5)

$= \Pr\left(\frac{12.5-16}{3.098} < Z < \frac{19.5-16}{3.098}\right) = \Pr(-1.13 < Z < 1.13)$

$= \Phi(1.13) - \Phi(-1.13) = (.8708 - .1292) = (.7416)$.

The exact value is: $\sum_{x=13}^{19} \binom{40}{x} (.4)^x (.6)^{40-x} = (.7417)$.

Very close agreement indeed!

The Central Limit Theorem (CLT)

The Central Limit Theorem (CLT) is one of the most important theorems in the subject of Probability and Statistics. It is the basis for the techniques of large sample statistical inference that we will discuss in some of the remaining chapters of this book.

We already know from the Law of Large Numbers that for any random variable X, the distribution of $\bar{X}$ becomes highly-peaked with decreasing variability as n gets larger. We know that the $E(\bar{X}) = \mu$, and that the $\text{Var}(\bar{X}) = \frac{\sigma^2}{n}$, where μ and σ^2 are the mean and the variance of the parent distribution that we are sampling from. It turns out that we know even more about the distribution of $\bar{X}$ for large samples, which comes from the so-called Central Limit Theorem. At this time, we want to state that we will write $(n \to \infty)$ to mean "as n gets larger beyond all bounds." The proof of the CLT requires techniques of advanced calculus which are beyond the scope of this book, so we will state the CLT without proof.

The Central Limit Theorem:

(Part 1) For any distribution X, the distribution of $\bar{X}$ for large samples is approximately normally distributed with mean μ and variance $\frac{\sigma^2}{n}$, and the approximation improves as $n \to \infty$.

Therefore, for large samples, $Z = \left(\dfrac{\bar{X} - \mu}{\left(\frac{\sigma}{\sqrt{n}}\right)} \right)$ has approximately a Normal(0,1) distribution, and the approximation improves as $n \to \infty$.

(Part 2) The sample standard deviation $S \approx \sigma$ for large samples. So, when we don't know σ we can substitute S for σ and we have the important result that for large samples, $Z = \left(\dfrac{\bar{X} - \mu}{\left(\frac{S}{\sqrt{n}}\right)}\right)$ has approximately a Normal(0,1) distribution, and the approximation improves as $n \to \infty$.

Calculations Concerning $\bar{X}$ for Large Samples

Example (1):

A physicist plans to measure the momentum of $n = 45$ bullets one second after they have been fired from a new type of rifle that the army wants to buy. The manufacturer claims that the momemtum X of a rifle bullet one second after firing has a mean $\mu = 375$ and a standard deviation $\sigma = 48$. Since we have a large sample ($n = 45$), the CLT says that the distribution of $\bar{X}$ is approximately normal with parameters $\mu = 375$, and standard deviation $\sigma = \dfrac{48}{\sqrt{45}}$.

We write that $\bar{X} \sim N(\mu = 375, \ \sigma = 7.15)$.

(A) What is the $\Pr(370 \leq \bar{X} \leq 380)$?

The $\Pr(370 \leq \bar{X} \leq 380) = \Pr\left(\left(\dfrac{370-375}{7.15}\right) \leq Z \leq \left(\dfrac{380-375}{7.15}\right)\right)$

$= \Pr(-.70 \leq Z \leq .70)$

$$= \Pr(Z \le .70) - \Pr(Z \le -.70)$$

$$= \Phi(.70) - \Phi(-.70) = (0.7580) - (0.2420)$$

$$= (0.516) = 51.6\%.$$

(B) What is the probability that $\bar{X} \ge 385$?

$$\text{The } \Pr(\bar{X} \ge 385) = \Pr\left(Z \ge \left(\frac{385-375}{7.15}\right)\right) = \Pr(Z \ge 1.40)$$

$$= 1 - \Pr(Z \le 1.40) = 1 - (0.9192)$$

$$= (0.0808) = 8.08\%.$$

In sampling n = 45 firings we would find that the sample average $\bar{X}$ is greater than 385 about 8.08% of the time. If this seems strange, since the mean for a single bullet is 375, it's only because of the high variability $\sigma = 48$ of the random variable X.

Example (2):
We flip a coin n = 400 times and define a success as "Heads." The statistic $\bar{p} = \dfrac{X}{n} = \dfrac{(\# \, Heads)}{(\# \, Tosses)}$ is an average of 400 $0's$ and $1's$. So, just like $\bar{X}$, the CLT applies to $\bar{p}$ as well because it's an average.

In fact, as n → ∞, $\bar{p}$ is distributed normally with mean $\mu = p$ and standard deviation $\sigma = \sqrt{\dfrac{p(1-p)}{n}}$. In this case, the mean is $p = \frac{1}{2} = (0.5)$, and the standard deviation is

$$\sqrt{\frac{(0.5)(0.5)}{400}} = (0.025).$$

(A) What is the $\Pr(0.47 \leq \bar{p} \leq 0.53)$?

The $\Pr(0.47 \leq \bar{p} \leq 0.53)$

$$= \Pr\left(\left(\frac{0.47-0.50}{0.025}\right) \leq Z \leq \left(\frac{0.53-0.50}{0.025}\right)\right)$$

$$= \Pr(-1.20 \leq Z \leq 1.20)$$

$$= \Pr(Z \leq 1.20) - \Pr(Z \leq -1.20)$$

$$= \Phi(1.20) - \Phi(-1.20) = (0.8849) - (0.1151)$$

$$= (0.7698) = 76.98\%.$$

(B) What is the probability that in 400 flips, the number of "Heads" is between 185 and 225 inclusive?

The $\Pr(185 \leq X \leq 225) = \Pr\left(\frac{185}{400} \leq \bar{p} \leq \frac{225}{400}\right)$

$$= \Pr(0.4625 \leq \bar{p} \leq 0.5625)$$

$$= \Pr\left(\left(\frac{0.4625-0.50}{0.025}\right) \leq Z \leq \left(\frac{0.5625-0.50}{0.025}\right)\right)$$

$= \Pr(-1.50 \leq Z \leq 2.50)$

$= \Pr(Z \leq 2.50) - Pr(Z \leq -1.50)$

$= \Phi(2.50) - \Phi(-1.50)$

$= (0.9938) - (0.0668)$

$= (0.9270) = 92.70\%$

Example (3):
We sample 100 adult Rattlesnakes and record their weight X. The random variable X has a mean $\mu = 2.17$ pounds with standard deviation $\sigma = 0.29$ pounds.

(A) What is the $\Pr(2.10 \leq \bar{X} \leq 2.20)$ when n = 100?

The $\Pr(2.10 \leq \bar{X} \leq 2.20)$

$= \Pr\left(\left(\dfrac{2.10-2.17}{\left(\frac{0.29}{\sqrt{100}}\right)}\right) \leq Z \leq \left(\dfrac{2.20-2.17}{\left(\frac{0.29}{\sqrt{100}}\right)}\right)\right)$

$= \Pr(-2.41 \leq Z \leq 1.03)$

$= \Pr(Z \leq 1.03) - \Pr(Z \leq -2.41)$

$= \Phi(1.03) - \Phi(-2.41)$

$\approx (0.8485) - (0.0080)$

$= (0.8405) = 84.05\%$

(B) What is the $Pr(\bar{X} \geq 2.30)$ when n = 100?

The $Pr(\bar{X} \geq 2.30) = Pr\left(Z \geq \left(\frac{2.30 - 2.17}{\left(\frac{0.29}{\sqrt{100}} \right)} \right) \right)$

$= Pr(Z \geq 4.48) = 1 - Pr(Z \leq 4.48)$

$= (1 - \Phi(4.48)) \approx (0.0000)$

There is practically no chance of sampling the weights of n = 100 adult Rattlesnakes and finding the sample average $\bar{X}$ to be greater than or equal to 2.30 pounds.

Chi-Square Distributions $\left(\chi^2_{(k)} \right)$

In the final chapters of this book, we will have some use for so-called Chi-Square distributions, which always have a parameter k called the degrees of freedom associated with them. We will sometimes denote a Chi-Square distribution with k degrees of freedom as $\chi^2_{(k)}$. See Figure (14) on the next page. There are several important ways in which this important family of distributions crops up. Some important cases are given on the following pages:

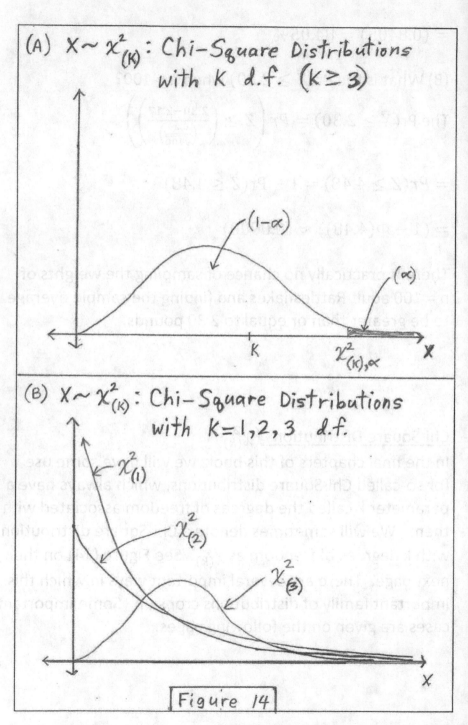

(A) $X \sim \chi^2_{(K)}$: Chi-Square Distributions with K d.f. (K ≥ 3)

$(1-\alpha)$

(α)

K

$\chi^2_{(K), \alpha}$

X

(B) $X \sim \chi^2_{(K)}$: Chi-Square Distributions with K = 1, 2, 3 d.f.

$\chi^2_{(1)}$

$\chi^2_{(2)}$

$\chi^2_{(3)}$

X

Figure 14

(a) If $X = Z^2$, where $Z \sim N(0,1)$, then X has a Chi-Square distribution with 1 degree of freedom.

(b) If we have a SRS of k normal(μ, σ) random variables, then the sum $X = \sum_{i=1}^{k}(Z_i)^2 = \sum_{i=1}^{k}\left(\frac{X_i-\mu}{\sigma}\right)^2$, has a Chi-Square distribution with k degrees of freedom.

(c) If we have a large SRS of n random variables with mean μ and standard deviation σ, then $X = \left(\frac{\bar{X}-\mu}{\left(\frac{\sigma}{\sqrt{n}}\right)}\right)^2$ has a Chi-Square distribution with 1 degree of freedom.

(d) If we have a large SRS of n normal(μ, σ) random variables, then $X = \sum_{i=1}^{n}\left(\frac{X_i-\bar{X}}{\sigma}\right)^2 = \frac{(n-1)s^2}{\sigma^2}$ has a Chi-Square distribution with (n − 1) degrees of freedom.

Since a Chi-square random variable X ($X \sim \chi^2_{(k)}$) is the sum of k independent squared standard normal random variables, it follows that X is always positive valued, that is, it takes values in the interval $[0, \infty)$. For k = 1, the curve comes down from $+\infty$ (asymptotic to the y-axis) and trails off toward 0 as $x \to \infty$. For k = 2, the curve comes downward from a certain intercept on the y-axis and trails off toward

0 as $x \to \infty$. For $k \geq 3$, X starts at the origin and is skewed right and mound-shaped with a mean $\mu = k$ and trails off toward 0 as $x \to \infty$. Therefore, all Chi-square curves are skewed to the right and get infinitesimally close to the x-axis as $x \to \infty$. They are pdf's and therefore the area under them all on the interval $[0,\infty)$ is 1.

We have tabulated certain percentiles of the Chi-square distribution, for degrees of freedom from 1 to 40, in the table labeled Appendix B.

****

For those that have studied calculus, the distributions of so-called exponential and chi-square random variables are special cases of Gamma(α, β) random variables. The pdf of a Gamma(α, β) random variable is:

$$f(x) = \left\{ \frac{x^{\alpha-1}e^{-\frac{x}{\beta}}}{\Gamma(\alpha)\ \beta^\alpha},\ \ x \geq 0 \right\}, \text{ where } \Gamma(\alpha) = \int_0^\infty x^{\alpha-1}e^{-x}dx.$$

(a) If $X \sim \chi^2_{(k)}$, then X has a special type of Gamma distribution where $\alpha = \dfrac{k}{2}$ and $\beta = 2$. The pdf of X is:

$$f(x) = \left\{ \frac{x^{\frac{k}{2}-1}e^{-\frac{x}{2}}}{\Gamma\left(\frac{k}{2}\right)(2)^{\frac{k}{2}}},\ \ x \geq 0 \right\}.$$

444

(b) For the special case where $\alpha = 1$, then:

$$f(x) = \left\{ \left(\frac{1}{\beta}\right) e^{-\frac{x}{\beta}}, \ x \geq 0 \right\}.$$

This is the pdf of an exponential random variable, which has a lot of applications in science and engineering.

****

Some of the topics in the next chapters on hypothesis testing and confidence intervals, which concern a variance σ^2 or a standard deviation σ, involve Chi-Square random variables (Chi-square distributions).

################# Exercises #################

(1) If X ~ Normal($\mu = 20, \sigma = 3$),
   (a) What is the $Pr(X \leq 18.5)$?
   (b) What is the $Pr(X \geq 22)$?
   (c) What is the $Pr(19 < X < 21)$?

(2) If X ~ Normal($\mu = 1000, \sigma = 50$),
   (a) What is the $Pr(X \leq 910)$?
   (b) What is the $Pr(X \geq 1080)$?
   (c) What is the $Pr(1000 < X < 1100)$?

(3) A scientist takes a SRS of size n = 60 from a random

variable X with mean $\mu = 105$ and $\sigma = 17$.
(a) What is the approximate distribution of $\bar{X}$?
(b) What is the $\Pr(\bar{X} \leq 106.5)$?
(c) What is the $\Pr(\bar{X} \geq 102)$?
(d) What is the $\Pr(103 < \bar{X} < 107)$?

(4) A scientist studies a random variable X that is believed
to be distributed Bernoulli(p = .38). He takes a SRS of
size n = 100 independent measurements and considers
the variable Y = $\sum_{i=1}^{100} X_i$.
(a) What is the distribution of Y?
(b) Why is the scientist justified in considering the
distribution of $\bar{p}$ to be approximately normally
distributed?
(c) What is the $\Pr(\bar{p} \geq .45)$?
(d) What is the $\Pr(.35 < \bar{p} < .41)$?
(e) What is the $\Pr(\bar{p} \leq .32)$

(9) <u>LARGE SAMPLE HYPOTHESIS TESTING</u>

(9.1) <u>Introduction</u>

We now turn our attention to some methods of statistical inference. Statistical inference is centered about two main topics, hypothesis testing and confidence interval estimation for distribution parameters.

In this chapter, we will discuss hypothesis testing for the parameters μ, p, and σ^2. Hypothesis testing is all about deciding between two competing statements concerning the true value of an unknown distribution parameter.

Hypothesis tests for a mean μ and for a proportion p are based directly on the Central Limit Theorem when n is large. The CLT says that there is no need to require sampling from normal distributions for large n because the distribution of $\bar{X}$ will always be approximately normal for large sample sizes regardless of the type of distribution we are sampling from, and the approximation gets better as n increases. As a rule of thumb, a large sample size is when n $\geq$ 30.

In contrast, hypothesis testing for the parameter σ^2 does not involve the CLT. When performing tests for variances, we will assume that we are sampling from normal distributions, and that there is no need for large sample sizes.

Test Procedure

Testing procedures generally consist of the steps outlined below. What we are testing for is the true but unknown value of a parameter associated with a certain distribution.

(1) Decide on the hypotheses that we want to test. This involves choosing a so-called null hypothesis H_0, based on so-called null values μ_0, p_0, or σ_0^2 for a parameter, and an associated alternative hypothesis H_a. The null hypothesis states that the null value is the true value of a parameter. We conduct a test to try to determine whether a parameter is equal to the null value, versus an alternative hypothesis that the parameter is less than, greater than, or not equal to the null value (either lesser or greater). These are the three forms that an alternative hypothesis H_a can take. The burden of proof for the experimenter is with the alternative hypothesis. We need to have significant evidence to reject the null hypothesis and accept the alternative hypothesis.

For a mean μ, the three scenarios are:

$$\begin{cases} H_o: \mu = \mu_0 \\ H_a: \mu < \mu_0 \end{cases} \qquad \begin{cases} H_o: \mu = \mu_0 \\ H_a: \mu > \mu_0 \end{cases} \qquad \begin{cases} H_o: \mu = \mu_0 \\ H_a: \mu \neq \mu_0 \end{cases}$$

For a proportion p, the three scenarios are:

$$\begin{Bmatrix} H_o: p = p_0 \\ H_a: p < p_0 \end{Bmatrix} \quad \begin{Bmatrix} H_o: p = p_0 \\ H_a: p > p_0 \end{Bmatrix} \quad \begin{Bmatrix} H_o: p = p_0 \\ H_a: p \neq p_0 \end{Bmatrix}$$

For a variance σ^2, the three scenarios are:

$$\begin{Bmatrix} H_o: \sigma^2 = \sigma_0^2 \\ H_a: \sigma^2 < \sigma_0^2 \end{Bmatrix} \quad \begin{Bmatrix} H_o: \sigma^2 = \sigma_0^2 \\ H_a: \sigma^2 > \sigma_0^2 \end{Bmatrix} \quad \begin{Bmatrix} H_o: \sigma^2 = \sigma_0^2 \\ H_a: \sigma^2 \neq \sigma_0^2 \end{Bmatrix}$$

(2) After you have conducted your experiment and collected your sample, calculate the statistics that are relevant to your testing situation, such as $\bar{X}$, $\bar{p}$, S^2, and S.

(3) Calculate your test statistic. In what we will we do here this will be either a Z-statistic Z_0 or a Chi-square statistic χ_0^2. This is the quantity which will be used to discriminate between the null hypothesis H_o and the alternative hypothesis H_a. Since there is always variability with sample data, we always run the risk of making an error. We can make one of two possible errors, a type I or a type II error.

A type I error is the rejection of a true null hypothesis and making the inference that the true parameter value is something different. This type of error we have control over and we say that it has probability α, which is chosen by the experimenter beforehand. We call α the level of significance for the test. We always construct our test statistic as if the null hypothesis is true, so we have control over when we will reject the null hypothesis. We reject the null hypothesis in such a way that makes the probability of a type I error equal to a certain chosen α.

A type II error is failing to reject the null hypothesis when it is false. We call the probability of a type II error β. This is more difficult to compute because we usually don't know what the true alternative value of the parameter is. We can compute β if we do in fact know the correct alternative value for the parameter. Tables for this are available in more advanced statistics books.

The Power of a test has the probability $(1 - \beta)$, it is the probability of rejecting a false null hypothesis. Sometimes this is computed for various alternative values of the parameter and tabulated. The smaller β is for our test, by being able to correctly decide for a true alternative value, the greater is the power $(1 - \beta)$ of the test. In this treatment of the subject, by considering only large samples, we greatly improve the power of our tests. The greater the power of our tests, the greater the chance that we make correct inferences from experimental data.

(4) Determine your p-value (to be discussed soon) and make your conclusion. If the p-value is less than α, then we reject the null hypothesis and conclude that the true but unknown parameter value is either less than, greater than, or simply different from the null value. If the p-value is greater than α, then we fail to reject the null hypothesis. The p-value

tells us how strongly we reject or fail to reject the null hypothesis.

The Test Statistic
Most of the time we make a test about a mean μ or a proportion p. In this case we will need to calculate a Z-statistic Z_0:

$$Z_0 = \left(\frac{\bar{X} - \mu_0}{\left(\frac{S}{\sqrt{n}}\right)}\right) \quad \text{or} \quad Z_0 = \left(\frac{\bar{p} - p_0}{\sqrt{\frac{p_0(1-p_0)}{n}}}\right), \text{ where } Z_0 \sim N(0,1).$$

If we want to make a test about a variance σ^2, we will need to calculate a χ^2-statistic χ_0^2:

$$\chi_0^2 = \left(\frac{(n-1)S^2}{\sigma_0^2}\right), \text{ where } \chi_0^2 \sim \chi_{(n-1)}^2.$$

The (p-value) for tests concerning a mean or proportion
A p-value is a probability and will be used to make a conclusion. If we are testing about a mean μ or about a proportion p, and we calculate a Z-statistic Z_0, then the p-value is one of these three probabilities:

(1) If your test is whether a mean μ is a null value μ_0 or if a proportion p is a null value p_0, versus the alternative that the true μ is less than μ_0 or the true p is less than p_0, then the p-value = $\Pr(Z \leq Z_0)$.

(2) If your test is whether a mean μ is a null value μ_0 or if a proportion p is a null value p_0, versus the alternative that the true μ is greater than μ_0 or the true p is greater than p_0, then the p-value = $\Pr (Z \geq Z_0)$.

(3) If your test is whether a mean μ is a null value μ_0 or if a proportion p is a null value p_0, versus the alternative that the true μ is different from μ_0 or the true p is different from p_0 (different meaning either lesser or greater), then the

p-value = $(\Pr(Z \leq -Z_0) + \Pr(Z \geq Z_0))$, (if $Z_0 > 0$), or the
p-value = $(\Pr(Z \leq Z_0) + \Pr(Z \geq -Z_0))$, (if $Z_0 < 0$).

The (p-value) for tests concerning variances
If we are testing about a variance σ^2 and we calculate a χ^2-statistic χ_0^2, then your p-value is one of these three possibilities:

(1) If your test is whether a variance σ^2 is a null value σ_0^2 or whether the true σ^2 is less than σ_0^2, then the p-value = $\Pr(\chi_{(n-1)}^2 \leq \chi_0^2)$.

(2) If your test is whether a variance σ^2 is a null value σ_0^2 or whether the true σ^2 is greater than σ_0^2, then the p-value = $\Pr(\chi_{(n-1)}^2 \geq \chi_0^2)$.

(3) If your test is whether a variance σ^2 is a null value σ_0^2 or whether the true σ^2 is simply different than σ_0^2, then the

p-value $= \Pr\left(\chi^2_{(n-1)} \leq \chi^2_0\right)$, if χ^2_0 is in the lower half of the $\chi^2_{(n-1)}$ distribution, or the

p-value $= \Pr\left(\chi^2_{(n-1)} \geq \chi^2_0\right)$, if χ^2_0 is in the upper half of the $\chi^2_{(n-1)}$ distribution.

(9.2) Tests for the Mean μ

Example (1):
A researcher conducts an experiment and is concerned with a random variable X. He wants to test the hypotheses at the $\alpha = .05$ level:

$$\begin{cases} H_o: \mu = 53 \\ H_a: \mu > 53 \end{cases}$$

He takes a SRS of size n = 64. He calculates $\bar{X} = 54.9$ and S = 5.61. Taking advantage of the CLT for large samples, he computes the test statistic:

$$Z_0 = \left(\frac{\bar{X} - \mu_0}{\frac{S}{\sqrt{n}}}\right) = \left(\frac{54.9 - 53}{\frac{5.61}{\sqrt{64}}}\right) = 2.71$$

The p-value is the $\Pr(Z \geq Z_0) = \Pr(Z \geq 2.71)$
$$= 1 - \Pr(Z \leq 2.71)$$
$$= (1 - 0.9966) = 0.0034 .$$

Since the p-value (0.0034) $\ll$ (.05), he can very confidently reject H_0 and accept $H_a: \mu > 53$.

Example (2):
An economist is conducting an experiment and is concerned with a random variable X. He wants to test the hypotheses at the $\alpha = .01$ level:

$$\begin{cases} H_o: \mu = 650 \\ H_a: \mu \neq 650 \end{cases}$$

He takes a SRS of size n = 100. He calculates $\bar{X}$ = 645 and S = 32.6. Taking advantage of the CLT for large samples, he computes the test statistic:

$$Z_0 = \left(\frac{\bar{X} - \mu_0}{\left(\frac{S}{\sqrt{n}}\right)}\right) = \left(\frac{645 - 650}{\left(\frac{32.6}{\sqrt{100}}\right)}\right) = -1.53$$

The p-value is the $\Pr(Z \leq -1.53) + \Pr(Z \geq 1.53)$
$$= 2 \cdot \Pr(Z \leq -1.53)$$
(because of the symmetry of normal distributions about their mean)
$$= (2 \cdot (0.0630)) = 0.1260 .$$

Since the p-value (0.1260) $\gg$ (.01), he fails to reject H_0 and accepts that the mean of his random variable X is 650.

Example (3):
A store manager conducts an experiment and is concerned with a random variable X. He wants to test the hypotheses at the $\alpha = .10$ level:

$$\begin{cases} H_0: \mu = 214 \\ H_a: \mu < 214 \end{cases}$$

He takes a SRS of size n = 90. He calculates $\bar{X}$ = 212.99 and S = 7.13. Taking advantage of the CLT for large samples, he computes the test statistic:

$$Z_0 = \left(\frac{\bar{X} - \mu_0}{\left(\frac{S}{\sqrt{n}}\right)}\right) = \left(\frac{212.99 - 214}{\left(\frac{7.13}{\sqrt{90}}\right)}\right) = -1.34$$

The p-value is the $Pr(Z \le Z_0) = Pr(Z \le -1.34)$
$$= (0.0901).$$

Since the p-value = (0.0901) < (.10), he rejects H_0 and accepts that $\mu < 214$. This result is statistically significant, but just barely. It would not have been significant at the $\alpha = .05$ level. The manager thinks that he should repeat the experiment again with an even larger sample size to see what happens in a follow-up experiment.

################# Exercises #################

(1) A researcher is trying to determine whether the mean of a random variable X is equal to 140, or whether it is simply different from 140, at the $\alpha = .01$ level.
   (a) What hypotheses is he trying to test?
   (b) He takes a simple random sample of size n = 70 and calculates $\bar{X} = 141.50$ and $S^2 = 8.13$. What is the value of his test statistic Z_0?
   (c) What is the distribution of Z_0 and why?

(d) What is the p-value and his conclusion?

(2) A researcher takes a SRS of size n = 36. He is interested in determining if the mean μ of a certain random variable X which he is studying is less than 52, at the $\alpha = .05$ level.

(a) What hypotheses is he trying to test?

(b) He calculates $\bar{X} = 51.5$ and $S^2 = 180.36$ from his data. What is the value of his test statistic Z_0?

(c) What is the distribution of Z_0 and why?

(d) What is the p-value and his conclusion?

(9.3) Tests for a Proportion p

Example (1):

A pollster is interested in the proportion of voters that are in favor of a certain candidate A (success) or not in favor of candidate A (failure). Of course, candidate A wins if he gets more than 50% of the votes. So, the pollster wants to test the following hypotheses at the $\alpha = .05$ level:

$$\begin{cases} H_o: p = .50 \\ H_a: p > .50 \end{cases}$$

The pollster counts the number of successes among a SRS of size n = 800 potential voters chosen at random from the electorate, and finds 432 in favor of candidate A. So,

$\bar{p} = \frac{432}{800} = (0.54)$. Under the null hypotheses, $p_0 = (0.50)$. We always set up our test statistic assuming that the null hypothesis is true. The CLT says that $\bar{p}$ is approximately normally distributed with mean $p = p_0$ and standard deviation $\sigma = \sqrt{\frac{(p_0)(1-p_0)}{n}}$. So, our test statistic is:

$$Z_0 = \left(\frac{\bar{p}-p_0}{\sqrt{\frac{(p_0)(1-p_0)}{n}}} \right) = \left(\frac{(.54)-(.50)}{\sqrt{\frac{(.50)(1-.50)}{800}}} \right) = 2.26$$

The p-value is the $\Pr(Z \geq Z_0) = \Pr(Z \geq 2.26)$
$$= (1 - \Pr(Z \leq 2.26))$$
$$= (1 - 0.9881) = 0.0119$$

Since the p-value (0.0119) $\ll$ .05, then we will reject the null hypothesis and accept the alternative that the true proportion of voters in favor of candidate A is greater than 50%. So, the pollster believes that candidate A is likely to win the election.

Example (2):
A quality control engineer is interested in the number of defective (success) products from a SRS of size n = 100 products chosen from an assembly line. A non-defective product is generically called (failure). There is a problem with the manufacturing process if the proportion of defectives is greater than 10%. He finds 12 defective products in the sample, so $\bar{p} = 0.12$. The engineer wants

to test the following hypotheses at the $\alpha = .05$ level:

$$\begin{cases} H_o: p = .10 \\ H_a: p > .10 \end{cases}$$

Under the null hypotheses, $p_0 = (0.10)$. We always set up our test statistic assuming that the null hypothesis is true. The CLT says that $\bar{p}$ is approximately normally distributed with mean $p = p_0$ and standard deviation $\sigma = \sqrt{\dfrac{(p_0)(1-p_0)}{n}}$.

So, our test statistic is:

$$Z_0 = \left(\frac{\bar{p}-p_0}{\sqrt{\frac{(p_0)(1-p_0)}{n}}} \right) = \left(\frac{(.12)-(.10)}{\sqrt{\frac{(.10)(1-.10)}{100}}} \right) = 0.67$$

The p-value is the $\Pr(Z \geq Z_0) = \Pr(Z \geq 0.67)$
$$= (1 - \Pr(Z \leq 0.67))$$
$$= (1 - 0.7486) = 0.2514$$

Since the p-value (0.2514) $\gg$ .05, then we fail to reject the null hypothesis. There is no significant evidence that would indicate that the true proportion of defectives is greater than 10%. Even though the proportion of defectives in this sample was 12%, that could easily be explained as just due to randomness in the manufacturing process.

Example (3):
An astronomer takes a SRS of size n = 346 stars and is interested in the proportion of them that are so-called MR39 variable stars. He suspects that the true proportion of MR39 variable stars in the Milky Way is less than 6%. In his sample he finds X = 11 of them. So, the sample proportion of such stars is $\bar{p} = \frac{11}{346} = (0.0318)$. The astronomer wants to test the following hypotheses at the $\alpha = .01$ level:

$$\begin{cases} H_o: p = .06 \\ H_a: p < .06 \end{cases}$$

Under the null hypotheses, $p_0 = (0.06)$. We always set up our test statistic assuming that the null hypothesis is true. The CLT says that $\bar{p}$ is approximately normally distributed with mean $p = p_0$ and standard deviation $\sigma = \sqrt{\frac{(p_0)(1 - p_0)}{n}}$.

So, our test statistic is:

$$Z_0 = \left(\frac{\bar{p} - p_0}{\sqrt{\frac{(p_0)(1 - p_0)}{n}}} \right) = \left(\frac{(.0318) - (.06)}{\sqrt{\frac{(.06)(1 - .06)}{346}}} \right) = -2.21$$

The p-value is the $\Pr(Z \le Z_0) = \Pr(Z \le -2.21)$
$$= (0.0136)$$

Since the p-value (0.0136) > .01, we fail to reject the null hypothesis that 6% of stars are MR39 variable stars. The proportion of MR39 variables found in his sample was

3.18%, but this is not considered significantly less than 6% at the .01 level. It would have been significantly less at the .05 level, but the astronomer wants to set the bar high for making such a conclusion. It could very well be that the astronomer's sample had a low percentage of MR39 variables just by chance.

################# Exercises #################

(1) A politician believes that the proportion of people that support a ballot initiative (the outcome "success" for random variable X) is greater than 0.50, and he wants to test this hypothesis at the $\alpha = .05$ level.
   (a) What are the hypotheses that he wants to test?
   (b) He hires a statistical consulting firm and they poll n = 600 people at random. The result of the poll is X = 327 successes, or $\bar{p} = (0.545)$.
   What is the value of his test statistic?
   (c) What is the p-value and his conclusion?

(2) A Casino manager suspects that a certain coin used in gambling is unfair. So he has somebody flip it n = 200 times and the number of successes (Heads) is X = 110 . So the sample proportion of Heads is $\bar{p} = \frac{110}{200} = 0.55$.
   He wants to test his hypothesis at the $\alpha = .05$ level.
   (a) What are his hypotheses?
   (b) What is the value of the test statistic?
   (c) What is the p-value for this test and what are his conclusions?

(9.4) Tests for a Variance σ^2

Example (1):
A scientist conducts an experiment and gets a SRS of n = 35 observations of a normally distributed random variable X. He calculates the sample variance S^2 to be 12.53. In the research literature the variance of this random variable is always reported as (10.00). He suspects that the true variance for X is greater than (10.00). So, he wants to test the following hypotheses at the $\alpha = .05$ level:

$$\begin{cases} H_0: \sigma^2 = 10.00 \\ H_a: \sigma^2 > 10.00 \end{cases}$$

The scientist uses the fact that $\frac{(n-1)S^2}{\sigma_0^2}$ is distributed $\chi^2_{(n-1)}$ when H_0 is true. So, the test statistic is:

$$\chi_0^2 = \frac{(34)(12.53)}{(10.00)} = (42.602) .$$

By examining the chi-square table in Appendix B, the p-value is greater than .05, Therefore, he fails to reject H_0. There is not significant evidence to conclude that $\sigma^2 > (10.00)$ at the $\alpha = .05$ level.

Example (2):
An engineer designs an automobile part to have a breaking strength X with a mean μ of 1056 pounds, and a variance σ^2

of less than 30 (pounds)^2. Note that in statistics, the units for variance is (units-squared), and the units for the mean and standard deviation is just (units). He also has good reason to believe that X is normally distributed. He wants to establish at the $\alpha = .01$ level that the true variance is less than 30 (pounds)^2. So, he conducts an experiment by gathering a SRS of 40 of these automobile parts and calculates the sample variance S^2 to be 20.19 (pounds)^2. His hypotheses are:

$$\begin{cases} H_0: \sigma^2 = 30.00 \\ H_a: \sigma^2 < 30.00 \end{cases}$$

The engineer uses the fact that $\frac{(n-1)S^2}{\sigma_0^2}$ is distributed $\chi^2_{(n-1)}$ when H_0 is true. So, the test statistic is:

$$\chi_0^2 = \frac{(39)(20.19)}{(30.00)} = (26.25) .$$

By examining the chi-square table in Appendix B, the p-value is much greater than .01, Therefore, he fails to reject H_0 at the .01 level. Believe it or not, there is not significant evidence to conclude that $\sigma^2 < (30.00)$ even at the .05 level of significance.

################# Exercises #################

(1) An airline executive believes that airplane inspection times are normally distributed with a variance that is less than 20.0. He has an engineer take a SRS of size

n = 30 inspection times and he calculates $S^2 = 18.96$.
He wants to test his belief at the $\alpha = .05$ level.
(a) What hypotheses is he testing?
(b) What is the value of his test statistic χ_0^2?
(c) What is the p-value for his test and his conclusions?

(2) A biologist has read in the literature that the variance
σ^2 for a specific random variable X is always taken to be
35.0. But this biologist believes that it is greater. So, he
wants to test his idea at the $\alpha = .05$ level. He must
assume that the random variable X is normally
distributed, and he has evidence to believe this is so.
(a) What hypotheses is he testing?
(b) From a SRS of size n = 39, he calculates $S^2 = 49.29$.
    What is the value of his test statistic χ_0^2?
(c) What is the p-value and what does he conclude?

(9.5) Two Sample Tests for the Difference of Two Means

We can also conduct a large sample test for the
hypothesized difference between two means.

A soil scientist performs an experiment in a large plot of
land, where he takes two independent SRS of sizes $n_x = n_y = 100$. He measures a random variable X on each of the
experimental units in one of the samples, and measures a

random variable Y on each of the experimental units in the other sample. He wants to try to determine if $\mu_X = \mu_Y$, or in other words if $\mu_X - \mu_Y = 0$, versus the alternative that $\mu_X - \mu_Y \neq 0$. So, he wants to test the hypotheses at the $\alpha = .05$ level:

$$\begin{cases} H_0: \mu_X - \mu_Y = 0 \\ H_a: \mu_X - \mu_Y \neq 0 \end{cases}$$

He calculates $\bar{X} = 25.1$, $\bar{Y} = 27.6$, $S_X^2 = 5.32$, and $S_Y^2 = 4.99$.

Since $\bar{X}$ and $\bar{Y}$ are independent random variables, the $Var(\bar{X} - \bar{Y}) = Var(\bar{X}) + Var(\bar{Y}) = \left(\frac{\sigma_X^2}{n_X} + \frac{\sigma_Y^2}{n_Y} \right)$. He doesn't know the values of these variances, so he substitutes the variance estimates S_X^2 and S_Y^2 for σ_X^2 and σ_Y^2 respectively, to get the approximation:

$Var(\bar{X} - \bar{Y}) \approx \left(\frac{S_X^2}{n_X} + \frac{S_Y^2}{n_Y} \right)$.   So, the standard deviation

of $(\bar{X} - \bar{Y}) \approx \sqrt{\frac{S_X^2}{n_X} + \frac{S_Y^2}{n_Y}}$.   From the CLT, the statistic:

$Z_0 = \frac{(\bar{X} - \bar{Y}) - (0)}{\sqrt{\frac{S_X^2}{n_X} + \frac{S_Y^2}{n_Y}}} \sim N(0,1)$, because n_X and n_Y are large.

So, the test statistic is $Z_0 = \left(\frac{(25.1 - 27.6) - (0)}{\sqrt{\frac{5.32}{100} + \frac{4.99}{100}}} \right) = (-7.79)$

464

The p-value is the $(\Pr(Z \leq -7.79) + \Pr(Z \geq 7.79))$ which is approximately (0.0000). Therefore, we strongly reject H_0 and conclude that $\mu_X \neq \mu_Y$. There is very strong evidence in this case that $\mu_X \neq \mu_Y$ because the p-value is exceedingly small, α was chosen to be .05. This points out that with large samples we can determine small but real differences between the means of two different random variables.

****

If we have good reason to believe that the variances of the two random variables X and Y are equal, then we can use a pooled estimate of the variance from the two samples. We denote the pooled estimate as S_p^2, which is calculated:

$$S_p^2 = \left(\frac{(n_X - 1)S_X^2 + (n_Y - 1)S_Y^2}{(n_X + n_Y - 2)} \right).$$

Then the standard deviation of $(\bar{X} - \bar{Y}) \approx \left(\sqrt{\frac{S_p^2}{n_x} + \frac{S_p^2}{n_y}} \right)$,

which equals $S_p \sqrt{\frac{1}{n_x} + \frac{1}{n_y}}$.

So, in the example above, since the researcher believes that the variances of X and Y are the same (and the data seems to suggest that) he uses a pooled estimate of their common variance:

$$S_p^2 = \left(\frac{(99)(5.32) + (99)(4.99)}{(198)} \right) = 5.155.$$

465

Our test statistic Z_0 is then $Z_0 = \left(\dfrac{(25.1 - 27.6) - (0)}{(\sqrt{5.155}) \cdot \sqrt{\frac{1}{100} + \frac{1}{100}}} \right)$, or

$Z_0 = (-7.79)$.

In this case we get the exact same value for the test statistic as above, rounded to two decimal places. The p-value is then exactly the same as it was before, that is, the p-value is essentially (0.0000). So, in this case our conclusion would be the same. We would reject H_0 that the means of X and Y are the same, and conclude that they are significantly different.

****

################ Exercises ##################

(1) A consumer testing organization wants to determine if the mean average tread life of brand X and brand Y tires are equal or simply different. They want to conduct a test at the $\alpha = .01$ level of significance.
  (a) What are the hypotheses to be tested?
  (b) A SRS of n = 50 tires of each brand is tested, resulting in $\bar{X} = 50{,}490$, $\bar{Y} = 50{,}510$, $S_X^2 = 3267$, and $S_Y^2 = 2946$. We will assume that the variances are equal. Compute a pooled estimate of their common variance S_p^2.
  (c) What is the value of the test statistic Z_0?
  (d) What is the p-value and what is the conclusion?

(2) A physicist wants to determine if the means of random
variables X and Y are equal or simply different. They
want to conduct a test at the $\alpha = .05$ level.
(a) What are the hypotheses to be tested?
(b) A SRS of n = 50 measurements of both random
variables X and Y results in $\bar{X} = 2100$, $\bar{Y} = 2180$,
$S_X^2 = 854$, and $S_Y^2 = 712$. We will not assume that
the variances are equal. What is the value of the test
statistic Z_0?
(c) What is the p-value and what is the conclusion?

(9.6) <u>Two Sample Tests for the</u>
<u>Difference of Two Proportions</u>

We can also conduct a large sample test for the
hypothesized difference between two proportions.

A demographer conducts an experiment with two
independent SRS of sizes $n_X = n_Y = 400$, and measures
two random variables X and Y. He measures the random
variable X on one of the groups of 400 people and the
random variable Y on the other group. Each of the two
random variables takes only the values 1 or 0 (success or
failure). He finds that the first sample has X = 300 successes
and the second sample has Y = 305 successes.
So, $\bar{p}_X = \frac{300}{400} = (0.75)$ and $\bar{p}_Y = \frac{305}{400} = (0.7625)$.

He wants to test at the $\alpha = .05$ level the hypotheses:

$$\begin{cases} H_0: p_X - p_Y = 0 \\ H_a: p_X - p_Y < 0 \end{cases} \text{ or equivalently } \begin{cases} H_0: p_X = p_Y \\ H_a: p_X < p_Y \end{cases}.$$

Since $\bar{p}_X$ and $\bar{p}_Y$ are independent random variables, the

$$Var(\bar{p}_X - \bar{p}_Y) = Var(\bar{p}_X) + Var(\bar{p}_Y)$$

$$= \left(\frac{(p_X)(1-p_X)}{n_X} \right) + \left(\frac{(p_Y)(1-p_Y)}{n_Y} \right)$$

Since we don't know p_X and p_Y, we estimate them with $\bar{p}_X$ and $\bar{p}_Y$. So, the estimated variance of $(\bar{p}_X - \bar{p}_Y)$ is:

$$Var(\bar{p}_X - \bar{p}_Y) \approx \left(\frac{(\bar{p}_X)(1-\bar{p}_X)}{n_X} \right) + \left(\frac{(\bar{p}_Y)(1-\bar{p}_Y)}{n_Y} \right).$$

The estimated standard deviation of $(\bar{p}_X - \bar{p}_Y)$ is then:

$$\sqrt{\left(\frac{(\bar{p}_X)(1-\bar{p}_X)}{n_X} \right) + \left(\frac{(\bar{p}_Y)(1-\bar{p}_Y)}{n_Y} \right)}.$$

Our test statistic, which is based on the CLT is:

$$Z_0 = \frac{(\bar{p}_X - \bar{p}_Y) - (0)}{\sqrt{\left(\frac{(\bar{p}_X)(1-\bar{p}_X)}{n_X} \right) + \left(\frac{(\bar{p}_Y)(1-\bar{p}_Y)}{n_Y} \right)}} \sim N(0,1) \text{, which equals}$$

$$Z_0 = \frac{(0.75 - 0.7625) - (0)}{\sqrt{\left(\frac{(0.75)(0.25)}{400} \right) + \left(\frac{(0.7625)(0.2375)}{400} \right)}} = (-0.41).$$

The p-value is the $\Pr(Z \le Z_0) = \Pr(Z \le -0.41) = (0.3409)$.

Since the p-value (0.3409) $\gg$ (0.05), we fail to reject H_0, and we conclude that there is no difference between the two proportions p_X and p_Y.

****

Under H_0, since we assume that $p_X = p_Y$, we should use a pooled estimate of the common proportion, and perhaps get a more powerful test. The pooled proportion is:

$$\bar{p}_p = \left(\frac{(n_X)(\bar{p}_X) + (n_Y)(\bar{p}_Y)}{(n_X + n_Y)} \right).$$

So, we should use $\sqrt{(\bar{p}_p)(1 - \bar{p}_p)\left(\frac{1}{n_x} + \frac{1}{n_y}\right)}$ instead of $\sqrt{\left(\frac{(\bar{p}_X)\,(1-\bar{p}_X)}{n_X}\right) + \left(\frac{(\bar{p}_Y)\,(1-\bar{p}_Y)}{n_Y}\right)}$ as our estimate of the standard deviation of $(\bar{p}_X - \bar{p}_Y)$.

We calculate $\bar{p}_p = \left(\frac{(400)(0.75) + (400)(0.7625)}{(800)} \right) = \left(\frac{300 + 305}{800} \right)$

$$= \left(\frac{605}{800} \right) = (0.75625).$$

Then our test statistic $Z_0 = \dfrac{(0.75 - 0.7625) - (0)}{\sqrt{(.75625)\,(1 - .75625)\left(\frac{1}{400} + \frac{1}{400}\right)}}$

$$= -(0.41).$$

The value of the test statistic is exactly the same as above, so our conclusions are exactly the same.

################# Exercises #################

(1) Two independent polls of voter preference for a senate seat result in rather different sample proportions of the electorate that intend to vote for the Republican candidate. Poll X says that $\bar{p} = 0.536$ and poll Y says that $\bar{p} = 0.581$ based on $n_X = 750$ and $n_Y = 810$ people. The Republican candidate has his staff test for a difference at the $\alpha = .05$ level of significance.
   (a) What are the hypotheses to be tested?
   (b) Under H_0, $p_X = p_Y$. What is the pooled estimate $\bar{p}_p$ of p_X and p_Y?
   (c) What is the value of the test statistic Z_0?
   (d) What is the p-value and what is the Republican Candidate's conclusion?

(2) Two independent polls, one from France and one from Germany, of consumer preference for a brand of Ice Cream results in two sample proportions of approval. The poll of random variable X, approval from French consumers, results in an approval (success) proportion of $\bar{p} = 0.410$. The poll of random variable Y, approval from German consumers, results in an approval (success) proportion of $\bar{p} = 0.395$. These polls were based on $n_X = 75$ French and $n_Y = 81$ Germans. The Ice Cream manufacturer wants to test whether these proportions are equal or not, at the $\alpha = .05$ level of significance.

470

(a) What are the hypotheses to be tested?

(b) Under H_0, $p_X = p_Y$. What is the pooled estimate $\bar{p}_p$ of p_X and p_Y?

(c) What is the value of the test statistic Z_0?

(d) What is the p-value and what is the Ice Cream company's conclusion?

(10) <u>LARGE SAMPLE</u>
<u>CONFIDENCE</u>
<u>INTERVALS</u>

(10.1) <u>Introduction</u>

The statistics $\bar{X}$ and S^2, as we have explained in Chapter 7, are point estimators of distribution parameters. With these two it is μ and σ^2 respectively. We will now discuss interval estimates for μ and σ^2. An interval estimate of a parameter is an interval of numbers on the real line, in which we have a certain amount of confidence that it contains the true value of the parameter. We choose α in order to have a certain amount of confidence that the interval will capture the true but unknown parameter value, and to give us an idea of the precision that we have in how well we have estimated the parameter, from the data in a sample.

We will see that the CLT also plays a role in interval estimation as well, since when constructing an interval estimate of μ, we will rely on the statistic:

$Z = \left(\dfrac{\bar{X} - \mu}{\left(\frac{S}{\sqrt{n}} \right)} \right)$ which we know is distributed N(0,1) for large samples. We will also use the statistic:

$$Z = \left(\frac{\bar{p} - p}{\sqrt{\frac{(\bar{p})(1-\bar{p})}{n}}} \right) \text{ to construct interval estimates of p.}$$

Since $\bar{p}$ is the average of a sample of 0's and 1's, we have the CLT for large samples that tells us that it is distributed N(0,1) also.

The interval estimates that we will construct are called confidence intervals. We decide on the amount of confidence that we want beforehand through our choice of α. The amount of confidence is usually 90%, 95%, or 99%.

(10.2) Confidence Intervals for Means

We know that for a large sample taken from a distribution with mean μ and variance σ^2, and for a chosen α, the CLT says that:

The $\Pr\left(-Z\alpha_{/2} \leq \left(\frac{\bar{X} - \mu}{\left(\frac{S}{\sqrt{n}}\right)} \right) \leq Z\alpha_{/2} \right) = (1 - \alpha)$.

The notation $\left(Z\alpha_{/2} \right)$ is the value of the standard normal distribution such that the $Pr\left(Z \geq Z\alpha_{/2} \right) = \alpha_{/2}$. This $\left(Z\alpha_{/2} \right)$ can be found from Appendix A. So, the

$$\Pr\left(\left(-Z_{\alpha/2}\right)\left(\tfrac{S}{\sqrt{n}}\right) \le (\bar{X} - \mu) \le \left(Z_{\alpha/2}\right)\left(\tfrac{S}{\sqrt{n}}\right)\right) = (1 - \alpha).$$

Subtracting $\bar{X}$ from the three parts of the inequality gives:

$$\Pr\left(\left(-Z_{\alpha/2}\right)\left(\tfrac{S}{\sqrt{n}}\right) - \bar{X} \le -\mu \le \left(Z_{\alpha/2}\right)\left(\tfrac{S}{\sqrt{n}}\right) - \bar{X}\right) = (1 - \alpha).$$

Multiplying through by (-1) isolates μ in the middle of the inequality, so that we have:

$$\Pr\left(\bar{X} - \left(Z_{\alpha/2}\right)\left(\tfrac{S}{\sqrt{n}}\right) \le \mu \le \bar{X} + \left(Z_{\alpha/2}\right)\left(\tfrac{S}{\sqrt{n}}\right)\right) = (1 - \alpha).$$

So, we say that we have $(1 - \alpha) \cdot 100\%$ confidence that the interval on the real line:

$$\left(\bar{X} - \left(Z_{\alpha/2}\right)\left(\tfrac{S}{\sqrt{n}}\right), \bar{X} + \left(Z_{\alpha/2}\right)\left(\tfrac{S}{\sqrt{n}}\right)\right)$$

contains the true value of μ, and we call it a $(1 - \alpha) \cdot 100\%$ confidence interval for μ.

Example (1):
We have a SRS of size n = 48 for some random variable X.
From the sample, we calculate $\bar{X} = 112.3$ and S = (3.57).
Construct a 90%, 95%, and 99% confidence interval for μ.
The formula is:

$$\left(\bar{X} - \left(Z_{\alpha/2}\right)\left(\tfrac{S}{\sqrt{n}}\right), \bar{X} + \left(Z_{\alpha/2}\right)\left(\tfrac{S}{\sqrt{n}}\right)\right).$$

The 90% confidence interval for μ is:

$$\left(112.3 - (1.645)\left(\frac{3.57}{\sqrt{48}}\right), 112.3 + (1.645)\left(\frac{3.57}{\sqrt{48}}\right)\right)$$

$= (111.45, 113.15)$.

The 95% confidence interval for μ is:

$$\left(112.3 - (1.96)\left(\frac{3.57}{\sqrt{48}}\right), 112.3 + (1.96)\left(\frac{3.57}{\sqrt{48}}\right)\right)$$

$= (111.29, 113.31)$.

The 99% confidence interval for μ is:

$$\left(112.3 - (2.575)\left(\frac{3.57}{\sqrt{48}}\right), 112.3 + (2.575)\left(\frac{3.57}{\sqrt{48}}\right)\right)$$

$= (110.97, 113.63)$.

The price we pay for higher confidence is a wider interval, that is, less precision in our estimate. But note that the width of the interval will be less (for a given level of confidence) when we increase the sample size n.

Example (2):
A government statistician takes a SRS of n = 75 U.S. counties and gets a value for the random variable X, where X = the number of building permits issued for the county. He calculates from the sample $\bar{X} = 1290$ and S = 179.5. Calculate a 95% confidence interval for the mean number μ of building permits issued per county. The formula is:

$$\left(\bar{X} - \left(Z_{\alpha/2}\right)\left(\frac{S}{\sqrt{n}}\right), \; \bar{X} + \left(Z_{\alpha/2}\right)\left(\frac{S}{\sqrt{n}}\right)\right).$$

The 95% confidence interval for μ is:

$$\left(1290 - (1.96)\left(\frac{179.5}{\sqrt{75}}\right), 1290 + (1.96)\left(\frac{179.5}{\sqrt{75}}\right)\right)$$

$= (1249, 1331)$.

****

Once we substitute numbers into a confidence interval formula, we get a specific interval that is no longer random. The true value of μ would be in the specific interval with probability 0 or 1. This does not really say anything useful about the specific interval. So, once we plug the numbers into the formula and get an interval, we do not say that the specific interval has a probability of $(1 - \alpha)$ of containing μ. The correct interpretation is that in repeated sampling, i.e. repeating the entire experiment again and again an infinite number of times to consider all possible samples, then $(1 - \alpha) \cdot 100\%$ of the time the specific interval we get contains the true value of μ. So, we say that we have $(1 - \alpha) \cdot 100\%$ confidence that the specific interval that we get contains the true value of μ. We call this a $(1 - \alpha) \cdot 100\%$ confidence interval for the true value of μ.

We will see that a confidence interval for μ or p consists of these two parts:
(1) the point estimate of the parameter, denoted $\hat{\beta}$, and
(2) a margin of error E.

The confidence interval has the form: $\left(\hat{\beta} - E, \ \hat{\beta} + E\right)$.
From above, the Confidence Interval for the mean μ uses
the point estimator $\hat{\beta} = \bar{X}$, and the margin of error E which
is $\left(Z_{\alpha/2}\right)\left(\frac{S}{\sqrt{n}}\right)$. It is built entirely from the sample statistics
$\bar{X}$ and S, and a Z value depending on the choice of the
confidence level that we want (from our choice of α). The
common choices of α and the associated confidence levels
are:

$\alpha = .10$ $\leftrightarrow$ confidence level = 90%
$\alpha = .05$ $\leftrightarrow$ confidence level = 95%
$\alpha = .01$ $\leftrightarrow$ confidence level = 99%

With $\alpha = .10$, $Z_{\alpha/2} = Z_{.05} = 1.645$.
With $\alpha = .05$, $Z_{\alpha/2} = Z_{.025} = 1.96$.
With $\alpha = .01$, $Z_{\alpha/2} = Z_{.005} = 2.575$.

The margin of error E is a multiple of estimated standard
deviations that we want the interval to extend on either side
of the point estimator $\hat{\beta}$. For the confidence interval given
above for μ, $\left(\frac{S}{\sqrt{n}}\right)$ provides the estimated standard deviation
of $\bar{X}$, and $\left(Z_{\alpha/2}\right)$ is the desired number of estimated
standard deviations. So, E is the product of the two, the
product $\left(Z_{\alpha/2}\right) \cdot \left(\frac{S}{\sqrt{n}}\right)$.

****

478

(1) For a SRS of size n = 65, construct a 95% confidence interval for the mean μ of a random variable X, if we calculate that $\bar{X} = 27.5$ and S = 3.2.

(2) For a SRS of size n = 34, construct a 99% confidence interval for the mean μ of a random variable X, if we calculate that $\bar{X} = 103$ and S = 5.5.

(10.3) <u>Confidence Intervals for Proportions</u>

We take a large SRS from a random variable X (that takes outcomes 0 or 1) and compute $\bar{p}$. Since we don't know the true p, we substitute $\bar{p}$ for it. This generally works well.

A $(1 - \alpha) \cdot 100\%$ confidence interval for p is found from the probability statement (based on the CLT):

The $\Pr\left(-Z\alpha/_2 \leq \left(\dfrac{\bar{p}-p}{\sqrt{\dfrac{(\bar{p})(1-\bar{p})}{n}}} \right) \leq Z\alpha/_2 \right) = (1-\alpha).$

We could rearrange the inequality as we did for the confidence interval for μ, but let's just note that the point estimator of p is $\bar{p}$ and the estimated standard deviation of $\bar{p}$ is $\sqrt{\dfrac{(\bar{p})(1-\bar{p})}{n}}$. So, for a chosen α, the margin of error is

$E = \left(Z\alpha_{/2}\right)\left(\sqrt{\dfrac{(\bar{p})(1-\bar{p})}{n}}\right)$. Therefore, $(\bar{p} - E,\ \bar{p} + E)$ yields

the $(1 - \alpha) \cdot 100\%$ Confidence Interval for p:

$$\left(\bar{p} - \left(Z\alpha_{/2}\right)\left(\sqrt{\dfrac{(\bar{p})(1-\bar{p})}{n}}\right),\ \bar{p} + \left(Z\alpha_{/2}\right)\left(\sqrt{\dfrac{(\bar{p})(1-\bar{p})}{n}}\right)\right).$$

Example (1):

A coin is tossed 33 times. So we say that we have a SRS of size n = 33. We find 14 Heads (successes) and 19 Tails (failures). So, $\bar{p} = (.4242)$. Find a 99% confidence interval for p, the true proportion of Heads. The formula is:

$$\left(\bar{p} - \left(Z\alpha_{/2}\right)\left(\sqrt{\dfrac{(\bar{p})(1-\bar{p})}{n}}\right),\ \bar{p} + \left(Z\alpha_{/2}\right)\left(\sqrt{\dfrac{(\bar{p})(1-\bar{p})}{n}}\right)\right)$$

$$= \left((0.4242) \pm (2.575)\sqrt{\dfrac{(0.4242)(.5758)}{33}}\right)$$

$$= (0.2027, 0.6457)$$

With the high confidence level, we cannot say that the coin is not fair, even though there were only 14 Heads in 33 tosses. We say this since the computed confidence interval contains 0.50. Apparently, this result is still well within the bounds of likely outcomes.

Example (2):
A California lawyer hires a company to do an opinion poll for a ballot initiative which will be voted on by California citizens. They take a SRS of size n = 650 of the electorate, and find that 334 favor the initiative. Calculate a 95% confidence Interval for p = the true proportion of supporters of the initiative.

The sample proportion of supporters (our point estimate) is:
$$\bar{p} = \frac{334}{650} = (0.5138).$$

The margin of error is $(1.96)\left(\sqrt{\frac{(0.5138)(0.4862)}{650}}\right) = (0.0384).$

The 95% confidence interval for p is: (0.4754, 0.5522).

This is sometimes stated: $(51.38\% \pm 3.84\%)$. The outcome of the vote is still up in the air, it could go either way.

################ Exercises ################

(1) For a SRS of size n = 80, construct a 95% confidence interval for the proportion p of a binomial random variable X, if we calculate that $\bar{p} = 0.62$.

(2) For a SRS of size n = 59, construct a 99% confidence interval for the proportion p of a binomial random variable X, if we calculate that $\bar{p} = 0.23$.

(3) (Worst Case Scenario):

Recall the discussion about parabolas in Section (2.7).

(a) Show that $f(p) = p(1 - p) = p - p^2$ is maximized When $p = 0.50$.

(b) Then show that the margin of error:

$$E = \left(Z_{\frac{\alpha}{2}}\right) \cdot \left(\sqrt{\frac{(0.5)(0.5)}{n}}\right)$$

associated with a 95.4% confidence interval for the true value of p when we use $\bar{p} = (0.50)$ is $E = \frac{1}{\sqrt{n}}$.

Therefore, we have a quick conservative 95% confidence interval estimate of p:

$$\left(\bar{p} - \frac{1}{\sqrt{n}}, \bar{p} + \frac{1}{\sqrt{n}}\right).$$

(10.4) <u>Confidence Intervals for Variances σ^2</u>

Recall that $\left(\frac{(n-1)S^2}{\sigma^2}\right) \sim \chi^2_{(n-1)}$, when we are sampling from normal(μ, σ) distributions. For $X \sim \chi^2_{(k)}$, $\chi^2_{(k),\alpha/2}$ is the outcome such that $Pr\left(X \geq \chi^2_{(k),\alpha/2}\right) = \alpha/2$, and $\chi^2_{(k),1-\alpha/2}$ is the outcome such that $Pr\left(X \geq \chi^2_{(k),1-\alpha/2}\right) = 1 - \alpha/2$.

Then, the $Pr\left(\chi^2_{(k),1-\alpha/2} \leq \left(\frac{(n-1)S^2}{\sigma^2}\right) \leq \chi^2_{(k),\alpha/2}\right) = (1 - \alpha)$

Then, the $\Pr\left(\frac{1}{\chi^2_{(k),\alpha/2}} \le \left(\frac{\sigma^2}{(n-1)S^2}\right) \le \frac{1}{\chi^2_{(k),1-\alpha/2}}\right) = (1-\alpha)$

Then, the $\Pr\left(\left(\frac{(n-1)S^2}{\chi^2_{(k),\alpha/2}}\right) \le \sigma^2 \le \left(\frac{(n-1)S^2}{\chi^2_{(k),1-\alpha/2}}\right)\right) = (1-\alpha)$.

This says that a $(1-\alpha) \cdot 100\%$ confidence interval for σ^2

is: $\left[\left(\frac{(n-1)S^2}{\chi^2_{(k),\alpha/2}}\right), \left(\frac{(n-1)S^2}{\chi^2_{(k),1-\alpha/2}}\right)\right]$.

Note that this does not have the form $[\hat{\beta} - E, \hat{\beta} + E]$ that we had in the confidence intervals for μ and p. This interval is not based on the CLT, but we did assume that we were sampling from a normal distribution.

As an example, suppose that we are sampling from $N(\mu, \sigma)$ distributions and we choose $\alpha = .05$. We take a SRS of size n = 32, and we calculate S^2 to be 10.83. From the formula above, and noting that from Appendix B:

$\chi^2_{(31),0.025} = 48.231$, and $\chi^2_{(31),0.975} = 17.538$,

our 95% confidence interval for σ^2 is:

$\left[\left(\frac{(31)(10.83)}{48.231}\right), \left(\frac{(31)(10.83)}{17.538}\right)\right] = (6.961, 19.143)$.

################# Exercises #################

(1) Assume that we are sampling from normal distributions. For a SRS of size n = 40, construct a 95% confidence interval for σ^2 if $S^2 = (50.6)$.

(2) (a) Assume that we are sampling from normal distributions. For a SRS of size n = 30, construct a 90% confidence interval for σ^2 if $S^2 = (150.9)$.
(b) What would be a 90% confidence interval for σ?

(10.5) <u>Two Sample Confidence Interval for a Difference of Two Means</u>

Now we will consider confidence intervals for the difference of the means of two different independent random variables X and Y. Suppose we have a large SRS of size n_X containing measurements of the random variable X, and a large SRS of size n_Y containing measurements of the random variable Y. We will have no assumption of equal variances, because confidence intervals depend entirely on the data, with as few assumptions as possible. From the data, we calculate $\bar{X}$, $\bar{Y}$, S_X^2, and S_Y^2. We want a

$(1 - \alpha) \cdot 100\%$ confidence interval for $\mu_X - \mu_Y$.

Note that the following argument is based on the CLT and large samples. The statistic $(\bar{X} - \bar{Y})$ is the point estimator of $\mu_X - \mu_Y$ that is commonly used.

From section (9.5) we know that the estimated standard deviation of $(\bar{X} - \bar{Y})$ is $\sqrt{\frac{S_X^2}{n_X} + \frac{S_Y^2}{n_Y}}$. So, the margin of error $E = \left(Z_{\alpha/2}\right)\left(\sqrt{\frac{S_X^2}{n_X} + \frac{S_Y^2}{n_Y}}\right)$. Therefore, a $(1 - \alpha) \cdot 100\%$ confidence interval for $\mu_X - \mu_Y$ is:

$$(\bar{X} - \bar{Y}) \pm \left(Z_{\alpha/2}\right)\left(\sqrt{\frac{S_X^2}{n_X} + \frac{S_Y^2}{n_Y}}\right).$$

As an example, suppose we take two independent samples: a SRS of size n_X = 110 economy cars and a SRS of size n_Y = 92 midsize cars from the New York City area, and we take gas mileage measurements X from the economy cars, and gas mileage measurements Y from the midsize cars. Each of these random variables has a mean, and we want to measure the difference in these two means $\mu_X - \mu_Y$.

We calculate a 99% confidence interval for $\mu_X - \mu_Y$ with the sample statistics $\bar{X} = 38.1$, $\bar{Y} = 33.6$, $S_X^2 = (9.6)$, and $S_Y^2 = (11.2)$.

The point estimate is: $(\bar{X} - \bar{Y}) = (38.1 - 33.6) = (4.5)$,

and the margin of error is:

$$E = \left(Z_{\alpha/2}\right)\left(\sqrt{\frac{S_X^2}{n_X} + \frac{S_Y^2}{n_Y}}\right) = (2.575)\left(\sqrt{\frac{9.6}{110} + \frac{11.2}{92}}\right)$$
$$= (2.575) \cdot \left(\sqrt{0.08727 + 0.12174}\right) = (1.1772)$$

Then, our 99% confidence interval for $\mu_X - \mu_Y$ is:

$(4.5 \pm 1.1772) = (3.3228, 5.6772)$.

It appears that there is a real difference in the average gas mileage of the two types of cars because the 99% confidence interval does not even nearly contain 0.

################# Exercises ##################

(1) For two independent SRS of sizes $n_X = 50$ and $n_Y = 60$, where we measure random variables X and Y respectively, calculate a 95% confidence interval for $(\mu_X - \mu_Y)$ if $\bar{X} = 32.6$, $\bar{Y} = 34.7$, $S_X^2 = 5.6$, and $S_Y^2 = 4.9$.

(2) For two independent SRS of sizes $n_X = 150$ and $n_Y = 160$, where we measure random variables X and Y respectively, calculate a 99% confidence interval for $(\mu_X - \mu_Y)$ if $\bar{X} = 45.6$, $\bar{Y} = 50.1$, $S_X^2 = 13.6$, and $S_Y^2 = 14.1$.

(10.6) Two Sample Confidence Interval for a Difference of Two Proportions

Now we will consider large sample confidence intervals for the difference in the proportion of successes for two different independent random variables X and Y (where both X and Y have the outcomes 0 or 1). Suppose we have a large SRS of size n_X containing measurements of the random variable X, and a large SRS of size n_Y containing measurements of the random variable Y. We calculate $\bar{p}_X$ and $\bar{p}_Y$ for the two samples. We want a $(1 - \alpha) \cdot 100\%$ confidence interval for $p_X - p_Y$.

Note that the following argument is based on the CLT and large samples.

The statistic $(\bar{p}_X - \bar{p}_Y)$ is the point estimator of $p_X - p_Y$.

From section (9.6) we know that the estimated standard deviation of the point estimator $(\bar{p}_X - \bar{p}_Y)$ is:

$$\sqrt{\frac{(\bar{p}_X)(1-\bar{p}_X)}{n_X} + \frac{(\bar{p}_Y)(1-\bar{p}_Y)}{n_Y}}.$$

We don't use a pooled estimate as we did before in section (9.6) since we have no assumption (under a null hypothesis) that the two proportions are equal.

So, $E = \left(Z_{\alpha/2}\right)\left(\sqrt{\frac{(\bar{p}_X)(1-\bar{p}_X)}{n_X} + \frac{(\bar{p}_Y)(1-\bar{p}_Y)}{n_Y}}\right).$

Therefore, a $(1 - \alpha) \cdot 100\%$ confidence interval for $p_X - p_Y$ is:

$$(\bar{p}_X - \bar{p}_Y) \pm \left(Z_{\alpha/2}\right)\left(\sqrt{\frac{(\bar{p}_X)(1-\bar{p}_X)}{n_X} + \frac{(\bar{p}_Y)(1-\bar{p}_Y)}{n_Y}}\right).$$

487

As an example, suppose we take two independent samples. We take a SRS of size n_X = 530 Vermont voters and a SRS of size n_Y = 610 Virginia voters from these areas, and we take presidential preferences X from the Vermont voters, and presidential preferences Y from the Virginia voters, where there are only two possible outcomes for each of X and Y. We consider the Democratic candidate a success and the Republican candidate a failure (generic terms). Each of these random variables has a proportion associated with it, and we want to estimate the difference in these two proportions $(p_X - p_Y)$.

From the samples we have X = 346 successes in Vermont and Y = 325 successes in Virginia. We calculate a 99% confidence interval for $(p_X - p_Y)$ with the sample statistics

$$\bar{p}_X = \frac{346}{530} = (0.653), \text{ and } \bar{p}_Y = \frac{325}{610} = (0.533).$$

The point estimate is $(\bar{p}_X - \bar{p}_Y) = (0.653 - 0.533) = 0.120$, and the margin of error is

$$E = \left(Z_{\alpha/2}\right)\left(\sqrt{\frac{(\bar{p}_X)(1-\bar{p}_X)}{n_X} + \frac{(\bar{p}_Y)(1-\bar{p}_Y)}{n_Y}}\right)$$

$$= (2.575)\left(\sqrt{\frac{(0.653)(0.347)}{530} + \frac{(0.533)(0.467)}{610}}\right)$$

$$= (2.575) \cdot \left(\sqrt{0.00042753 + 0.000408051}\right)$$

$= (0.0744)$.

Then, our 99% confidence interval for $p_X - p_Y$ is:
$(0.120 \pm 0.0744) = (0.0456, 0.1944) = (4.56\%, 19.44\%)$.

Therefore, it seems that there is a highly variable but real difference in the proportion of voters for the Democratic candidate in Vermont versus that of Virginia.

############### Exercises ###############

(1) For two independent SRS of sizes $n_X = 35$ and $n_Y = 37$, where we make measurements of the binomial random variables X and Y respectively, calculate a 99% confidence interval for $(p_X - p_Y)$ if $\bar{p}_X = 0.743$ and $\bar{p}_Y = 0.757$.

(2) For two independent SRS of sizes $n_X = 85$ and $n_Y = 87$, where we make measurements of the binomial random variables X and Y, calculate a 95% confidence interval for $(p_X - p_Y)$ if $\bar{p}_X = 0.644$ and $\bar{p}_Y = 0..668$.

(11) <u>SMALL SAMPLE INFERENCE</u>

(11.1) <u>Introduction</u>

Unfortunately, for researchers that want to use statistical methods, they may have to deal with a small sample where the sample size n < 30. To use the methods of this chapter, where we are not able to invoke the CLT, we must assume that we are sampling from normal distributions. Many times the researcher may have to perform some kind of test to determine whether or not the distributions are normal. We will not discuss such tests in this book.

Instead of using a standard normal random variable (Z) for the test statistic in a hypothesis test, or as a basis for the construction of a confidence interval for μ, we are forced to use a so-called T statistic. Unlike the Z test statistic, there is a separate T distribution for different sample sizes.

Let's now discuss the T distribution. Suppose we have a random variable U which has a N(0,1) distribution, and a random variable V which has a $\chi^2_{(k)}$ (chi-square distribution with k degrees of freedom). Furthermore, assume that random variables U and V are independent. Then it follows that:

$T = \dfrac{U}{\sqrt{\frac{V}{k}}}$ has a $T_{(k)}$ distribution. $T_{(k)}$ stands for a t-distribution with k degrees of freedom (d.f.). The calculation of the pdf for t-distributions with k d.f. depends on advanced methods of calculus, which will not concern us in this book. See Figure (15) on the next page.

Let $U = \left(\dfrac{\bar{X} - \mu}{\frac{\sigma}{\sqrt{n}}} \right)$. We know that for normal distributions, this is distributed N(0,1). We also know that if we are sampling from normal distributions, $V = \dfrac{(n-1)S^2}{\sigma^2}$ is a $\chi^2_{(n-1)}$ random variable. The fact that U and V are independent random variables follows from the fact that when sampling from normal distributions, $\bar{X}$ and S^2 are independent random variables (though we will not prove it). So, it follows that:

$$T = \left(\dfrac{\frac{\bar{X} - \mu}{\frac{\sigma}{\sqrt{n}}}}{\sqrt{\frac{\left(\frac{(n-1)S^2}{\sigma^2} \right)}{(n-1)}}} \right) = \left(\dfrac{\bar{X} - \mu}{\frac{\sigma}{\sqrt{n}}} \right) \cdot \left(\dfrac{\sigma}{S} \right) = \left(\dfrac{\bar{X} - \mu}{\frac{S}{\sqrt{n}}} \right) \sim T_{(n-1)} .$$

This is exactly the same test statistic as the Z test statistic that we have been using in the previous two chapters. With large sample sizes, the CLT says that it has an approximate N(0,1) distribution. However, with small sample sizes, even when sampling from normal distributions, it is not nearly distributed N(0,1). Instead, it turns out that it follows a

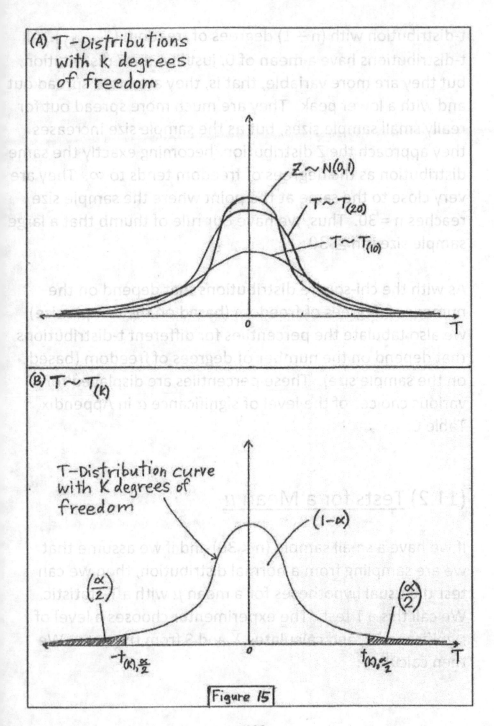

(A) T-Distributions with k degrees of freedom

$Z \sim N(0,1)$

$T \sim T_{(20)}$

$T \sim T_{(10)}$

0

T

(B) $T \sim T_{(k)}$

T-Distribution curve with K degrees of freedom

$(1-\alpha)$

$\left(\dfrac{\alpha}{2}\right)$

$\left(\dfrac{\alpha}{2}\right)$

$-t_{(k),\frac{\alpha}{2}}$

0

$t_{(k),\frac{\alpha}{2}}$

T

Figure 15

t-distribution with (n − 1) degrees of freedom $\left(T_{(n-1)}\right)$. All t-distributions have a mean of 0, just like the Z distribution, but they are more variable, that is, they are more spread out and with a lower peak. They are much more spread out for really small sample sizes, but as the sample size increases they approach the Z distribution, becoming exactly the same distribution as the degrees of freedom tends to ∞. They are very close to the same at the point where the sample size reaches n = 30. Thus, we have our rule of thumb that a large sample size is n ≥ 30.

As with the chi-square distributions that depend on the number of degrees of freedom (based on the sample size), we also tabulate the percentiles for different t-distributions that depend on the number of degrees of freedom (based on the sample size). These percentiles are displayed for various choices of the level of significance α in Appendix Table C.

(11.2) Tests for a Mean μ

If we have a small sample (n < 30) and if we assume that we are sampling from a normal distribution, then we can test the usual hypotheses for a mean μ with a T-statistic. We call this a T test. The experimenter chooses a level of significance α, and calculates $\bar{X}$ and S from the data. We then calculate:

$$T_0 = \left(\frac{\bar{X} - \mu_0}{\frac{S}{\sqrt{n}}} \right). \quad T_0 \text{ has a t-distribution with } (n-1) \text{ d.f.}$$

Just as with the large sample tests using the statistic Z_0, we reject or fail to reject $H_0: \mu = \mu_0$ and make a conclusion based on the p-value, according to whichever of the three alternatives we are interested in. We find the p-value by consulting the table in Appendix C:

If the alternative hypothesis is $H_a: \mu \neq \mu_0$, then the p-value is $2 \cdot (\Pr(T_{(n-1)} > |T_0|))$.

If the alternative hypothesis is $H_a: \mu < \mu_0$, then the p-value is $(\Pr(T_{(n-1)} < T_0))$.

If the alternative hypothesis is $H_a: \mu > \mu_0$, then the p-value is $(\Pr(T_{(n-1)} > T_0))$.

Example (1):
A chemist has a SRS of size n = 16 measurements of a random variable X, the number of grams of copper precipitate after a certain amount of two chemicals are mixed. He assumes that X is normally distributed. He wants to test the hypothesis that the amount of Copper produced in each repetition of the experiment is greater than 16.5 grams. So, he wants to test the hypotheses:

$$\begin{cases} H_0: \mu = 16.5 \\ H_a: \mu > 16.5 \end{cases}$$

at the $\alpha = .05$ level. From his SRS, he computes the two statistics $\bar{X} = 18.1$, and S = 4.7. Since he has a small sample, he performs a T-test. He calculates the test statistic:

$$T_0 = \left(\frac{\bar{X}-\mu_0}{\frac{S}{\sqrt{n}}}\right) = \left(\frac{18.1-16.5}{\frac{4.7}{\sqrt{16}}}\right) = 1.362 \, .$$

His test statistic T_0 has (n − 1) = 15 degrees of freedom. From a table of t-distributions, he figures that the p-value is between .05 and .10. So he fails to reject H_0, and concludes that the amount of Copper precipitate is not greater than 16.5 at the .05 level.

Example (2):
A SRS of size n = 22 customers at a bank reveals that the amount of cash withheld from their paycheck deposit is $\bar{X} = \$80.00$, with S = \$15.00. The bank manager assumes that the amount of cash withheld, a random variable X, is normally distributed. He wants to test the hypotheses:

$$\begin{cases} H_0: \mu = 60 \\ H_a: \mu \neq 60 \end{cases} \text{ at the } \alpha = .01 \text{ level.}$$

Because he has a small sample, his test statistic is:

$$T_0 = \left(\frac{\bar{X}-\mu_0}{\frac{S}{\sqrt{n}}}\right) = \left(\frac{80-60}{\frac{15}{\sqrt{22}}}\right) = 6.254.$$

T_0 has (n − 1) = 21 degrees of freedom. He determines that the p-value $\ll .01$. Therefore, he rejects H_0 and concludes that the mean amount of cash withheld is not equal to $60.00. In fact, he thinks that it is probably more by looking at the value of $\bar{X}$ from his sample.

################# Exercises #################

(1) (a) A geologist has a SRS of size n = 18 measurements of the hardness X (in units of Sterlings) of a certain type of Limestone. What must he assume about X?
(b) He's interested in determining whether the mean hardness is greater than 235 Sterlings, at level $\alpha = .05$. Formulate the hypotheses that he wants to test.
(c) From his sample, he calculates $\bar{X} = 248$ and S = 13.2. What is the value of his test statistic T_0?
(d) What is the distribution of T_0 if the null hypothesis and his assumptions are true?
(e) What is the p-value for this test and his conclusions?

(2) (a) A Biologist has a SRS of size n = 11 measurements of the Blood Pressure X of Australian Gray Lizards. What must he assume about X?
(b) He's interested in determining whether the mean Blood Pressure is 125 units or not, at level $\alpha = .01$. Formulate the hypotheses that he wants to test.
(c) From his sample, he calculates $\bar{X} = 128$ and S = 3.73. What is the value of his test statistic T_0?
(d) What is the distribution of T_0 if the null hypothesis and his assumptions are true?
(e) What is the p-value for this test and his conclusions?

(11.3) Confidence Intervals for a Mean μ

With a chosen α, to construct a $(1 - \alpha) \cdot 100\%$ confidence interval for the mean μ, start with the following:

The $\Pr\left(-t_{(n-1),\alpha/2} < \left(\frac{\bar{X}-\mu}{\frac{S}{\sqrt{n}}}\right) < t_{(n-1),\alpha/2}\right) = (1 - \alpha)$

After rearranging this inequality, we find that the

$$\Pr\left(\bar{X} - \left(t_{(n-1),\alpha/2}\right)\left(\frac{S}{\sqrt{n}}\right) < \mu < \bar{X} + \left(t_{(n-1),\alpha/2}\right)\left(\frac{S}{\sqrt{n}}\right)\right)$$

equals $(1 - \alpha)$.

So, a $(1 - \alpha) \cdot 100\%$ confidence interval for the mean μ based on a small sample of size n is:

$$\left(\bar{X} - \left(t_{(n-1),\alpha/2}\right)\left(\frac{S}{\sqrt{n}}\right), \bar{X} + \left(t_{(n-1),\alpha/2}\right)\left(\frac{S}{\sqrt{n}}\right)\right)$$

Example (1):
A SRS of size n = 25 measurements of a random variable X , yields an average $\bar{X} = 53.2$ and S = 5.7. It is assumed that the distribution being sampled from is normally distributed. The researcher wants a 90% confidence interval for the true but unknown mean μ. Since he has a small sample, his confidence interval is:

$$\left(\bar{X} - \left(t_{(24),.05}\right)\left(\frac{S}{\sqrt{n}}\right), \bar{X} + \left(t_{(24),.05}\right)\left(\frac{S}{\sqrt{n}}\right)\right), \text{ which is:}$$

$$\left(53.2 - (1.711)\left(\frac{5.7}{\sqrt{25}}\right), 53.2 + (1.711)\left(\frac{5.7}{\sqrt{25}}\right)\right), \text{ which is:}$$

$(51.249, 55.151)$.

Example (2):
A SRS of size n = 9 from an experiment, consisting of measurements of a random variable X (assumed to be normally distributed) is taken, the manager calculates $\bar{X} = 101$, and S = 12. He wants a 99% confidence interval for the true but unknown mean μ from this small sample. He calculates:

$$\left(\bar{X} - \left(t_{(8),.005}\right)\left(\frac{S}{\sqrt{n}}\right), \bar{X} + \left(t_{(8),.005}\right)\left(\frac{S}{\sqrt{n}}\right)\right), \text{ which is:}$$

$$\left(101 \pm (3.355)\left(\frac{12}{\sqrt{9}}\right)\right), \text{ which is } (87.58, 114.42)$$

################# Exercises #################

(1) A SRS of size n = 13 measurements from a normally distributed random variable X yields $\bar{X} = 84.7$ and $S^2 = 103.6$. Calculate a 95% confidence interval for the true but unknown mean μ of the distribution of X.

(2) A SRS of size n = 29 measurements of a normally distributed random variable X yields $\bar{X} = 17.85$ and $S = 2.38$. Calculate a 99% confidence interval for the true but unknown mean μ of the distribution of X.

(11.4) <u>Two Sample Tests for the Difference of Two Means</u>

If we have two independent small SRS of sizes n_x and n_y measuring random variables X and Y respectively, then in addition to assuming that X and Y are normally distributed, we need to assume that the random variables X and Y have equal variances. From the samples we calculate $\bar{X}$, $\bar{Y}$, S_X^2, and S_y^2. Since we assume that the variances are equal, we will use a pooled variance:

$$S_p^2 = \frac{(n_x-1)S_x^2+(n_y-1)S_y^2}{n_x+n_y-2}.$$

We will test the hypotheses: $\begin{cases} H_0: \mu_X - \mu_Y = 0 \\ H_a: \mu_X - \mu_Y \neq 0 \end{cases}$ at level α.

We use the test statistic $T_0 = \left(\dfrac{(\bar{X}-\bar{Y})}{S_p\sqrt{\frac{1}{n_x}+\frac{1}{n_y}}} \right)$, with

$(n_x + n_y - 2)$ degrees of freedom. Then we determine the p-value according to the percentiles found in the Appendix C table, and make our conclusions accordingly.

Example (1):
Two small SRS of sizes $n_x = 10$ and $n_y = 10$, for two normally distributed random variables X and Y, yields $\bar{X} = 12.7$, $\bar{Y} = 13.9$, $S_x^2 = 1.7$, and $S_y^2 = 2.1$.

The researcher wants to test $\begin{cases} H_0: \mu_x - \mu_y = 0 \\ H_a: \mu_x - \mu_y \neq 0 \end{cases}$ at the

$\alpha = .05$ level. Since we must assume that the random variables X and Y have equal variances, we will pool the variances:

$$S_p^2 = \frac{(n_x-1)S_x^2+(n_y-1)S_y^2}{n_x+n_y-2} = \frac{(9)(1.7)+(9)(2.1)}{18} = 1.9 \, .$$

The test statistic is:

$$T_0 = \left(\frac{(\bar{X}-\bar{Y})}{S_p\sqrt{\frac{1}{n_x}+\frac{1}{n_y}}} \right) = \left(\frac{12.7-13.9}{(\sqrt{1.9})\sqrt{\frac{1}{10}+\frac{1}{10}}} \right) = -1.947 \, .$$

This test statistic has a t-distribution with 18 degrees of freedom. The $.05 <$ p-value $< .10$. Therefore, we fail to reject H_o, and conclude that the means of the random variables X and Y are not different.

################# Exercises #################

(1) (a) Two small SRS of sizes $n_x = 14$ and $n_y = 15$, for two normally distributed random variables X and Y, yields $\bar{X} = 75.1$, $\bar{Y} = 76.6$, $S_x^2 = 3.35$, $S_y^2 = 4.15$. The researcher wants to test $\begin{cases} H_0: \mu_x - \mu_y = 0 \\ H_a: \mu_x - \mu_y \neq 0 \end{cases}$ at

the $\alpha = .05$ level. Since we must assume that the

random variables X and Y have equal variances, we will pool the variances. Calculate the pooled variance.

(b) Calculate the test statistic T_0 that is used to test the hypotheses above.

(c) How many degrees of freedom are associated with this statistic T_0?

(d) What is the p-value and what conclusion does the researcher make?

(11.5) Two Sample Confidence Interval for Difference of Two Means

Let X and Y be two independent normally distributed random variables, with equal variances. From two small SRS of sizes n_x and n_y, we can construct a $(1 - \alpha) \cdot 100\%$ confidence interval for $\mu_x - \mu_y$, using the t-distribution.

Suppose from our samples we calculate $\bar{X}$, $\bar{Y}$, S_x^2, and S_y^2. Since we have equal variances, we pool them:

$$S_p^2 = \frac{(n_x-1)S_x^2+(n_y-1)S_y^2}{n_x+n_y-2}.$$

The estimator of $\mu_x - \mu_y$ is $(\bar{X} - \bar{Y})$ and its standard deviation is $\sqrt{\frac{\text{Var(X)}}{n_x} + \frac{\text{Var(Y)}}{n_y}} = \sqrt{S_p^2 \left(\frac{1}{n_x} + \frac{1}{n_y}\right)} = S_p\sqrt{\frac{1}{n_x} + \frac{1}{n_y}}.$

So, a $(1 - \alpha) \cdot 100\%$ confidence interval for the true but unknown difference of means $\mu_x - \mu_y$ is:

$$(\bar{X} - \bar{Y}) \pm \left(t_{(n_x+n_y-2),\frac{\alpha}{2}}\right) S_p \sqrt{\frac{1}{n_x} + \frac{1}{n_y}}.$$

As an example, suppose we have two independent small SRS from normally distributed random variables X and Y, with equal variances. We want a 90% confidence interval for the difference of means $\mu_x - \mu_y$. From our SRS of sizes $n_x = n_y = 20$, we calculate $\bar{X} = 20$, $\bar{Y} = 22.4$, $S_x^2 = 5.6$, and $S_y^2 = 6.1$. We pool the variances:

$$S_p^2 = \frac{(19)(5.6)+(19)(6.1)}{38} = 5.85$$

Then a 90% confidence interval for $\mu_x - \mu_y$ is:

$$(20 - 22.4) \pm \left(t_{(38),.05}\right)(2.4187)\left(\sqrt{\frac{1}{20} + \frac{1}{20}}\right), \text{ which is:}$$

$(-2.4) \pm (1.686)(2.4187)(0.31623)$, which is:

$(-3.690, -1.110)$.

################# Exercises #################

(1) (a) Suppose we have two independent small SRS from normally distributed random variables X and Y, with equal variances. We want a 99% confidence interval for the difference of means $\mu_x - \mu_y$. From our samples of sizes $n_x = n_y = 15$, we calculate $\bar{X} = 212$,

$\bar{Y} = 224$, $S_x^2 = 15.6$, and $S_y^2 = 16.1$. What is the pooled variance?

(b) What is the confidence interval for the difference in means?

(c) What are we doing differently in this section that we stated in section (10.5) that we should not do with confidence intervals involving two samples?

(12) <u>LINEAR REGRESSION AND CORRELATION</u>

(12.1) The Least Squares Line

Suppose we are doing some experiment and we collect data on two variables X and Y, and we wish to fit a linear relationship to these variables. We will try to find the best fitting line to the set of data points $\{(x_i, y_i),\ i = 1, 2, \ldots, n\}$ in the X-Y plane, that is, a best fitting line to the resulting scatter plot of data in the X-Y plane. The line will have the form:

$y = \beta_0 + \beta_1 x$,   where β_0 and β_1 are real numbers.

The parameter β_0 is the y-intercept and β_1 is the slope. We need to find estimates $\hat{\beta}_0$ and $\hat{\beta}_1$ of β_0 and β_1 respectively. We consider X to be the independent variable and the set of x-values are considered to be non-random. We assume that we choose the set of x-values that we want to have in our experiment. For a given value of x, we have a y-value that is considered to be random. We have the model for the y-values:

$y_i = \beta_0 + \beta_1 x_i + \varepsilon_i$ (where $\varepsilon_i \sim N(0, \sigma^2)$, $(i = 1, 2, \ldots, n)$, across the entire range of x-values. When we fit the line to the data, we have decided that the criterion that we will make use of to find the estimators of β_0 and β_1 is a least squares criterion. We want to find estimators that minimize the sum of squares SS $= \sum_{i=1}^{n} \varepsilon_i^2$. The estimates of β_0 and β_1 from the data that minimizes SS are called the least squares estimators of β_0 and β_1, and these are the estimators $\hat{\beta}_0$ and $\hat{\beta}_1$ that we will use. From our model for y_i, we can solve for ε_i to get:

$$\varepsilon_i = y_i - \beta_0 - \beta_1 x_i . \quad \text{Then, } \varepsilon_i^2 = (y_i - \beta_0 - \beta_1 x_i)^2.$$

$$\text{So, SS} = \sum_{i=1}^{n} \varepsilon_i^2 = \sum_{i=1}^{n}(y_i - \beta_0 - \beta_1 x_i)^2$$

is a function of β_0 and β_1 that we need to minimize.

****

To derive the least squares estimators of β_0 and β_1, we will take an excursion into calculus. For those readers that have not studied calculus, they can skip this part of the discussion and simply use the results that we will derive.

We take the derivative of SS with respect to β_0 and β_1, set the two derivatives equal to 0, and solve for β_0 and β_1. So,

$$\frac{d(SS)}{d\beta_0} = (-2)[\sum_{i=1}^{n}(y_i - \beta_0 - \beta_1 x_i)]$$
$$= (-2)\sum_{i=1}^{n} y_i + (2\beta_0)\sum_{i=1}^{n}(1) + (2\beta_1)\sum_{i=1}^{n} x_i$$

$$\frac{d(SS)}{d\beta_1} = (-2)[\Sigma_{i=1}^{n}(y_i - \beta_0 - \beta_1 x_i)] \cdot (x_i)$$
$$= (-2)\Sigma_{i=1}^{n}(x_i y_i) + (2\beta_0)\Sigma_{i=1}^{n} x_i + (2\beta_1)\Sigma_{i=1}^{n} x_i^2$$

Setting the $\frac{d(SS)}{d\beta_0}$ and the $\frac{d(SS)}{d\beta_1}$ equal to 0, we get the so-called normal equations:

$$\begin{cases} (n)\beta_0 & + & (\Sigma_{i=1}^{n} x_i)\beta_1 & = & \Sigma_{i=1}^{n} y_i \\ (\Sigma_{i=1}^{n} x_i)\beta_0 & + & (\Sigma_{i=1}^{n} x_i^2)\beta_1 & = & \Sigma_{i=1}^{n} x_i y_i \end{cases}.$$

Using Cramer's Rule for finding the solution of a system of equations:

$$\hat{\beta}_1 = \frac{\begin{vmatrix} n & \Sigma_{i=1}^{n} y_i \\ \Sigma_{i=1}^{n} x_i & \Sigma_{i=1}^{n} x_i y_i \end{vmatrix}}{\begin{vmatrix} n & \Sigma_{i=1}^{n} x_i \\ \Sigma_{i=1}^{n} x_i & \Sigma_{i=1}^{n} x_i^2 \end{vmatrix}} = \frac{(n)(\Sigma_{i=1}^{n} x_i y_i) - (\Sigma_{i=1}^{n} x_i)(\Sigma_{i=1}^{n} y_i)}{(n)(\Sigma_{i=1}^{n} x_i^2) - (\Sigma_{i=1}^{n} x_i)^2}$$

$$= \frac{(n)(\Sigma_{i=1}^{n} x_i y_i) - (n^2)(\bar{x})(\bar{y})}{(n)(\Sigma_{i=1}^{n} x_i^2) - (n^2)(\bar{x})^2}$$

$$= \frac{(\Sigma_{i=1}^{n} x_i y_i) - n(\bar{x})(\bar{y})}{(\Sigma_{i=1}^{n} x_i^2) - n(\bar{x})^2}.$$

Dividing the first of the normal equations by n, and substituting $\hat{\beta}_1$ for β_1, we get:

$$\hat{\beta}_0 = \bar{y} - \hat{\beta}_1 \bar{x}$$

****

So, the best fitting line $Y = \hat{\beta}_0 + \hat{\beta}_1 X$ to the data $\{(x_i, y_i), \ i = 1, 2, \ldots, n\}$ uses the least squares estimators of β_0 and β_1:

$$\hat{\beta}_0 = \bar{y} - \hat{\beta}_1 \bar{x} \quad \text{and} \quad \hat{\beta}_1 = \frac{(\sum_{i=1}^{n} x_i y_i) - n(\bar{x})(\bar{y})}{(\sum_{i=1}^{n} x_i^2) - n(\bar{x})^2}.$$

Example (1):

Suppose an agricultural researcher has n = 30 small plots of land, and on each he takes two measurements X and Y. X is the percentage of an artificial fertilizer in the top 6 inches of the soil, and Y is the yield in pounds of New Guinea Grape Tomatoes on the plot. From an examination of the data scatterplot, he believes that there is a linear relationship between X and Y. X varies over the range 5.0% to 10.0%. He assumes that for every x-value in the range [.05, .10], that Y is normally distributed with mean equal to the y-value of the fitted line, and with variance σ^2 vertically about the line. So, he wants a line:

$$Y = \beta_0 + \beta_1 X$$

which will fit the data: $\{(x_i, y_i), \ i = 1, 2, \ldots, 30\}$ in the least squares sense.

(A) The researcher has the statistics:
$\sum_{i=1}^{30} x_i = 2.41$, $\sum_{i=1}^{30} y_i = 30.67$,

$\sum_{i=1}^{30} x_i^2 = 1.094$, $\sum_{i=1}^{30} y_i^2 = 57.22$, and

$\sum_{i=1}^{30} x_i y_i = 4.463.$

From these statistics: $\bar{X} = 0.08$, and $\bar{Y} = 1.022$.

So, $\hat{\beta}_1 = \dfrac{(\sum_{i=1}^{n} x_i y_i) - n(\bar{x})(\bar{y})}{(\sum_{i=1}^{n} x_i^2) - n(\bar{x})^2} = \dfrac{(4.463 - (30)(0.08)(1.022))}{(1.094 - (30)(0.08)^2)}$

$= \dfrac{2.0102}{0.902} = (2.2286).$

$\hat{\beta}_0 = \bar{y} - \hat{\beta}_1 \bar{x} = (1.022 - (2.2286)(0.08)) = (0.8437).$

Therefore, the best fitting line is:
$Y = (0.8437) + (2.2286)X$

(B) The researcher calculates from this line that if x = 0.09, or 9% of the top 6 inches of the soil is the artificial fertilizer, then the expected yield Y is 1.0443 pounds of New Guinea Grape Tomatoes.

(C) Anybody that uses a regression analysis like this must be careful about extrapolating a y-value for an x-value that is outside the range of the x-values, outside of [.05, .10] in this example. Beyond this range, the relationship between X and Y may no longer be linear.

There are many different types of linear regression analyses. For example, we may try to fit the model:

$$Y = \beta_0 + \beta_1 X_1 + \beta_2 X_2 + \beta_3 X_3$$

or we may try to fit a completely different model to the data, such as one of these:

$$Y = \beta_0 + \beta_1 X + \beta_2 X^2$$

or, $$Y = \beta_0 + \beta_1 X + \beta_2 X^2 + \beta_3 X^3$$

or, $$Y = \beta_0 + \beta_1 X_1 + \beta_2 X_2 + \beta_3 X_1 X_2$$

These are all called linear models because they are linear in the parameters (the $\beta_i's$). Once we get beyond the case of simple linear regression, the level of complexity in finding and solving the normal equations for the parameters $\{\beta_i\}$ rises very quickly. Fortunately, there exists techniques of matrix algebra in conjunction with high speed computers that are very useful in formulating and solving regression problems.

(12.2) Correlation Between X and Y

For a bivariate distribution between two random variables X and Y, the distribution correlation coefficient ρ is defined:

$$\rho = \frac{Cov(X,Y)}{\sigma_X \cdot \sigma_Y} = \frac{E(X-\mu_X)(Y-\mu_Y)}{\sqrt{E(X-\mu_X)^2} \cdot \sqrt{E(Y-\mu_Y)^2}}.$$

The value of ρ measures the degree of linear association between the two random variables X and Y: $-1 \le \rho \le 1$.

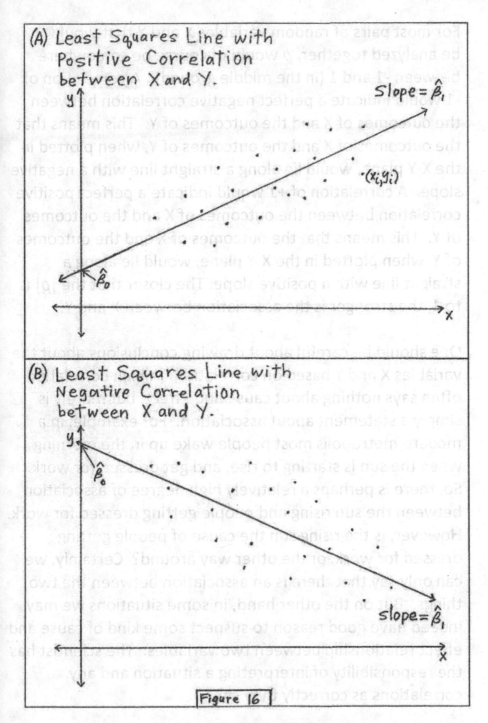

(A) Least Squares Line with Positive Correlation between X and Y.

Slope = $\hat{\beta}_1$

y

$\cdot(x_i, y_i)$

$\hat{\beta}_0$

x

(B) Least Squares Line with Negative Correlation between X and Y.

y

$\hat{\beta}_0$

slope = $\hat{\beta}_1$

x

Figure 16

For most pairs of random variables X and Y that would be analyzed together, ρ would of course be somewhere between -1 and 1 (in the middle ground). A correlation of -1 would indicate a perfect negative correlation between the outcomes of X and the outcomes of Y. This means that the outcomes of X and the outcomes of Y, when plotted in the X-Y plane, would lie along a straight line with a negative slope. A correlation of +1 would indicate a perfect positive correlation between the outcomes of X and the outcomes of Y. This means that the outcomes of X and the outcomes of Y, when plotted in the X-Y plane, would lie along a straight line with a positive slope. The closer that the $|\rho|$ is to 1, the stronger is the association between X and Y.

One should be careful about drawing conclusions about the variables X and Y based on correlation. A high correlation often says nothing about cause and effect, but merely is simply a statement about association. For example, in a modern metropolis most people wake up in the morning when the sun is starting to rise, and get dressed for work. So, there is perhaps a relatively high degree of association between the sun rising and people getting dressed for work. However, is the rising sun the cause of people getting dressed for work, or the other way around? Certainly, we can only say that there is an association between the two things. But on the other hand, in some situations we may indeed have good reason to suspect some kind of cause and effect relationship between two variables. The scientist has the responsibility of interpreting a situation and any correlations as correctly they can.

From data such as we studied in the previous section on simple linear regression: $\{(x_i, y_i), \ i = 1, 2, \ldots, n\}$ we can get an estimate of ρ with the sample correlation coefficient, which we call r:

$$r = \frac{\sum_{i=1}^{n}(x_i y_i) - n(\bar{X})(\bar{Y})}{\sqrt{\sum_{i=1}^{n}(x_i - \bar{x})^2} \cdot \sqrt{\sum_{i=1}^{n}(y_i - \bar{y})^2}}.$$

We would have $-1 \leq r \leq 1$.

The sample correlation coefficient r gives us a measure of the strength of the linear relationship between X and Y. "r" is not the slope of the least squares line, but a measure of how tightly the data fit the line.

Example (1):
(A) Using example (1) of section (12.1), another way to calculate r given that we have already calculated $\hat{\beta}_1$ is:

$$r = (\hat{\beta}_1) \cdot \left(\frac{\sqrt{\sum_{i=1}^{n}(x_i - \bar{x})^2}}{\sqrt{\sum_{i=1}^{n}(y_i - \bar{y})^2}} \right) = (\hat{\beta}_1) \cdot \left(\frac{\sqrt{\sum_{i=1}^{n} x_i^2 - n(\bar{X})^2}}{\sqrt{\sum_{i=1}^{n} y_i^2 - n(\bar{Y})^2}} \right)$$

$$= (2.2286) \cdot \left(\frac{0.949737}{5.08777} \right) = (0.416).$$

As a general rule, $0 \leq |r| \leq 0.5$ are low correlations, and $0.5 \leq |r| \leq 0.8$ are moderate correlations, and $0.8 \leq |r| \leq 1.0$ are high correlations.

(B) In simple linear regression, the statistic r^2 is called the Coefficient of Determination. For this example:

$$r^2 = (\text{sample correlation})^2 = (0.416)^2 = 0.173 = 17.3\%.$$

The statistic r^2 tells us the amount of the variability in Y that is explained by fitting a line to the data. The correlation r = 0.416 is a low correlation. As a result, $r^2 = 0.173$ is low too. It tells us that only 17.3% of the variability in Y is explained by fitting the line to the data, or put another way, about 82.7% of the variability in the data is random error about the line.

We have assumed that the x-values in a simple linear regression are fixed values that an experimenter can choose. Now, let's make the assumption that both X and Y are normal random variables and that they have a joint bivariate normal distribution.

The normal distribution is bell-shaped on the domain of the real numbers. Similarly, the bivariate normal distribution is a bell-shaped surface over the X-Y plane, with a joint pdf f(x,y). We will not concern ourselves with this pdf f(x,y) since it is rather complicated and beyond the scope of what we are doing here. Suffice it to say that the bivariate normal distribution is a natural extension of the normal distribution to two dimensions. Its marginal distributions are normal distributions in the X and Y directions with parameters μ_X, σ_X and μ_Y, σ_Y respectively. The bivariate normal distribution has the additional parameter ρ to take into account the tendency for there to be a linear component in the bivariate distribution's shape above the X-Y plane, that is, a linear component to the shape of the joint pdf surface f(x,y).

Statisticians have determined that if we assume (in a simple linear regression situation) that the two variables X and Y have a bivariate normal distribution, then we can test the null hypothesis $H_0: \rho = \rho_0$ versus one of the usual three alternatives (where $-1 \le \rho_0 \le 1$). Testing these hypotheses depends on recognizing that the random variable W:

$W = \frac{1}{2} \ln \left(\frac{1+r}{1-r} \right)$, for large n, is approximately normally distributed with mean $\mu_W = \frac{1}{2} \ln \left(\frac{1+\rho}{1-\rho} \right)$ and variance $\sigma_W^2 = \left(\frac{1}{n-3} \right)$.

To test the null hypothesis $H_0: \rho = \rho_0$, the test statistic:

$Z = \dfrac{\frac{1}{2}\ln\left(\frac{1+r}{1-r}\right) - \frac{1}{2}\ln\left(\frac{1+\rho_0}{1-\rho_0}\right)}{\left(\frac{1}{\sqrt{n-3}}\right)}$ is approximately Normal(0,1) in

distribution. The p-value can be determined for whichever of the three alternatives the experimenter decides upon ahead of time (with level of significance α).

Example (2):

(A) For the sample correlation calculated in example (1) of this section, test $H_0: \rho = 0.5$ versus $H_a: \rho < 0.5$, at the level $\alpha = .05$. The test statistic is calculated to be:

$$Z_0 = \frac{\frac{1}{2} ln\left(\frac{1 + 0.416}{1 - 0.416}\right) - \frac{1}{2} ln\left(\frac{1 + 0.5}{1 - 0.5}\right)}{\left(\frac{1}{\sqrt{27}}\right)} = \frac{\frac{1}{2} ln\left(\frac{1.416}{0.584}\right) - \frac{1}{2} ln\left(\frac{1.5}{0.5}\right)}{(0.19245)}$$

$$= \frac{0.4428 - 0.5493}{0.19245} = -(0.55).$$

The p-value is the $\Pr(Z \leq Z_0)$
$$= \Pr(Z \leq -0.55) = (0.2912)$$

Therefore, since the p-value is much greater than $\alpha = .05$, we fail to reject H_0, and we conclude that there is no significant evidence in this sample that the true value of ρ is less than (0.5).

(B) To test that the correlation is significantly greater than 0, at the $\alpha = .05$ level, we test $H_0: \rho = 0$ versus $H_a: \rho > 0$. The test statistic is calculated to be:

$$Z_0 = \frac{\frac{1}{2} ln\left(\frac{1 + 0.416}{1 - 0.416}\right) - \frac{1}{2} ln\left(\frac{1 + 0}{1 - 0}\right)}{\left(\frac{1}{\sqrt{27}}\right)} = \frac{\frac{1}{2} ln\left(\frac{1.416}{0.584}\right)}{(0.19245)} = 2.30 .$$

The p-value is the $\Pr(Z \geq Z_0) = 1 - \Pr(Z \leq Z_0)$
$$= 1 - \Pr(Z \leq 2.30)$$
$$= 1 - (0.9893) = (0.0107)$$

Since the p-value is much less than α, we reject H_0 and conclude that there is significant evidence that the correlation between X and Y is greater than 0, even though it is not a large correlation.

################## Exercises ##################

(1) A scientist is studying lifespan Y (in years) versus diet quality X (measured on a scale of 1 to 10) for Rhesus monkeys in the Metropolitan Zoo. She has collected the following data on n = 20 monkeys and wants to fit a line through the data. The righthand column is the product xy.

(x_i, y_i)	$(x_i)(y_i)$
(1, 7)	7
(1, 12)	12
(2, 5)	10
(2, 8)	16
(2, 6)	12
(3, 10)	30
(4, 13)	52
(4, 14)	56
(5, 17)	85
(5, 16)	80
(6, 18)	108
(6, 15)	90
(7, 21)	147
(7, 18)	126
(8, 18)	144
(8, 20)	160
(9, 17)	153
(9, 21)	189
(10, 22)	220
(10, 17)	170

(a) Calculate: $\sum_{i=1}^{20} x_i$, $\sum_{i=1}^{20} x_i^2$, $\sum_{i=1}^{20} y_i$, $\sum_{i=1}^{20} y_i^2$

(b) Calculate: $\bar{X}$, $\bar{Y}$, $\sum_{i=1}^{20} x_i y_i$

(c) Calculate: $\hat{\beta}_0$, $\hat{\beta}_1$

(d) What is the least squares line through the scatterplot?

(e) Calculate r and r^2 and state your conclusions about the meaning of the Correlation and the Coefficient of Determination in this case.

(2) Before the experiment, the scientist had believed that the correlation between X and Y would be greater than 0.75. Test the Hypotheses $\begin{cases} H_0: \rho = .75 \\ H_a: \rho > .75 \end{cases}$ at the .05 level. What are your results?

APPENDICES

519

Appendix A: Standard Normal Curve Areas: Table value is Pr $(Z \leq z)$.

z	.00	.01	.02	.03	.04	.05	.06	.07	.08	.09
-3.4	.0003	.0003	.0003	.0003	.0003	.0003	.0003	.0003	.0003	.0002
-3.3	.0005	.0005	.0005	.0004	.0004	.0004	.0004	.0004	.0004	.0003
-3.2	.0007	.0007	.0006	.0006	.0006	.0006	.0006	.0005	.0005	.0005
-3.1	.0010	.0009	.0009	.0009	.0008	.0008	.0008	.0008	.0007	.0007
-3.0	.0013	.0013	.0013	.0012	.0012	.0011	.0011	.0011	.0010	.0010
-2.9	.0019	.0018	.0017	.0017	.0016	.0016	.0015	.0015	.0014	.0014
-2.8	.0026	.0025	.0024	.0023	.0023	.0022	.0021	.0021	.0020	.0019
-2.7	.0035	.0034	.0033	.0032	.0031	.0030	.0029	.0028	.0027	.0026
-2.6	.0047	.0045	.0044	.0043	.0041	.0040	.0039	.0038	.0037	.0036
-2.5	.0062	.0060	.0059	.0057	.0055	.0054	.0052	.0051	.0049	.0038
-2.4	.0082	.0080	.0078	.0075	.0073	.0071	.0069	.0068	.0066	.0064
-2.3	.0107	.0104	.0102	.0099	.0096	.0094	.0091	.0089	.0087	.0084
-2.2	.0139	.0136	.0132	.0129	.0125	.0122	.0119	.0116	.0113	.0110
-2.1	.0179	.0174	.0170	.0166	.0162	.0158	.0154	.0150	.0146	.0143
-2.0	.0228	.0222	.0217	.0212	.0207	.0202	.0197	.0192	.0188	.0183
-1.9	.0287	.0281	.0274	.0268	.0262	.0256	.0250	.0244	.0239	.0233
-1.8	.0359	.0352	.0344	.0336	.0329	.0322	.0314	.0307	.0301	.0294
-1.7	.0446	.0436	.0427	.0418	.0409	.0401	.0392	.0384	.0375	.0367
-1.6	.0548	.0537	.0526	.0516	.0505	.0495	.0485	.0475	.0465	.0455
-1.5	.0668	.0655	.0643	.0630	.0618	.0606	.0594	.0582	.0571	.0559
-1.4	.0808	.0793	.0778	.0764	.0749	.0735	.0722	.0708	.0694	.0681
-1.3	.0968	.0951	.0934	.0918	.0901	.0885	.0869	.0853	.0838	.0823
-1.2	.1151	.1131	.1112	.1093	.1075	.1056	.1038	.1020	.1003	.0985
-1.1	.1357	.1335	.1314	.1292	.1271	.1251	.1230	.1210	.1190	.1170
-1.0	.1587	.1562	.1539	.1515	.1492	.1469	.1446	.1423	.1401	.1379
-0.9	.1841	.1814	.1788	.1762	.1736	.1711	.1685	.1660	.1635	.1611
-0.8	.2119	.2090	.2061	.2033	.2005	.1977	.1949	.1922	.1894	.1867
-0.7	.2420	.2389	.2358	.2327	.2296	.2266	.2236	.2206	.2177	.2148
-0.6	.2743	.2709	.2676	.2643	.2611	.2578	.2546	.2514	.2483	.2451
-0.5	.3085	.3050	.3015	.2981	.2946	.2912	.2877	.2843	.2810	.2776
-0.4	.3446	.3409	.3372	.3336	.3300	.3264	.3228	.3192	.3156	.3121
-0.3	.3821	.3783	.3745	.3707	.3669	.3632	.3594	.3557	.3520	.3482
-0.2	.4207	.4168	.4129	.4090	.4052	.4013	.3974	.3936	.3897	.3859
-0.1	.4602	.4562	.4522	.4483	.4443	.4404	.4364	.4325	.4286	.4247
-0.0	.5000	.4960	.4920	.4880	.4840	.4801	.4761	.4721	.4681	.4641

z	.00	.01	.02	.03	.04	.05	.06	.07	.08	.09
0.0	.5000	.5040	.5080	.5120	.5160	.5199	.5239	.5279	.5319	.5359
0.1	.5398	.5438	.5478	.5517	.5557	.5596	.5636	.5675	.5714	.5753
0.2	.5793	.5832	.5871	.5910	.5948	.5987	.6026	.6064	.6103	.6141
0.3	.6179	.6217	.6255	.6293	.6331	.6368	.6406	.6443	.6480	.6517
0.4	.6554	.6591	.6628	.6664	.6700	.6736	.6772	.6808	.6844	.6879
0.5	.6915	.6950	.6985	.7019	.7054	.7088	.7123	.7157	.7190	.7224
0.6	.7257	.7291	.7324	.7357	.7389	.7422	.7454	.7486	.7517	.7549
0.7	.7580	.7611	.7642	.7673	.7704	.7734	.7764	.7794	.7823	.7852
0.8	.7881	.7910	.7939	.7967	.7995	.8023	.8051	.8078	.8106	.8133
0.9	.8159	.8186	.8212	.8238	.8264	.8289	.8315	.8340	.8365	.8389
1.0	.8413	.8438	.8461	.8485	.8508	.8531	.8554	.8577	.8599	.8621
1.1	.8643	.8665	.8686	.8708	.8729	.8749	.8770	.8790	.8810	.8830
1.2	.8849	.8869	.8888	.8907	.8925	.8944	.8962	.8980	.8997	.9015
1.3	.9032	.9049	.9066	.9082	.9099	.9115	.9131	.9147	.9162	.9177
1.4	.9192	.9207	.9222	.9236	.9251	.9265	.9278	.9292	.9306	.9319
1.5	.9332	.9345	.9357	.9370	.9382	.9394	.9406	.9418	.9429	.9441
1.6	.9452	.9463	.9474	.9484	.9495	.9505	.9515	.9525	.9535	.9545
1.7	.9554	.9564	.9573	.9582	.9591	.9599	.9608	.9616	.9625	.9633
1.8	.9641	.9649	.9656	.9664	.9671	.9678	.9686	.9693	.9699	.9706
1.9	.9713	.9719	.9726	.9732	.9738	.9744	.9750	.9756	.9761	.9767
2.0	.9772	.9778	.9783	.9788	.9793	.9798	.9803	.9808	.9812	.9817
2.1	.9821	.9826	.9830	.9834	.9838	.9842	.9846	.9850	.9854	.9857
2.2	.9861	.9864	.9868	.9871	.9875	.9878	.9881	.9884	.9887	.9890
2.3	.9893	.9896	.9898	.9901	.9904	.9906	.9909	.9911	.9913	.9916
2.4	.9918	.9920	.9922	.9925	.9927	.9929	.9931	.9932	.9934	.9936
2.5	.9938	.9940	.9941	.9943	.9945	.9946	.9948	.9949	.9951	.9952
2.6	.9953	.9955	.9956	.9957	.9959	.9960	.9961	.9962	.9963	.9964
2.7	.9965	.9966	.9967	.9968	.9969	.9970	.9971	.9972	.9973	.9974
2.8	.9974	.9975	.9976	.9977	.9977	.9978	.9979	.9979	.9980	.9981
2.9	.9981	.9982	.9982	.9983	.9984	.9984	.9985	.9985	.9986	.9986
3.0	.9987	.9987	.9987	.9988	.9988	.9989	.9989	.9989	.9990	.9990
3.1	.9990	.9991	.9991	.9991	.9992	.9992	.9992	.9992	.9993	.9993
3.2	.9993	.9993	.9994	.9994	.9994	.9994	.9994	.9995	.9995	.9995
3.3	.9995	.9995	.9995	.9996	.9996	.9996	.9996	.9996	.9996	.9997
3.4	.9997	.9997	.9997	.9997	.9997	.9997	.9997	.9997	.9997	.9998

Appendix B: Percentiles for Chi-Square Distributions:

If $X \sim \chi^2_{(k)}$, $Pr(X > (\text{table value})) = \alpha$.

k	.995	.975	.95	.05	.025	.005
1	0.000	0.001	0.004	3.843	5.025	7.882
2	0.010	0.051	0.103	5.992	7.378	10.597
3	0.072	0.216	0.352	7.815	9.348	12.837
4	0.207	0.484	0.711	9.488	11.143	14.860
5	0.412	0.831	1.145	11.070	12.832	16.748
6	0.676	1.237	1.635	12.592	14.440	18.548
7	0.989	1.690	2.167	14.067	16.012	20.276
8	1.344	2.180	2.733	15.507	17.534	21.594
9	1.735	2.700	3.325	16.919	19.022	23.587
10	2.156	3.247	3.940	18.307	20.483	25.188
11	2.603	3.816	4.575	19.675	21.920	26.755
12	3.074	4.404	5.226	21.026	23.337	28.300
13	3.565	5.009	5.892	22.362	24.735	29.817
14	4.075	5.629	6.571	23.685	26.119	31.319
15	4.600	6.262	7.261	24.996	27.488	32.799
16	5.142	6.908	7.962	26.296	28.845	34.267
17	5.697	7.564	8.682	27.587	30.190	35.716
18	6.265	8.231	9.390	28.869	31.526	37.156
19	6.843	8.906	10.117	30.143	32.852	38.580
20	7.434	9.591	10.851	31.410	34.170	39.997

(continued)

k	.995	.975	.95	.05	.025	.005
21	8.033	10.283	11.591	32.670	35.478	41.399
22	8.643	10.982	12.338	33.924	36.781	42.796
23	9.260	11.688	13.091	35.172	38.075	44.179
24	9.886	12.401	13.848	36.415	39.364	45.558
25	10.519	13.120	14.611	37.652	40.646	46.925
26	11.160	13.844	15.379	38.885	41.923	48.290
27	11.807	14.573	16.151	40.113	43.194	49.642
28	12.461	15.308	16.928	41.337	44.461	50.993
29	13.120	16.147	17.708	42.557	45.772	52.333
30	13.787	16.791	18.493	43.773	46.979	53.672
31	14.457	17.538	19.280	44.985	48.231	55.000
32	15.134	18.291	20.072	46.194	49.480	56.328
33	15.814	19.046	20.866	47.400	50.724	57.646
34	16.501	19.806	21.664	48.602	51.966	58.964
35	17.191	20.569	22.465	49.802	53.203	60.272
36	17.887	21.336	23.269	50.998	54.437	61.581
37	18.584	22.105	24.075	52.192	55.677	62.880
38	19.289	22.878	24.884	53.384	56.896	64.181
39	19.994	23.654	25.695	54.572	58.119	65.473
40	20.706	24.433	26.509	55.758	59.342	66.766

(continued)

Appendix Table C: Percentiles of $T_{(k)}$ Distributions.

If X ~ $T_{(k)}$, The Pr(X > (table value)) = α

| | | | α | | |
k	.10	.05	.025	.01	.005
1	3.078	6.314	12.706	31.821	63.657
2	1.886	2.920	4.303	6.965	9.925
3	1.638	2.353	3.182	4.541	5.841
4	1.533	2.132	2.776	3.747	4.604
5	1.476	2.015	2.571	3.365	4.032
6	1.440	1.943	2.447	3.143	3.707
7	1.415	1.895	2.365	2.998	3.499
8	1.397	1.860	2.306	2.896	3.355
9	1.383	1.833	2.262	2.821	3.250
10	1.372	1.812	2.228	2.764	3.169
11	1.363	1.796	2.201	2.718	3.106
12	1.356	1.782	2.179	2.681	3.055
13	1.350	1.771	2.160	2.650	3.012
14	1.345	1.761	2.145	2.624	2.977
15	1.341	1.753	2.131	2.602	2.947
16	1.337	1.746	2.120	2.583	2.921
17	1.333	1.740	2.110	2.567	2.898
18	1.330	1.734	2.101	2.552	2.878
19	1.328	1.729	2.093	2.539	2.861
20	1.325	1.725	2.086	2.528	2.845
21	1.323	1.721	2.080	2.518	2.831
22	1.321	1.717	2.074	2.508	2.819
23	1.319	1.714	2.069	2.500	2.807
24	1.318	1.711	2.064	2.492	2.797
25	1.316	1.708	2.060	2.485	2.787
26	1.315	1.706	2.056	2.479	2.779
27	1.314	1.703	2.052	2.473	2.771
28	1.313	1.701	2.048	2.467	2.763
29	1.311	1.699	2.045	2.462	2.756
30	1.310	1.697	2.042	2.457	2.750

(continued)

k	.10	.05	.025	.01	.005
32	1.309	1.694	2.037	2.449	2.738
34	1.307	1.691	2.032	2.441	2.728
36	1.306	1.688	2.028	2.434	2.719
38	1.304	1.686	2.024	2.429	2.712
40	1.303	1.684	2.021	2.423	2.704
50	1.299	1.676	2.009	2.403	2.678
60	1.296	1.671	2.000	2.390	2.660
120	1.289	1.658	1.980	2.358	2.617
∞	1.282	1.645	1.960	2.326	2.576

Appendix D: Answers to Exercises.

CHAPTER 1: LOGIC
Section (1.2)

(1) Let r = (p ∧ (~q))

p	q	~q	r	~r
T	T	F	F	T
T	F	T	T	F
F	T	F	F	T
F	F	T	F	T

(2) Let r = (p ∧ q), s = ((~p) ∧ (~q))

p	q	~p	~q	r	~r	s	~r → s
T	T	F	F	T	F	F	T
T	F	F	T	F	T	F	F
F	T	T	F	F	T	F	F
F	F	T	T	F	T	T	T

(3): DeMorgan's Laws for Logic:

(a) Let r = (p ∨ q), s = ((~p) ∧ (~q))

p	q	~p	~q	r	~r	s	~r ↔ s
T	T	F	F	T	F	F	T
T	F	F	T	T	F	F	T
F	T	T	F	T	F	F	T
F	F	T	T	F	T	T	T

Yes, ~(p ∨ q) ↔ ((~p) ∧ (~q)) is always a true statement.

(b): Let r = (p ∧ q), s = ((~p) ∨ (~q))

p	q	~p	~q	r	~r	s	~r ↔ s
T	T	F	F	T	F	F	T
T	F	F	T	F	T	T	T
F	T	T	F	F	T	T	T
F	F	T	T	F	T	T	T

Yes, ~(p ∧ q) ↔ ((~p) ∨ (~q)) is always a true statement.

CHAPTER 2: SETS AND NUMBERS

Section (2.3)

(1) (a) Yes, because every element of C is in A, and every element of A is in B.

(b) Yes, C is a proper subset of A, because A contains elements w and h which are not in C.

(c) No, A is not a proper subset of B, because there are no elements of B which are not in A, (A = B).

(2) (a) C is a subset of D, but D is not a subset of A, so the statement as a whole is not true.

(b) Yes, C is a proper subset of D because there are the elements w and r which are in D, but not in C.

(c) D is not a subset of A, proper or improper, since D contains element r which is not in A.

Section (2.4)

(1) (a) {0,2,4,6,7,8,9,10,11,12,13}

(b) {0,2,4,6,8,10,12}
(c) {0,2,4,6,8,10,15,17,20}
(d) {0,2,4,6,7,8,9,10,13,18,19}
(e) {4,6,7,8,9,10,11,12,13}
(f) {6,7,8,9,10,11,12,13,15,17,20}
(g) {0,6,7,8,9,10,11,12,13,18,19}
(h) {4,6,8,12,15,17,20}
(i) {0,4,6,7,8,9,12,13,18,19}
(j) {0,7,9,13,15,17,18,19,20}
(k) {0,4,6,7,8,9,12,13,15,17,18,19,20}

(2) (a) {6,8,10}   (b) {4,6,8}   (c) ⊘
    (d) {0}   (e) {6,8,12}   (f) ⊘
    (g) {7,9,13}   (h) ⊘   (i) ⊘
    (j) ⊘

(3) (a) {1,3,5,7,9,11,12,13,14,15,16,17,18,19,20}
    (b) {0,1,2,3,4,5,14,15,16,17,18,19,20}
    (c) {0,1,2,3,5,7,9,10,11,13,14,15,16,17,18,19,20}
    (d) {0,1,2,3,4,5,6,7,8,9,10,11,12,13,14,16,18,19}
    (e) {1,2,3,4,5,6,8,10,11,12,14,15,16,17,20}

(4) A and D, B and D, C and D, C and E, D and E.

(5) (a) No, (b) No   (6) (a) {1,3,5,14,16}, (b) Yes

(7) (a) Yes, (b) No   (8) Yes

(9) (a) {0,9,10,11}, (b) {6,7} (c) {9,10}
    (d) {6,7,9,10}, which is the set E.

528

(1) $\left(19 - \frac{3}{4}i\right)$ (2) $\left(17 + \frac{1}{9}i\right)$ (3) $\left(-1 - \frac{48}{7}i\right)$ (4) 2i

(5) $(22 - 4i)$ (6) $(33 - 9i)$ (7) $\left(-\frac{2}{5} - \frac{3}{10}i\right)$ (8) $-\frac{7}{18}$

(9) $-\frac{1}{7}i$

CHAPTER 3: FUNDAMENTALS OF ALGEBRA & GEOMETRY

Section (3.2)

(1) $4 < x < 22$ (2) $-30 \le x \le 30$ (3) $x \le -2$ or $x \ge 6$

(4) $x < (5 - h)$ or $x > (5 + h)$

(5) (Think of $|x + 20|$ as $|x - (-20)|$),

   So, $x < (-20 - \varepsilon)$ or $x > (-20 + \varepsilon)$

(6) $x \le 8$ or $x \ge 14$ (7) $15 < x < 35$

(8) $x^6 y^2$ (9) $\frac{x^4}{y^5}$ (10) $x^{\frac{3}{5}} y^{\frac{8}{5}} z^{\frac{4}{5}}$ (11) $\frac{x^{\frac{2}{3}}}{y^3}$ (12) x^6

(13) $x^6 y^9 z^{12}$ (14) $\frac{z^2}{xy^2}$ (15) $2x$ (16) $2x$ (17) x^2

(18) $2^{(4x^2 + x + 8)}$ (19) $\ln(60x^3)$ (20) $\log_{10}(5x)$

(21) $7 \cdot \log_4 x$ (22) $3 \cdot \ln(x)$

(23) $x = \pm 3$ (24) $x = \pm 5$ (25) $x = \pm\sqrt{27} = \pm 3\sqrt{3}$

(26) $x = -5$ (27) $x = 2$ (28) $x = \pm 3$

(29) (a) 9   (b) 100   (c) 2   (d) 30

   (e) 4   (f) $\sqrt[7]{13} \approx (1.4426)$

Section (3.3)

(1) $4x^2 - 16x + 16$ (2) $x^2 + 24x + 144$

(3) $9x^2 + 18x + 9$ (4) $16x^2 - 80x + 100$

(5) $3x^2 - 2xy + 4xz$ (6) $2x^2 + 15x - 1$

(7) $-x^3 + 9x^2 - 3x + 27$ (8) $x^2 - 9$
(9) $x^4 - x^6 + x^7 - x^9$ (10) $15x^2 + 41x + 28$
(11) $(x - 3)(x + 4)$ (12) $(x - 9)(x - 7)$
(13) $(x)(x + 5)(x + 6)$ (14) $(x^2)(x - 8)(x - 2)$
(15) $(x - 100)(x + 100)$ (16) $(4x - 2y)(4x + 2y)$
(17) $25(x + 1)(x - 6)$ (18) $(x + 1)(x - 1)^2$

Section (3.4)

(1) $\dfrac{3}{4x}$ (2) $\dfrac{-2x + 4}{x^2 - 1}$ (3) $2x + 1 + \dfrac{2}{x}$

(4) $2x - 5$ (5) $\dfrac{-3x + 14}{5x}$ (6) $7x^4$

(7) -4   (8) $(-1 - \sqrt{2}i), (-1 + \sqrt{2}i)$ (9) 0, -4

(10) $(-1 - \sqrt{721}), (-1 + \sqrt{721})$ (11) -3

(12) $\dfrac{1 - \sqrt{33}}{2}, \dfrac{1 + \sqrt{33}}{2}$

Section (3.6)

(1) $\dfrac{1}{34}$ (2) $\dfrac{7}{10}$ (3) $\dfrac{7}{16}$ (4) $-\dfrac{6}{7}$ (5) $\dfrac{6}{11}$

(6)(a) $y = \dfrac{5}{6}x + \dfrac{2}{3}$ (b) m = $\dfrac{5}{6}$, b = $\dfrac{2}{3}$, positively sloped line.

(7)(a) $y = -\dfrac{8}{3}x - \dfrac{14}{3}$ (b) m = $-\dfrac{8}{3}$, b = $-\dfrac{14}{3}$, a negatively
sloped line.

(8)(a) $z = -\frac{1}{7}x + \frac{4}{7}y - \frac{3}{7}$ (b) $\left(1, 2, \frac{4}{7}\right)$

Section (3.7)

(1) 4, -2   (2) $\dfrac{5 - \sqrt{21}}{2}, \dfrac{5 + \sqrt{21}}{2}$ (3) $-2 \pm \sqrt{5}$

(4) $\dfrac{1 - \sqrt{5}}{2}, \dfrac{1 + \sqrt{5}}{2}$ (5) $\dfrac{7 - \sqrt{41}}{4}, \dfrac{7 + \sqrt{41}}{4}$ (6) -1, 10

(7) -5, -1   (8) $\dfrac{-11 - \sqrt{117}}{2}, \dfrac{-11 + \sqrt{117}}{2}$

(9) y-intercept = 2, x-intercepts = $6 - \dfrac{3}{2}\sqrt{\dfrac{40}{3}}, 6 + \dfrac{3}{2}\sqrt{\dfrac{40}{3}}$,
vertex is at $(6, -10)$, parabola opens upward.

(10) y-intercept = 1, x-intercepts = -1, -1 (repeated root),
vertex is at $(-1, 0)$, parabola opens upward.

Section (3.8)

(1) (a) x = 5   (b) As $x \to -\infty, y \to -\infty$,
                      As $x \to +\infty, y \to +\infty$.

(2) (a) x = -3,0,2   (b) As $x \to -\infty, y \to +\infty$,
                      As $x \to +\infty, y \to +\infty$.

(3) (a) x = -3,3   (b) As $x \to -\infty, y \to -\infty$,
                      As $x \to +\infty, y \to -\infty$.

(4) (a) x = 5,2,8   (b) As $x \to -\infty, y \to +\infty$,
                      As $x \to +\infty, y \to -\infty$.

Section (3.9)

(1) $\dfrac{5}{6}$ (2) 2   (3) 6   (4) $-\dfrac{15}{46}$

(5) $\dfrac{13}{2} - \dfrac{\sqrt{61}}{2}$, $\dfrac{13}{2} + \dfrac{\sqrt{61}}{2}$ (6) -5

Section (3.10)

(1) $\left[\dfrac{1}{4}, \infty\right)$ (2) $\left(\dfrac{1}{2}, \infty\right)$ (3) $\left(\dfrac{4}{3}, \infty\right)$ (4) $\left(-\infty, \dfrac{35}{39}\right]$

(5) $\left(-\infty, \dfrac{29}{3}\right)$ (6) $(-5,5)$ (7) $(-\infty, -2] \cup [2, \infty)$

(8) $\left(\dfrac{7}{2}, \dfrac{9}{2}\right)$ (9) $\left[-\dfrac{1}{3}, \dfrac{1}{3}\right]$ (10) $(-2,3)$ (11) $\left(-\dfrac{1}{23}, \dfrac{12}{23}\right)$

(12) $\left(-\dfrac{7}{4}, -\dfrac{1}{2}\right)$ (13) $\left[-\sqrt{14}, \sqrt{14}\right]$ (14) $(-3, -1) \cup (1,3)$

Section (3.11)

(1) The point $(x,y) = \left(\dfrac{3}{2}, \dfrac{1}{2}\right)$.   (2) All points (x,y) on the line $2x - 7y = 12$.

(3) The point $(x,y) = \left(-\dfrac{25}{2}, 8\right)$.   (4) The point $(x,y) = (1,1)$.

Section (3.12)

(1) 147

(3) Both double sums equal 54.

(4) (a) 2750   (b) 5250   (c) 225

CHAPTER 4: FUNDAMENTALS OF TRIGONOMETRY
        AND THE CALCULUS

Section (4.4)

(1) (a) $\sin(\theta) = \frac{3}{5}$, $\cos(\theta) = \frac{4}{5}$, $\tan(\theta) = \frac{3}{4}$

$\csc(\theta) = \frac{5}{3}$, $\sec(\theta) = \frac{5}{4}$, $\cot(\theta) = \frac{4}{3}$

(b) $\theta = 36.87°$, or $\theta = 0.64$ radians   (c) 53.13°

(2) r = 11.51 , $\theta = 50.64°$

(3) (a) 10.63   (b) $|\vec{F_x}| = 8$, $|\vec{F_y}| = 7$ (c) -41.19°

(6) (a) angle B is $\approx 55.38°$ (b) angle C is $\approx 88.62°$

(c) side c is approximately of length 8.504.

(7) (a) 3.13   (b) 10.00   (c) 13.53

(d) $c^2 = a^2 + b^2$, The Pythagorean Theorem.

Section (4.6)

(1) 24   (2) 0   (3) $-\infty$ (4) ∞ (5) $-\infty$ (6) 0
(7) 2   (8) 0   (9) 0   (10) 0   (11) 0   (12) 0
(13) -1 (14) -9 (15) -1 (16) 1   (17) No   (18) A = 2

Section (4.7)

(1) (a) $14x - 4$ (b) $6x^2 + 1$

(2) (a) $f'(x) = -\frac{2}{3}x + 2$

(b) $f'(1) = \frac{4}{3}$. The function is increasing.

$f'(3) = 0$. The function is neither increasing or decreasing when x = 3. When x = 3, the function has reached a point where the slope of the tangent line is 0. Since this a parabola that opens downward, the function has reached its maximum value of f(3) = 1.

$f'(6) = -2$. The function is decreasing.

Section (4.8)

(1) $\frac{dy}{dx} = 12x^3 + 2x$ (2) $\frac{dy}{dx} = -32x^{-9} - 15x^2$

(3) $\frac{dy}{dx} = 4x + 9x^2 + 16x^3$ (4) $\frac{dy}{dx} = 1 + 2x + 4x^3$

(5) $\frac{dy}{dx} = -3x^{-4} - 4x^{-5} - 6x^{-7}$ (6) $\frac{dy}{dx} = 40x^3 + 300x^2$

Section (4.9)

(1) $\frac{dy}{dx} = (15 + 40x)(3x + 4x^2)^4$

(2) $\frac{dy}{dx} = 315x^2(15x^3 - 1)^6$

(3) $\frac{dy}{dx} = 6(2x + 1)^2 - (130x + 5)(13x^2 + x - 1)^4$

(4) $\frac{dy}{dx} = -72x^5(3 - 4x^6)^2 - 16x(10 - 2x^2)^3$

(5) $\frac{dy}{dx} = 2 + 18x - 192x^2$

(6) $\frac{dy}{dx} = 3(x - 1)^2 + 9(3x - 1)^2$

Section (4.10)

(1) $\frac{dy}{dx} = \frac{80x^4 - 3x^2}{3y^2}$ (2) $\frac{dy}{dx} = \left(\frac{1}{9y^2 + 4y + 1}\right)$

(3) $\frac{dy}{dx} = \left(\frac{5}{(4y+2)(2y^2+2y)^4}\right)$ (4) $\frac{dy}{dx} = \frac{75x^2 - 50x}{3y^2}$

(5) $\frac{dy}{dx} = \left(\frac{-2}{20y^4 - 17}\right)$ (6) $\frac{dy}{dx} = \frac{(x-1)}{4(y-3)}$

Section (4.11)

(1) $\frac{dy}{dx} = x^{-\frac{1}{3}} + 40$ (2) $\frac{dy}{dx} = \frac{21}{8}x^{-\frac{1}{8}} + x^{-\frac{3}{4}}$

(3) $\frac{dy}{dx} = 8x^{-\frac{3}{5}} + (x - 2)(x^2 - 4x)^{-\frac{1}{2}}$

(4) $\dfrac{dy}{dx} = 2(4x+2)^{-\frac{1}{2}} - \dfrac{2}{3}(2x-3)^{-\frac{2}{3}}$

(5) $\dfrac{dy}{dx} = \left(\dfrac{165}{2}\right)x^{\frac{9}{2}} - \left(\dfrac{130}{7}\right)x^{\frac{6}{7}}$ (6) $\dfrac{dy}{dx} = x^{-\frac{4}{5}} - 4x^{-\frac{1}{3}} + \dfrac{7}{6}x^{-\frac{5}{6}}$

Section (4.12)

(1) $\dfrac{dy}{dx} = \dfrac{(2x-7)^8}{2\sqrt{x}} + 16\sqrt{x}(2x-7)^7$

(2) $\dfrac{dy}{dx} = (3x^2-1)^3(2x-2)^4(78x^2-48x-10)$

(3) $\dfrac{dy}{dx} = (8x)(x-1)^3(x+1)^3$

(4) $\dfrac{dy}{dx} = \dfrac{(x-7)^2(9x^3-84x^2+6)}{(2-x^3)^5}$

(5) $\dfrac{dy}{dx} = \dfrac{(x^2+x+1)^2(6x^2+9x-6)}{(3x+2)^5}$

(6) $\dfrac{dy}{dx} = \dfrac{8(x-1)^3}{(x+1)^5}$

Section (4.13)

(1) 5.001   (2) 315,604.398   (3) $-37.515G$ Newtons

Section (4.14)

(1) (a) $y = 10x+3$, $x = \dfrac{1}{10}(y-3)$

(b) $\dfrac{dy}{dx} = 10,\ \dfrac{dx}{dy} = \dfrac{1}{10}$

(2) (a) $y = \dfrac{1}{8}x^2 + 2$, $x = \sqrt{8(y-2)}$

(b) $\dfrac{dy}{dx} = \dfrac{x}{4},\ \dfrac{dx}{dy} = \dfrac{1}{2}\left(\dfrac{1}{\sqrt{8(y-2)}}\right) \cdot (8) = \dfrac{4}{\sqrt{8\left(\left(\frac{1}{8}x^2+2\right)-2\right)}} = \dfrac{4}{x}$

(3) (a) $y = (x-3)^3$, $x = \sqrt[3]{y}+3$

(b) $\dfrac{dy}{dx} = 3(x-3)^2$,

535

$$\frac{dx}{dy} = \frac{1}{3}\left(\frac{1}{(\sqrt[3]{y})^2}\right) = \frac{1}{3}\frac{1}{(\sqrt[3]{(x-3)^3})^2} = \frac{1}{3(x-3)^2}.$$

Section (4.15)

(1) $c = \frac{5}{2} = 2.5$ (2) $c = \sqrt{\frac{1}{27}} \approx 0.19245$

Section (4.16)

(1) $\frac{5}{3}$ (2) 0   (3) ∞ (4) 0   (5) 0

(6) $\frac{3}{8}$ (7) $\frac{2}{3}$ (8) ∞

Section (4.17)

(1) $\frac{dy}{dx} = \cos x - \sec^2(x)$

(2) $\frac{dy}{dx} = -2\sin(2x) - 2\csc^2(2x)$

(3) $\frac{dy}{dx} = (2x)\sec(x^2)\tan(x^2)$

(4) $\frac{dy}{dx} = (9x^2 - 4)\sec^2(3x^3 - 4x)$

(5) $\frac{dy}{dx} = \cos(\cos(x^2)) \cdot (-\sin(x^2)) \cdot (2x)$

(6) $\frac{dy}{dx} = (-3x + 1)e^{(-3x^2 + 2x + 1)}$ (7) $\frac{dy}{dx} = \frac{(-2x - 1)}{(14 - x - x^2)}$

(8) $\frac{dy}{dy} = 2^{\sin x} \cdot \cos x \cdot \ln 2 + 3\csc(3x)\cot(3x)$

(9) $\frac{dy}{dx} = \left(\frac{5}{1 + 25x^2}\right)$ (10) $\frac{dy}{dx} = \left(\frac{20x}{\sqrt{1 - 100x^4}}\right)$

(11) (a) $L(x) = 1 + x$ (b) 1.003   (c) 1.003004505

(12) (a) $L(x) = x - 1$ (b) $-(0.003)$ (c) -0.003004509

(13) (a) $L(x) = x$ (b) 0.0001   (c) 0.000001745

Section (4.18)

(1) (a) $P(x) = -\frac{1}{2}x^2 + 44x$, $P'(x) = -x + 44$, the critical

536

value is x = 44.

(b) x = 44, y = 22 .   (c) 968

(d) P(x) is a parabola that opens downward. Therefore it has a maximum at its vertex (44, 968).

(e) x = 0, x = 88, the x-intercepts. These x-intercepts correspond to the absolute minimum and maximum values that x can take since x has to be non-negative.

(2) (a) Nowhere

(b) $f'(x) = \left(\dfrac{-1}{\sqrt{2\pi \, (a^3)}}\right)(x-b)e^{-\frac{1}{2}\left(\frac{x-b}{a}\right)^2}$, the critical value is x = b.

(c) $f''(x) = \left(\dfrac{-1}{\sqrt{2\pi \, (a^3)}}\right)\left[e^{-\frac{1}{2}\left(\frac{x-b}{a}\right)^2}\left(\left(\dfrac{-(x-b)^2}{a^2}\right)+1\right)\right]$, the inflection values are x = (b − a), and x = (b + a).

(d) 0, 0

(e) f(x) is increasing on (−∞, b) and f(x) is decreasing on (b, ∞).

(f) f(x) is concave up on (−∞, b − a) ∪ (b + a, ∞), f(x) is concave down on (b − a, b + a) .

(g) It has a relative maximum at $\left(b, \dfrac{1}{\sqrt{2\pi(a)}}\right)$. This is also an absolute maximum. It does not have an absolute minimum because the curve in its tails decreases towards zero, but never reaches zero.

(h) It should look like the graph of the normal curve shown in the first few pages of section (8.5), with $\mu = b, \sigma = a$.

(3) $\dfrac{dh}{dt} = \dfrac{44}{\sqrt{149}} \approx 3.605\ ^{ft}/_{sec.}$ (4) $1\ ^{ft}/_{sec.}$

Section (4.19)

(1) $\frac{x^4}{4} + \frac{x^3}{3} + \frac{x^2}{2} + x + C$ (2) $x + C$ (3) $48x + C$

(4) $-cosx + \ln|cotx - cscx| + C$

(5) $sinx - \ln|cosx| + C$ (6) $\frac{1}{4}x - \frac{1}{4}sinx + C$

(7) $\frac{1}{7}x - \frac{1}{7}cosx + C$ (8) $\ln|x| + C$ (9) $secx - cscx + C$

(10) $tanx - cotx + C$ (11) $13sin^{-1}x + C$

(12) $25tan^{-1}x + C$ (13) $\frac{x^3}{3} + sec^{-1}x + C$

(14) $\frac{x^2}{2} - cosx + (3)\ln|secx + tanx| + C$

Section (4.20)

(1) 24   (2) 2   (3) 0   (4) 560   (5) $(e^3 - 1) \approx 19.0855$

(6) $\ln(10) \approx 2.3026$ (7) $\frac{\pi}{2}$ (8) $\frac{\pi}{6}$ (9) ≈ 2.732

(10) $\ln|1 + \sqrt{2}| \approx 0.8814$ (11) $(1 - e^{-3}) \approx 0.9502$

(12) $\frac{3}{4}$

Section (4.21)

(1) 70.5 miles.   (2) 2.43474 meters^3 of fluid.

Section (4.22)

(1) $(x - 2)e^x - e^x + C$ (2) $\frac{(x^2 + x + 4)^5}{10} + C$

(3) $-\frac{1}{2}cos(x^2) + C$ (4) $x + \ln|x| + C$

(5) $\frac{1}{6}sec(3x^2) + C$ (6) $-\frac{1}{13}\ln|cos(13x + 3)| + C$

(7) $\frac{1}{4}tan^{-1}(4x) + C$ (8) $\frac{1}{5}sin^{-1}(x) + C$

(9) $\frac{1}{8}sin(4x^2 + 8x) + C$ (10) $(x)sin(x) + cos(x) + C$

(11) $(x)\ln(x) - x + C$

(12) $\frac{15}{2}\ln|\sec(2x^2) + \tan(2x^2)| + C$

(13) $(x + 3)e^{x+3} - e^{x+3} + C$

(14) $6\ln|x| - 3\ln|x^2 + 1| + C$

(15) e^3 (16) ∞ (17) $-\infty$

CHAPTER 5: COMBINATORICS
Section (5.1)

(1) 151,200   (2) 132   (3) 1680   (4) 15

(5) 120   (6) 330   (7) 2520   (8) 210,210

(9) 12,612,600   (10) 21   (11) 15   (12) 1

(13) 4096

(14) (a) 8   (b) { ⌀, {a}, {b}, {c}, {a,b}, {a,c}, {b,c}, {a,b,c} }

(15) 120

CHAPTER 6: FUNDAMENTALS OF PROBABILITY
Section (6.1)

(1) (a) $N(S) = 2^5 = 32$

   (b) A = {HHTTT, HTHTT, HTTHT, HTTTH, THHTT,
      THTHT, THTTH, TTHHT, TTHTH, TTTHH}

   (c) $\frac{5}{16}$

   (d) B = {HTTTT, THTTT, TTHTT, TTTHT, TTTTH}

   (e) $\frac{5}{32}$

(2) (a) S = {Whole numbers from 000 to 999} ,
      Card(S) = 1000

   (b) A = {505, 515, 525, 535, 545, 555, 565, 575, 585, 595}

(c) $\frac{1}{100}$ (d) $\frac{1}{500}$ (e) $\frac{9}{100}$

(3) $\frac{1}{30}$ (4) (a) $\frac{3}{5}$ (b) 120   (c) 12, $\frac{1}{10}$ (d) $\frac{1}{120}$

(5) (a) 1.599416×10^{10} (b) 405,600, 0.000025359

     (c) 0.000028571   (d) $1.250456417 \times 10^{-10}$

Section (6.2)

(1) (a) S = {(1,2), (1,3), (1,4), (2,1), (2,3), (2,4),
         (3,1), (3,2), (3,4), (4,1), (4,2), (4,3)}

   (b) $A_1 = \{(3,4), (4,3)\}$
      $A_2 = \{(1,3), (1,4), (2,3), (2,4),$
            $(3,1), (3,2), (4,1), (4,2)\}$
      $A_3 = \{(1,2), (2,1)\}$

   (c) Events $\{A_1, A_2, A_3\}$ are a mutually exclusive and exhaustive set of events from S.
      $\Pr(A_1) = \frac{1}{6}$. $\Pr(A_2) = \frac{2}{3}$. $\Pr(A_3) = \frac{1}{6}$.

   (d) $S = \{0, 1, 2\}$
      $\Pr(X = 0) = \frac{1}{6}$. $\Pr(X = 1) = \frac{2}{3}$. $\Pr(X = 2) = \frac{1}{6}$.
      These probabilities add to 1.

(2) (a) S = {Whole numbers from (00) to (99)}

   (b) Card(S) = 100.

   (c) $A_1 = \{03, 06, 09, 12, \ldots, 93, 96, 99\}$
      Card$(A_1) = 33$.

   (d) $A_2 = \{$ Whole numbers from (00) to (99), except for those that are evenly divisible by 3}
      Card$(A_2) = 67$.

   (e) Events $\{A_1, A_2\}$ are a mutually exclusive and exhaustive set of events from S.

(f) $Pr(A_1) = \frac{33}{100}$, $Pr(A_2) = \frac{67}{100}$

(g) $S = \{0,1\}$, $Pr(X = 0) = \frac{33}{100}$, $Pr(X = 1) = \frac{67}{100}$.

(3) (a) S = {Whole numbers from (00) to (99)}

(b) Card(S) = 100.

(c) A_1 = {00, 11, 22, 33, 44, 55, 66, 77, 88, 99}

Card(A_1) = 10.

(d) A_2 = { Whole numbers from (00) to (99), except for those where both digits are equal}

Card(A_2) = 90.

(e) Events $\{A_1, A_2\}$ are a mutually exclusive and exhaustive set of events from S.

(f) $Pr(A_1) = \frac{1}{10}$, $Pr(A_2) = \frac{9}{10}$

(g) $S = \{0,1\}$, $Pr(X = 0) = \frac{1}{10}$, $Pr(X = 1) = \frac{9}{10}$.

Section (6.3)

(1) (a) $\frac{5}{21}$ (b) $\frac{8}{21}$ (c) $\frac{3}{21}$ (d) $\frac{10}{21}$

(2) (a) $(A \cap C) = (A \cap D) = (C \cap D) = \emptyset$ (b) $\frac{4}{21}$

(c) $\frac{3}{21}$ (d) $\frac{12}{21}$

(3) (a) $\frac{2}{21}$ (b) $\frac{1}{4}$ (c) $\frac{1}{2}$

(4) (a) $\frac{1}{21}$ (b) $\frac{1}{3}$ (c) $\frac{1}{8}$ (5) 0.000000005

(6) (a) 0.000005802   (b) 0.14, 0.02

(7) (a) 0.21955   (b) 0   (8) 0.036

(9) $Pr(A) \leq 0.77$ (10) (a) 0.12   (b) 0.55

(11) (a) 0   (b) 0.5

(12) (a) 0.515

(b) $\{Pr(A|E) = 0.3883, Pr(A^c|E) = 0.6117\}$

(c) The data seems to suggest that customers feel less

favorably about male drivers and more favorably
about female drivers than was assumed before the
experiment was done.

Section (6.4)

$$(1) \quad p_Z(z) = \begin{cases} 1/8, & \text{for } z = 0 \\ 3/8, & \text{for } z = 1 \\ 3/8, & \text{for } z = 2 \\ 1/8, & \text{for } z = 3 \end{cases}$$

$$(2) \quad p_X(x) = \begin{cases} 2/3, & \text{for } x = 1 \\ 1/3, & \text{for } x = 2 \end{cases}$$

$$(3) \quad p_X(x) = \begin{cases} 1/6, & \text{for } x = 0 \\ 2/3, & \text{for } x = 1 \\ 1/6, & \text{for } x = 2 \end{cases}$$

$$(4) \quad p_X(x) = \begin{cases} 33/100, & \text{for } x = 0 \\ 67/100, & \text{for } x = 1 \end{cases}$$

$$(5) \quad p_X(x) = \begin{cases} 1/10, & \text{for } x = 0 \\ 9/10, & \text{for } x = 1 \end{cases}$$

Section (6.5)

(1) (a) $p_X(x) = \left\{ \frac{1}{7}, \text{ for } x = 7, 9, 15, 20, 25, 31, 33 \right\}$

(b) $\mu = 20$, $\sigma^2 = 90$, $\sigma = 9.49$

(2) (a) $p_X(x) = \begin{cases} \frac{1}{13}, & \text{for } x = -4, -3, -2, -1, 0, \\ & \quad 1, 2, 3, 4, 5, 6, 7, 8 \end{cases}$

(b) $\mu = 2$, $\sigma^2 = 14$, $\sigma = 3.74$

(3) (b) $\mu = 25$, $\sigma^2 = 140$, $\sigma = 11.83$

CHAPTER 7: SAMPLES AND STATISTICS
Section (7.3)
(1) n = 11, $\bar{X} = 15.3636$, $S^2 = 12.2545$, $S = 3.5006$
(2) n = 19, $\bar{X} = 0.5789$, $S^2 = 0.2573$, $S = 0.5073$
(3) n = 10, $\bar{X} = 100.62$, $S^2 = 2.4396$, $S = 1.5619$

CHAPTER 8: COMMON PROBABILITY DISTRIBUTIONS
Section (8.2)
(1) (a) 0.4, 0.0894   (b) 2.236
     (c) 18
(2) (a) 15   (b) 2.74
     (c) (9.52, 20.48)   (d) 0.5
     (e) 0.0913   (f) (0.3174, 0.6826)

Section (8.3)
(1) (a) 0.80085   (b) 0.0498

(1) (a) 0.3085   (b) 0.2514   (c) 0.2586
(2) (a) 0.0359   (b) 0.0548   (c) 0.4772
(3) (a) According to the CLT, with a large sample size of
     $n = 60$, $\bar{X} \sim$ Normal($\mu = 105, \sigma = 2.19$)
    (b) 0.7517   (c) 0.9147   (d) 0.6372
(4) (a) Y is distributed Binomial($n = 100, p = .38$)
    (b) $\bar{p}$ is the average of 100 $0's$ and $1's$, therefore, it is an
       average like $\bar{X}$. So, with the large sample size of
       n = 100, the CLT says that it will be approximately
       normally distributed.
    (c) 0.0749
    (d) 0.4648
    (e) 0.1075

CHAPTER 9: LARGE SAMPLE HYPOTHESIS TESTING
Section (9.2)

(1) (a) $\begin{cases} H_0: \mu = 140 \\ H_a: \mu \neq 140 \end{cases}$ (b) 4.40
    (c) Z_0 is distributed approx. N(0,1) if H_0 is true and we
       know this from the CLT for large sample sizes.
    (d) The p-value is $\ll .01$. He rejects H_0 and concludes
       that the true mean μ is different from 140.
(2) (a) $\begin{cases} H_0: \mu = 52 \\ H_a: \mu < 52 \end{cases}$ (b) -0.22
    (c) Z_0 is distributed approx. N(0,1) if H_0 is true according
       to the CLT because the sample size n is large.
    (d) The p-value = 0.4129. Since the p-value is $\gg .05$, he

fails to reject H_0 and concludes that there is not significant evidence that $\mu < 52$.

Section (9.3)

(1) (a) $\left\{ \begin{matrix} H_0: p = 0.50 \\ H_a: p > 0.50 \end{matrix} \right\}$ (b) $Z_0 = 2.20$

(c) The p-value is 0.0139, and since this is less than $\alpha = .05$, he rejects H_0 and concludes that the true value of p is greater than (0.50). This leads him to conclude that the ballot initiative will be adopted.

(2) (a) $\left\{ \begin{matrix} H_0: p = 0.50 \\ H_a: p > 0.50 \end{matrix} \right.$ (b) 1.41

(c) The p-value is 0.0793. Since this is greater than $\alpha = .05$, he must fail to reject H_0 and conclude that that there is not significant evidence that the coin is unfair.

Section (9.4)

(1) (a) $\left\{ \begin{matrix} H_0: \sigma^2 = 20.0 \\ H_a: \sigma^2 < 20.0 \end{matrix} \right.$ (b) 27.49

(c) The p-value is greater than $\alpha = .05$. So, the engineer fails to reject H_0 and reports that there is not significant evidence to conclude that the variance is less than 20.0.

(2) (a) $\left\{ \begin{matrix} H_0: \sigma^2 = 35.0 \\ H_a: \sigma^2 > 35.0 \end{matrix} \right.$ (b) 53.51

(c) The p-value is less than .05. So, he rejects H_0 and concludes that the true variance $\sigma^2 > 35.0$.

<u>Section (9.5)</u>

(1) (a) $\begin{cases} H_0: \mu_X - \mu_Y = 0 \\ H_a: \mu_X - \mu_Y \neq 0 \end{cases}$ (b) $S_p^2 = 3106.5$

(c) -1.79

(d) The p-value $= 0.0734 > .01$. At the $\alpha = .01$ level, we fail to reject H_0 and conclude that the average tread life of Brand X and Brand Y tires is the same.

(2) (a) $\begin{cases} H_0: \mu_X - \mu_Y = 0 \\ H_a: \mu_X - \mu_Y \neq 0 \end{cases}$ (b) -14.29

(c) The p-value is clearly $\approx (0.0000)$. So, we reject H_0 and conclude that there is a significant difference in the means of the random variables X and Y.

<u>Section (9.6)</u>

(1) (a) $\begin{cases} H_0: p_X - p_Y = 0 \\ H_a: p_X - p_Y \neq 0 \end{cases}$ (b) $\bar{p}_p = 0.5596$

(c) -1.79

(d) The p-value is 0.0734 which is $> (.05)$. Therefore, the Republican candidate fails to reject H_0, and concludes that the two polling proportions are not significantly different.

(2) (a) $\begin{cases} H_0: p_X - p_Y = 0 \\ H_a: p_X - p_Y \neq 0 \end{cases}$ (b) $\bar{p}_p = 0.4022$

(c) $Z_0 = 0.19$

(d) The p-value $= 0.8494$ which is $\gg .05$. So, the Ice Cream company fails to reject H_0 that the proportions are equal. The Ice Cream company is delighted that their product is liked equally in both countries, but wishes that the common proportion of approval was higher.

CHAPTER 10: LARGE SAMPLE CONFIDENCE INTERVALS

Section (10.2)
(1) (26.722, 28.278)   (2) (100.571, 105.429)

Section (10.3)
(1) (0.514, 0.726)   (2) (0.089, 0.371)

Section (10.4)
(1) (33.95, 83.43)

(2) (a) (102.83, 247.13)   (b) (10.14, 15.72)

Section (10.5)
(1) (−2.963, −1.237)

(2) (−5.5889, −3.4111)

Section (10.6)
(1) (−0.277, 0.249)

(2) (−0.166, 0.118)

CHAPTER 11: SMALL SAMPLE INFERENCE

Section (11.2)
(1) (a) X is normally distributed.

(b) $\begin{cases} H_0: \mu = 235 \\ H_a: \mu > 235 \end{cases}$ (c) $T_0 = 4.178$

(d) T_0 has a t-distribution with 17 degrees of freedom.

(e) The p-value $\ll$ .05. He rejects H_0 and concludes that the hardness is greater than 235 Sterlings.

(2) (a) X is normally distributed.

(b) $\begin{cases} H_0: \mu = 125 \\ H_a: \mu \neq 125 \end{cases}$ (c) $T_0 = 2.668$

(d) T_0 has a t-distribution with 10 degrees of freedom.

(e) .01 < p-value < .025. Therefore, he fails to reject H_0. Australian Gray Lizard Blood Pressure is not significantly different from 125 units at the $\alpha = .01$ level.

Section (11.3)

(1) (78.549, 90.851)   (2) (16.629, 19.071)

Section (11.4)

(1) (a) 3.765   (b) $T_0 = -2.081$ (c) 27

(d) .02 < p-value < .05. The researcher rejects H_0 at the $\alpha = .05$ level and concludes that the means for the random variables X and Y are different.

Section (11.5)

(1) (a) 15.85   (b) (-16.016, -7.984)

(c) We are pooling the variances! We can perhaps get away with it here since we are assuming equal variances.

CHAPTER 12: LINEAR REGRESSION AND CORRELATION

Section (12.2)

(1) (a) 109, 765, 295, 4869

(b) 5.45, 14.75, 1867

(c) 6.485, 1.517

(d) Y = (6.485) + (1.517)X

(e) r = 0.8714, $r^2 = 0.7593$

This is a high correlation. It seems that quality of diet is a good predictor of longevity for Rhesus monkeys. More than 75% of the variability in the variable Y is explained by fitting a line to the data.

(2) The value of the test statistic is $Z_0 = 1.51$. Since the p-value = .0655, then at the $\alpha = .05$ level we fail to reject H_0 and cannot conclude that the true correlation is greater than 0.75.